THE LONGMAN GUIDE TO STYLE AND WRITING ON THE INTERNET

Martha C. Sammons
Wright State University

New York • Boston • San Francisco
London • Toronto • Sydney • Tokyo • Singapore • Madrid
Mexico City • Munich • Paris • Cape Town • Hong Kong • Montreal

ISBN-13: 978-0-205-57629-6
ISBN-10: 0-205-57629-x

1 2 3 4 5 6 7 8 9 10—DOH—10 09 08 07

▷ Contents

▷ Topical Guide

Web Page Design

Types of Web Writing

Web Page Writing Techniques

Web Page Style

Editing

▷ Preface

What This Book Is About

This book is a reference guide for those who want to write and format effective Web documents. It focuses on writing and designing online hypertext documents. Many books explain how to create a Web page. While this book contains entries on HTML elements, layout, and other topics typically found in the books on the market, it emphasizes online writing. It includes topics such as designing and formatting Web text, using text in elements of a Web page, components to include in certain types of Web pages, effective online writing techniques, and the most common errors found in Web pages.

This book also focuses on what makes writing Web pages different from writing print documents. Writing online requires more consciousness about your readers, who are in a hurry and want to skim quickly. It also requires the ability to chunk and layer information and a more concise, direct writing style. Online writers also must be aware of the limitations of displaying text online and slow-loading pages over low-bandwidth connections. Thus, while it is helpful to know HTML codes and impressive Web page design, if your message is not effectively communicated, people will not return to your site.

This handbook is organized alphabetically, with entries that can be read in any order. Each entry contains definitions, concise tips and checklists, and numerous examples of good and bad writing techniques. The principles contained in this handbook can also be applied to other types of online hypertext documents, such as intranets and online documentation. In fact, information in this book is gleaned from the literature currently available on hypertext and online help systems.

All the examples in this book come from actual Web pages. These Web pages also provide a variety of ideas for effective techniques to use on your own Web pages. **The examples in this book come from the Internet in their original wording. While some are used as good examples of a particular technique, they may need to be improved or edited.**

The Scope of This Handbook

The emphasis of this handbook is online hypertext documents and the issues relevant to this type of writing. These issues result from the differences between online and print documents, such as screen resolution, file sizes, scrolling, and linking.

It does not explain how to code a Web page but does summarize HTML codes as they relate primarily to formatting techniques and Web page components. It does not describe how to use Gopher, Telnet, e-mail, browsers, etc. but does describe how to write Web documents in general, specific types of Web documents, and specific elements within Web pages. It also does not explain how to conduct research.

Most important, it is not meant to be a comprehensive writing handbook; there are many excellent books available that summarize grammar and punctuation rules. Instead, it reviews the most common errors committed on Web pages.

Who This Book Is For

This book is primarily aimed at the new writer who is unfamiliar with business and technical writing and has thus not written online documents. It is written for anyone who wants to write an effective Web page. The following types of writers will also find this handbook useful:

- Technical writing, business, journalism, and communication majors
- Students in writing classes who are studying Web writing
- Teachers
- Business and technical professionals
- Technical communicators

How This Book Is Organized

There are two parts. Part 1 (Web Writing and Design Principles) highlights key Web writing techniques and is written in a "readable," concise style. This section of the handbook discusses the most important principles of good writing style and contains extensive cross references to the most relevant handbook topics.

Part 2 (Handbook) contains over 300 entries arranged alphabetically from A-Z. Each entry contains a definition, HTML code, concise tips, and good and bad examples from real Web pages. These examples are based on observations of thousands of Web sites, usability studies, and a collection of techniques that "work." There are also "before and after" examples and simple visual diagrams that demonstrate writing techniques.

The appendix includes suggested reading, a glossary of Web terms, useful lists (e.g. wordy phrases), and a review checklist.

New to This Edition

The following are features new to this edition:

- Easier to read layout.
- New examples.
- Diagrams that illustrate writing techniques.
- Revised, expanded, and additional entries.
- New organization. The A-Z arrangement makes it convenient to look up a particular subject, as in a dictionary or encyclopedia. A new section summarizes key Web writing techniques and is written in a "readable," concise style rather than the list format found in the handbook. This section of the handbook discusses the most important principles of good writing style. Extensive cross references to the most relevant handbook topics are provided throughout.
- Expanded topical index showing related entries.

Conventions Used

Tips sections contain checklists of principles.

☑ **Examples** are good examples from actual Web sites. Please note that the wording and spelling in these examples are maintained as they were in the original sites.

☒ **Examples** are examples for poor writing techniques from actual Web sites.

☒ **Before** and ☑ **After** examples show original writing from actual Web sites with revisions.

Acknowledgements

I would like to thank the following people who contributed to this book:

- Eben Ludlow and the editors, proofreaders, and production staff at Longman.
- Martin C. Sammons for invaluable production assistance.
- Marci, Margi, and Mardi Sammons, and Dr. & Mrs. E. J. Cragoe, Jr., for your support.

Part 1:
Web Writing and Design Principles

▷ Comparing Online and Print Documents

To understand the guidelines for writing and designing Web pages, you should consider the differences between print and Web documents.

Web pages are
-Horizontal
-Scroll
-Linked
-Grainy
-Skimmed

Online Format: Screens are landscape orientation and about one-third of a normal page size. Because monitors vary in size and in the number of colors they display, you cannot be sure how your colors and graphics will display. Monitors also shine light in readers' eyes, unlike print documents that use reflected light. Thus it is more tiring to read online text. Less words, simpler sentences, more white space, and less punctuation are therefore appropriate.

Online Text: Text and graphics are grainy–about 50 to 100 dots per inch. Thus it is difficult to read long pages of text online. In fact, reading speed decreases by about 30%. Fonts appear different sizes in different resolutions. Web design is not WYSIWYG (What You See Is What You Get). A Web page will look different on every monitor and browser. Readers can change text size defaults in their browsers as well.

Web Features: Web documents do have many advantages over print documents: they allow you to use links, color, animation, interactivity and multimedia. Unfortunately, these elements can cause Web pages to load slowly and cause "information overload" if not used with discretion. The Web also allows you to distribute documents to a wide audience and update information frequently.

Online Readers: Web readers read differently–they scan and are usually in a hurry. So information must be brief, clear, and skimmable. Web readers also prefer bite-sized chunks of information. In addition, they must scroll or click to navigate. They cannot carry the documents with them or annotate them without printing them out. Print readers start at the upper left; on Web pages, readers see the entire screen as a whole. Thus positioning of important information is crucial.

All these differences affect decisions about how you arrange, write, and format Web documents.

▷ Understanding the Writing Process

Writing a Web page requires several stages. These may include any of the following tasks, and many steps may occur simultaneously or be repeated:

- Brainstorm topics
- Determine content and research material
- Consider copyright issues
- Determine the objectives and purpose
- Analyze the audience
- Organize material
- Plan the structure and navigation

- Create a concept document, flowchart, and/or storyboard to aid planning
- Design the page layout
- Select HTML software
- Plan file management
- Establish guidelines
- Write
- Create the pages
- Add graphics
- Edit
- Test and validate
- Conduct usability testing

These topics are described in more detail in the handbook section. The next sections focus on some of the most critical topics involved in writing and formatting Web documents.

RELATED HANDBOOK TOPICS

GUIDELINES
Cascading Style Sheets
Editorial Submission Guidelines
Privacy Policy
Style Guide
Style Sheets (Editorial)

FILE FORMATS & MANAGEMENT
Adobe Acrobat
Animation
Download Menu
File Management
Handheld Devices: Designing for. . .
HTML Editing Software
Multimedia
Printing
User Options

ISSUES
Accessibility
Copyright Issues
Deep Linking
Netiquette
Platform-Independent Terminology
Privacy/Personal Information

▷ Writing for Your Audience

Identifying Your Audience: One of the most important steps in writing a Web page is identifying your readers and the information they want. The target audience affects your entire site.

Accommodating Reader Levels: Although you should write for a target audience, you may still have readers who range from experts to novices. Experts include experienced Internet users and subject-matter experts. If you believe that many experts will be interested in your Web pages, try to accommodate their needs and interests. New users are unfamiliar with Web terminology, navigating, and configuring their browsers. They may also be unfamiliar with the subject. If you believe that many new users will be reading your Web pages, consider adapting your site for novices. One simple way to accommodate these varied types of readers is through layering information through links.

Using an Appropriate Style: Several writing techniques are used to explain technical terms, abbreviations, and acronyms that will be new to the user. Definitions help explain concepts and principles and accommodate novices. Glossary links allow readers to access definitions from any page. Examples and case studies can help explain abstract conceptual information or improve productivity. Analogies explain conceptual information or objects, help readers understand technical information, and elaborate on definitions. Site guides/help are also useful for novice users.

Using an Informal Style: Many Web pages tend to be more personal, direct, and conversational than most printed documents. Usability studies show that most Web readers prefer informal, conversational, down-to-earth writing and Web sites with personality rather than cold, impersonal sites. In addition, difficult writing slows readers down. Informal writing is achieved by a conversational sentence style and word choice (such as contractions). You should also avoid jargon and buzzwords and define acronyms. "You" orientation is another method of using informal writing and a personal tone. It emphasizes reader benefits and conveys a personal tone that is positive and friendly.

Providing Personalization and Community Features and Other Options: Readers also like to tailor the look and feel of the site to their own tastes. For example, they may select how they want to view your Web page, or the type of information to download. Allowing users to make choices makes your Web page reader-centered and independent of browser or hardware. Community features are elements you provide in your Web site for readers to interact. Providing community features allow readers to communicate with you and with each other, and to exchange information.

Making Your Site Accessible: Anyone should be able obtain information on your site regardless of platform, browser, devices, and disabilities. In designing an accessible site, you accommodate different audience levels, slow connections, users with text-only browsers, users in a hurry, users who want to print your page, and users with other handheld and other portable devices.

A number of accessibility guidelines are described in the handbook. The Web Accessibility Initiative (WAI) develops guidelines for accessibility of Web sites, browsers, and authoring tools. These guidelines benefit people both with and without disabilities and make it easier to use the Web. One of the most important guidelines is providing textual alternatives (ALT tags) for graphical elements on your page.

Through effective writing style and word choice, you also consider international readers. Even though you may have targeted a specific audience for your Web page, you may have many secondary readers. Because the Web is a global medium, your audience may reach readers around the world.

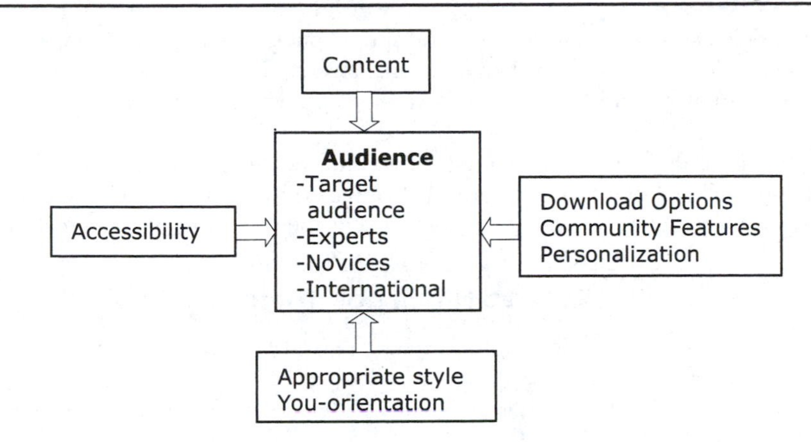

RELATED HANDBOOK TOPICS

WRITING FOR YOUR AUDIENCE
Accessibility
Acronyms
Analogies
Audience Analysis
Audience: Writing for Experts and Novices
Buzzwords
Children: Writing Web Pages for Kids
Definitions
Examples
Glossary
International Audience
Jargon
Readability
Site Guide/Help
"You" Orientation

INFORMAL WRITING STYLE
Active Voice
Contractions
Informal Writing
Personal Tone/Personalization
Tone
"You" Orientation

READER FEATURES
Blog
Community Features
Feedback
Interactivity
RSS Feed
User Options/Personalization

ACCESSIBILITY
Accessibility
Alternative (ALT) Text
Long Descriptions
Platform-Independent Terminology
User Options/Personalization
Validation

▷ Providing Content

Content
Archives
Background
FAQs
Interaction
Resources
Support

After you determine who your target audience is, determine the types of topics that are important to them. This is a key step in planning a Web site. Successful sites provide high-quality content. "Useless" Web sites have no purpose and are a waste of readers' time. Provide content that is relevant, useful, interesting, valuable, fresh, and original. Also strike a balance between providing too little and too much information. Provide links to FAQs, archives, background material, relevant sites, interactivity, and other support information.

However, don't take your existing content and simply put it online. Instead, rework it using the principles in this handbook, such as chunking, layering information, and making text skimmable. Also make sure your content is well-organized and well-written.

Finally, provide readers with new content (updated information) each time they visit (information, news, job postings, etc.). A well-maintained Web site encourages repeat visits and makes your site "sticky"–it attracts readers, keeps them on your site, and makes them return.

RELATED HANDBOOK TOPICS

Content
Stickiness
Updated information

▷ Chunking Information

After you determine your content, you begin to plan how to divide and organize it. Information "chunking" involves breaking information into small units. A chunk usually consists of one topic, idea or concept.

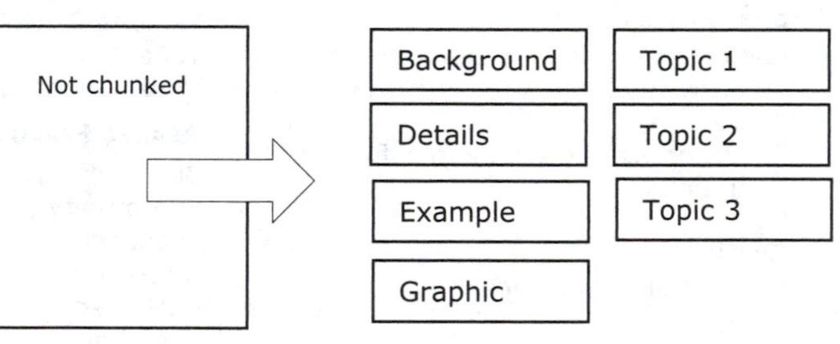

By chunking information into small units, you make large topics more manageable. Furthermore, chunking makes information easier to revise and update. Readers can also decide which topics they need. Presenting small chunks of information that link to more detailed or complex information makes optimal use of hypertext.

RELATED HANDBOOK TOPICS

Chunking

▷ Organizing Information

Once information is chunked, it must be organized into a structure appropriate for the site's goals and audience. Organizing information in a logical order helps users form a mental model of the structure and find information more quickly. Types of structure include linear/sequential, linear with alternative paths, hierarchical, web/network, and grid. On paper, the sequential arrangement of topics is often important. In online documents, however, content is more important than the topic order because readers can jump around.

You choose the appropriate organization for the type of document or information, such as a company or organization page; reference or educational information, FAQ, home page, business or service page; online journal, book, newsletter, or magazine; a list page; product or service information, sales and persuasive information, or a school Web site. You organize information at all levels, such as the content of the entire Web site, the order of topics in your menu or table of contents, the order of paragraphs within topics, and the order of points in a bulleted list. Types of organization include alphabetical, cause/effect, chronological, general to specific, hierarchical, most to least important, sequential, simple to complex, and topical.

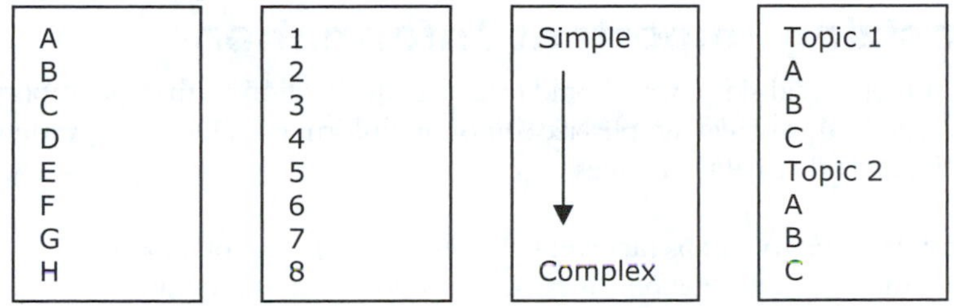

Related Handbook Topics

Organization

Next Links
Organization
Site Map
Structure
Structure: Sequential
Visual Cues

Types of Pages

Book: Online
Catalog Page
Company Page
Course Offering
Educational Information
FAQ
Home Page
Informational Page: Business or Service
Journal (Electronic)
List Page
Magazine or E-Zine
Newsletter
Organization Page
Personal Page
Press Release
Product or Service Information
Reference Information
Resume
Sales & Persuasive Sites
School Web Site

Sales

Action Steps
Attention
Blurbs
Credibility
Promotional Language
Sales & Persuasive Sites
Stickiness

Journalism

Blurbs
Headlines
Magazine or E-Zine
Newsletter
Publication Information

Technical Writing

FAQ
Instructions
Procedures
Process/How Things Work
Quick Reference
Support: Technical
Training/Tutorial
Troubleshooting
User Guide
White Paper

RELATED HANDBOOK TOPICS (CONTINUED)

OTHER TYPES OF ONLINE WRITING

E-Mail
E-Mail List Information
E-Mail Query
Usenet/Newsgroup Posting

▷ Emphasizing Important Information

Because readers are busy and skim, you should organize information so that the important information is easy to find. You can emphasize important information by using a variety of textual, organizational, and visual techniques.

Using Summaries: A summary or abstract condenses the main points of an article, document, or "bottom line" information. Because users skim sites, they read summaries to get the key information or determine what information they want to go to. "Blurbs" or annotations accompany headlines and may appear in a table of contents, "related topics," list of links, or e-newsletters. The description saves time by helping readers decide if they want to visit the page or know what to expect.

Using Introductions: The home page is the first page of a Web site–the page people start with and return to when navigating. It is the most important page on your site because it serves as an introduction that indicates the site's content, purpose, and scope. It also sets the tone and creates a first impression. The main menu/table of contents allows readers to find information.

Emphasize important information visually. Put important information first. Use summaries, introductions, and topic sentences.

Every Web page should tell readers who created the site, the sponsor, purpose, type of information, how often content is navigated, how to navigate, and how to contact the owners/authors.

Write a strong introduction to your Web site (introduction to site) and an introduction at the top of each Web page. Each introduction should explain what the page is about and summarize the contents. The introductory sentence should concisely state the purpose of the page and contain keywords. Because Web readers scan pages, the introduction can help them decide whether to continue reading.

Site Introduction
Who wrote the site?
What is the site about?
Where can I go?
When was it written?
Why is the site worth reading?
How will this site benefit me?

Using the Inverted Pyramid: The "inverted pyramid" method places the most important information first. It provides a bottom-line summary or conceptual overview. Secondary information is available later on the page or through a hyperlink. Information is presented in the following order:

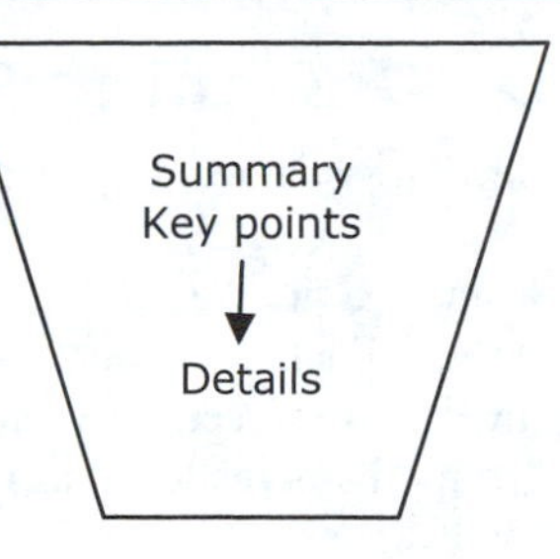

- Main point: who, what, where, when, why, and how
- Brief overview/preview of topics (advance organizer)
- Brief statement of the context
- Background information (or a link to background and conceptual information) needed to understand the topic.
- Link to details and examples

The following table illustrates the use of a hierarchical organization and inverted pyramid. It begins with "bottom line" or general principle and then gives details.

General principle or summary goes here.	
Heading	Details
Heading	Details
Heading	Details

Using Topic Sentences: Topic sentences are crucial in Web paragraphs because most readers skim Web pages. A topic sentence lets busy readers decide whether to read the paragraph. It also provides the context for the information, explains why the information is important, and previews the organization and contents.

Positioning Key Information: The important information is the message or main point of your Web page or site–what you want people to remember. It should appear early in your Web page, within the focal point, and above the "scroll line." If your home page contains numerous links, make it clear where to begin.

Related Handbook Topics

Important Information
Focal Point
Important Information
Introductions
Inverted Pyramid
Page Length
Scannability
Scrolling
Summary/Abstract
Topic Sentence

Summarizing Information
Blurb
Annotation
Summary

Site Introduction
Home Page
Introduction: Site
Purpose
Scope

▷ Designing Pages

Several visual techniques can help readers find key information and skim quickly.

Using a Grid: A grid divides and organizes your Web page into functional areas of the screen. Each area contains text and/or graphics and is devoted to a specific purpose. A grid divides the screen into categories and levels of information. Showing how information is grouped lets readers learn where to look for kinds of information on each page in your Web site.

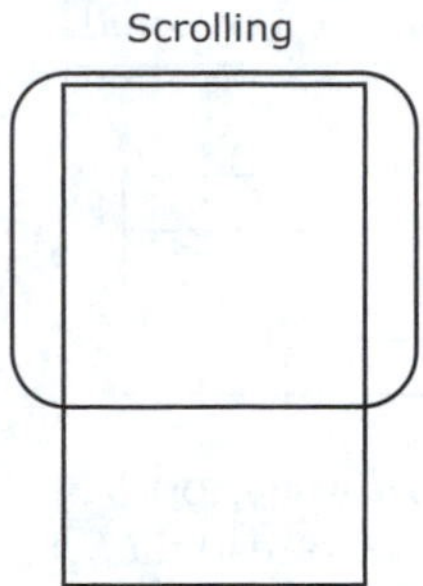

On a Web page, for example, zones might be a left navigation bar, title, body text, footer information.

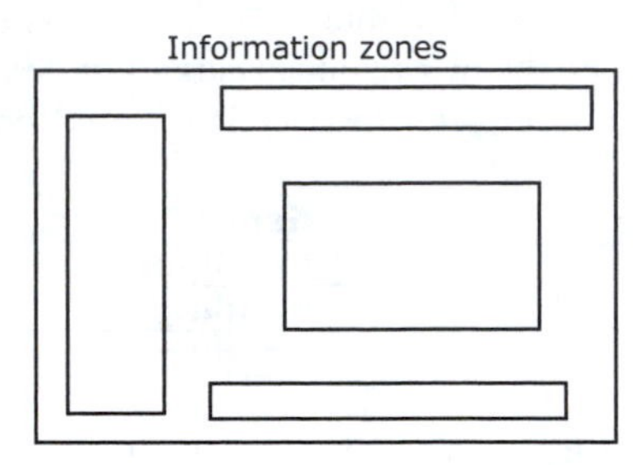

Readers should ideally not have to scroll. Short documents allow pages to load more quickly. Long documents are appropriate when you want readers to have both visual and conceptual continuity. Each option has pros and cons that you should consider.

Using Standard Elements: Every Web page should contain the following elements:

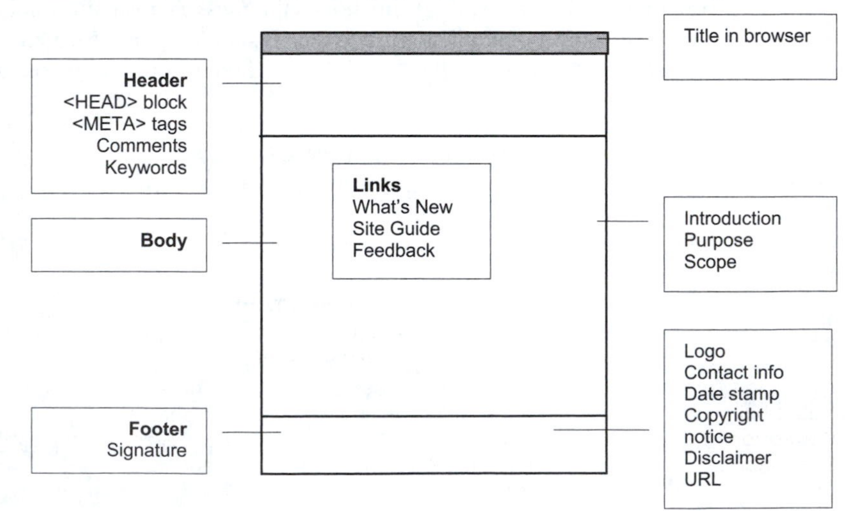

HTML is a structural markup language, not a page description language. When you use HTML tags, you are specifying the document's structure rather than its layout or formatting style. The layout will vary, depending on the browser.

Aiding Skimming: Make the structure visible by showing levels of importance, and design your page so it is easy to scan quickly. Visual cues help the reader determine what information is important.

Formatting techniques that help readers skim include the following:

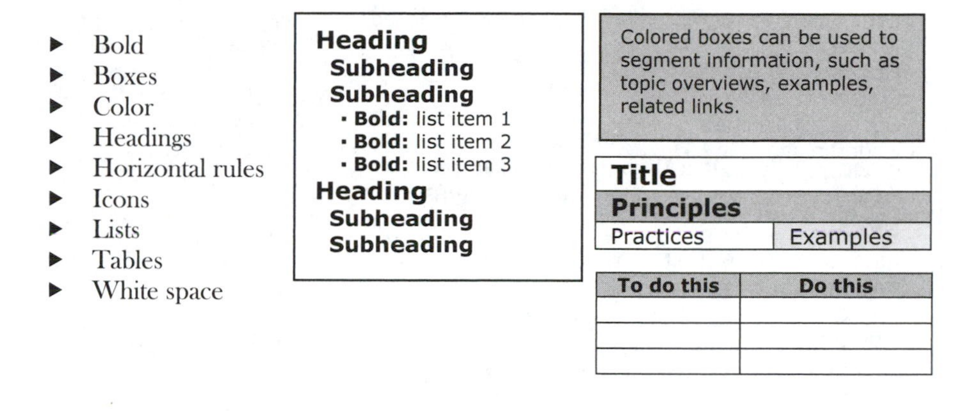

Using Good Page Design: Qualities of good Web page design are as follows:

- **Consistency & Predictability:** There is a common look and feel to every element; items are in the same location on every page.
- **Readability:** Fonts, emphasis, and justification are used for maximum effect.
- **Scannability:** The page can be quickly skimmed.
- **Simplicity:** Only necessary items are included. The site contains no large graphics that take time to load and uses devices of emphasis sparingly.

RELATED HANDBOOK TOPICS

QUALITIES: GOOD DESIGN
Consistency
Consistency: Layout
Consistency: Style
Patterned Information
Readability
Repetition
Scannability
Simplicity

PAGE LAYOUT
Balance
Focal Point
Grid
Grouping
Important Information
Page Size
Page Width
Position/Placement
Printing
Proximity
Scrolling
Visual Cues

DEVICES OF EMPHASIS
Bold
Boxes
Bulleted (Unordered) Lists
Contrast
Emphasis
Font Size
Headings
Numbered (Ordered) Lists
White Space

RELATED HANDBOOK TOPICS (CONTINUED)

TEXT FORMATTING
Bold
Capitalization
Font
Font Size
Italics
Monospaced Type
Preformatted Text
Special Characters
Underlining

LISTS
Bulleted (Unordered)
Lists
Lists
Numbered (Ordered)

LABELS
Callouts
Captions
Headings
Headlines
Labeling Information

SPACE
Columns
Line Length
Margins
Page Width
Readability
Scannability
Tables/Grids
Visual Cues
White Space

GRAPHICS
Alternative (ALT) Text
Callouts/Captions
Charts
Flowchart
Graphics
Icons
Image Map
Logos/Trademarks
Long Description
Tables/Grids
Thumbnail Graphics

COLOR
Background Color
Color
Colored Links
Colored Text

SPECIAL EFFECTS
Animation
Blinking Text
Informational Overload
Multimedia
Marquee
Noise

CODE
Alternative (ALT) Text
Body
Comments
Head
Keywords
Long Description
Meta Information
Search Engines: Writing For
Title

▷ Layering Information

Information should be prioritized by presenting minimal but important information "up front." Readers can then focus on a particular subtopic or more detailed elaboration. By keeping top-level pages simple, you reduce reader frustration inherent in information overload.

Layering is the technique of beginning with general information, then providing links to more details or supplementary information. This technique allows readers to reveal information only when they ask for it. Layering is a method of "progressive disclosure," or "zooming in" to details or explanation. It is essentially a method of "filtering" information as a way to separate primary from secondary information.

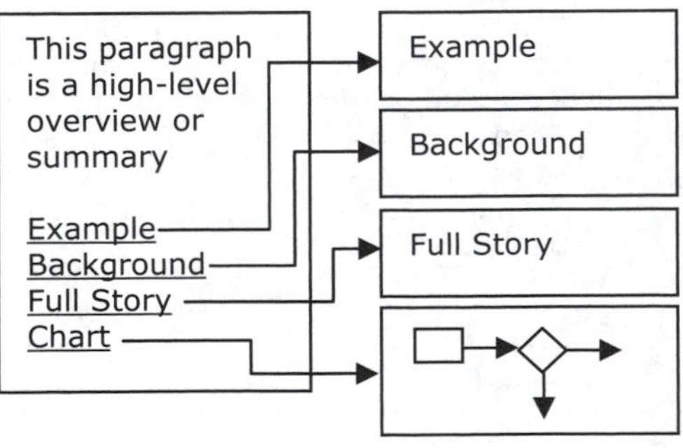

Layering can be used to let readers "drill down" to special sections, background, definitions, details, examples, external resources, a glossary, large graphics, help, multimedia, other document types, other topics, and cross references.

An example of layered information is a directory-structure site like Yahoo that lets readers make decisions in a hierarchical directory structure.

Layering information has many benefits. When skills or content areas are divided into discrete units and "layered," readers can select the amount and type of information they need on a topic. Content can be adapted to individual needs. Readers can also choose the types of material they want to view, including details, graphics, or multimedia.

More knowledgeable readers can easily skim the information that seems basic and move right on to more substantial information. Novices can view definitions, histories, and other supplemental information to the material, at their discretion. Readers can choose paths through the site. On the other hand, you cannot assume readers will navigate sequentially or will have viewed all material. In addition, readers may not return to the original location or may become lost unless sufficient guidance is provided. Orienting readers and providing clear navigation are thus crucial.

RELATED HANDBOOK TOPICS

LAYERING INFORMATION
Layering Information

SPECIAL SECTIONS
Archives
Biographical Information
Glossary
Mission Statement
News
Site Guide/Site Help
Site Map
Updated Information

▷ Orienting Readers

Using Advance Organizers: A means to orient the reader to the information in electronic format is essential, because materials are easily concealed, and users may feel lost or overwhelmed–"lost in hyperspace." Advance organizers are high-level summaries or general overviews and minimal instructions that can then link to details. These provide a mental model of how the information is organized. For example, advance organizers help orient readers to what they will see. They can also help them remember previous information they read.

In this section:
Topic 1
Topic 2
Topic 3

Topic 1

Topic 2

Topic 3

Advance organizers introduce the following types of information:

- Topics
- Sections in a topic
- Page contents
- Headings and subheadings
- Paragraphs

- Sentences
- Lists
- Examples

Advance organizers can be separated from the main text by using lines, boxes, lists, and other visual cues.

Using Contextual Clues: A danger of layering is that most readers get lost when there are more than three layers or paths through information. For example, if an online training course contains a main menu with links to lessons, then links to More Information, trainees will have already navigated through three layers of information. If more links are available, they will lose track of their location and the original context. Location identification helps orient readers and provides contextual clues. Always let users know their location: where they currently are, where they can go, and where they have been.

"Bread crumbs" are a common method of showing readers their location.

Topic 1>Topic 2>Topic 3

Contextual clues can also include the following:

- Color-coding
- Document and chapter names (path showing the location in a large document)
- Headings
- Identity graphic and logo on each page
- Menus and navigational aids
- Organization name, site sponsor, & contact info on each page
- Page number location/total number of pages
- Titles
- Content modules that open in another (smaller) browser window

Each page should provide information such as the site name, page topic, and navigational aids. A good Web page lets readers know what to do, how to do it, where to start, where to find content, how to navigate, where they are, how to return home, and what to expect.

Related Handbook Topics

Orienting Readers
Advance Organizers
Contextual Clues
Context Independence
Location Identification

▷ Directing Readers

Navigation is the pattern through which users move around your site. It provides a way for readers to jump from topic to topic and can indicate by link colors if they have seen everything.

Navigation is related to your site's structure. Navigational "levels" can provide visual cues of site organization. However, you must consider both the nature of the subject matter and the audience. For example, when site content provides facts, procedures, and rules, linear links assure that readers follow required paths through topics. However, for more complex knowledge levels, less structured hierarchical and referential linking is more appropriate.

Always provide more than one way to access the information, such as browsing options, index, menu, search engine, shortcuts, site map, and table of contents. Metaphors can also help readers know how to interact with, navigate, or find information. Examples of such metaphors include a book or hierarchical tree organization. Also place navigational aids in consistent locations and in a logical order.

RELATED HANDBOOK TOPICS

ACCESS TO INFORMATION

Browsing Options
Index
Menu
Search Engine
Shortcuts
Site Map
Table of Contents

▷ Connecting Ideas

Types of Links: Types of links include internal, intra-site, and external links.

Internal links (intra-page) connect locations inside one document. They are most often used on longer pages to link items in a menu or table of contents to a heading and then return to the menu. Use internal links to chunk information on long scrolling pages and as advance organizers about what will be covered. Make it clear that the link is internal rather than a link to another page.

Internal Links

This page discusses the following:
Topic 1
Topic 2
Topic 3

Topic 1
xxxxxxxxxxxxxx
Top

Topic 2
xxxxxxxxxxxxxx
Top

Topic 3
xxxxxxxxxxxxxx
Top

Intra-site links are cross references to pages within your own site.

External links jump to other Web sites. External links allow you to cross-reference other Web sites that are related to your topic. They also add credibility to your content. Make it obvious when links are external and provide useful annotation. However, avoid too many external links, or readers may not return.

Link Format: Links can be buttons, text, or both. Buttons are small graphics or icons used for hyperlinks and navigational aids. You can use simple buttons or combine them into *navigation bars*.

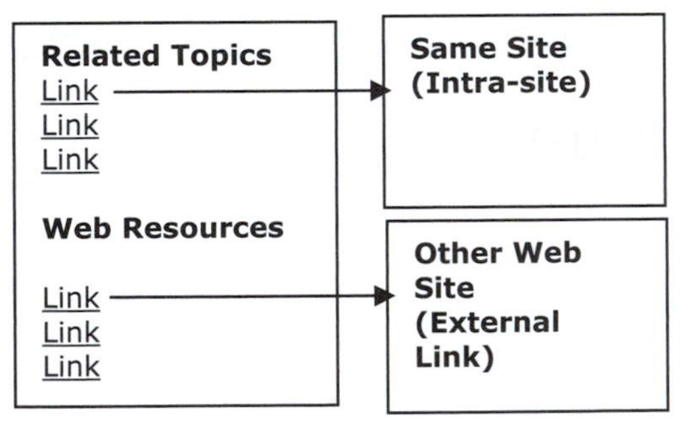

A *text link* is a word or phrase that jumps to an internal or external location. A text link provides a simple navigational aid or accompanies graphical links so your site is accessible. It also helps readers predict where the link leads and decide whether to click. One of the most important rules of linking is to always make it clear where a link goes. For example, links can be labeled to indicate the type of information they represent, such as Definition, Example, See Graphic. It is also beneficial to warn readers about what will happen when a link is clicked, such as that a link is offsite or that a link will open another browser window.

Choose between *embedded* text links (links within the paragraph), *link lists*, or a combination of both. Consider both types of links to accommodate readers who skim and search and readers who need the initial explanation. This technique is often useful on the home page.

Embedded Links

This is a link. This is another link. This is another link. This is another link.

Link List

Link
Link
Link

Embedded links can help preview organization. However, avoid too many embedded links within a paragraph. Make linking words or phrases part of meaningful sentences so readers have a clear understanding of where they are going. Use surrounding text/context to clarify the link text. Especially avoid links such as *Click here*.

Always arrange lists of links in a logical order. Also group links under descriptive headings to categorize related links. Avoid dead links by checking your links regularly.

RELATED HANDBOOK TOPICS

NAVIGATION SYSTEM
Browsing Options
Buttons/Navigation Bars
Colored Links
Cross-References
Deep Linking
Download Menu
Footnotes
Frames for Navigational Aids
Image Map
Index
Layering Information
Links
Links: Embedded Text
Links: External
Links: Internal
Links: Lists
Links: Text
List Page
Location Identification
Menu
Navigation
Navigation: Types of
Next Links
Page Numbers
Search Engine
Shortcuts
Site Map
Table of Contents

▷ Using Effective Writing Techniques

The writing style of a Web site includes the tone and word choices that reflect your goals and that will appeal to your audience. For example, your tone and style may be formal, professional, or casual.

The writing principles for print and online documents are the same. However, some principles are more difficult to implement online and are more crucial when writing online documents. A well-written Web page is geared to the audience by giving readers information quickly.

Skimmable Paragraphs

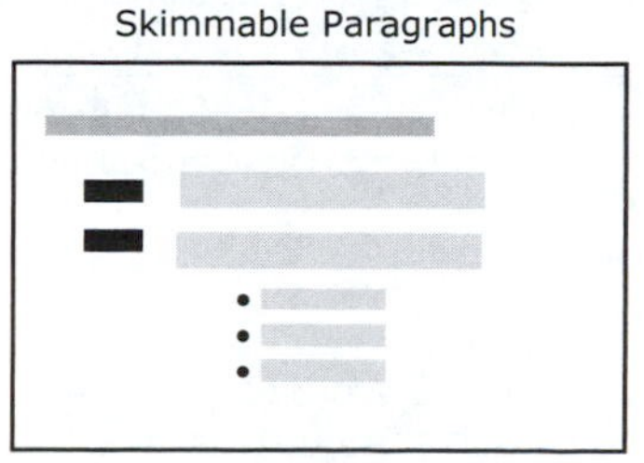

Web readers tend to scan documents, looking for a few sentences that contain the information they want. Web paragraphs should thus be short and focused. Text in short paragraphs is easier to read and skim quickly and adds more white space. Paragraphs formatted for easy skimming include headings, bold text, and lists.

Topic sentences in combination with specific headings also help readers skim. Consider emphasizing the topic sentence by using bold or indenting the paragraph text beneath it.

Sentences on the Web should be short and simple with the main point at the beginning of the sentence.

Your word choice affects clarity, tone, and how easily users can understand your Web pages. Use concrete, precise, and specific words appropriate to the audience. Be especially careful in choosing words for an international audience. Avoid vague, abstract, fuzzy words, buzzwords and acronyms.

RELATED HANDBOOK TOPICS

PARAGRAPHS
Paragraphs
Sentences
Topic Sentences
Transitions
Words to Watch For
Words: Errors to Watch For

WORDS
Buzzwords
Definitions
Glossary
Intensifying Words
Jargon
Usage: Computer and Internet
Verbs
Word Choice

▷ Cutting Excess

Conciseness is one of the most important qualities of good Web page style. Reading on screens is 30% slower than hard copy documents because online documents are more difficult to read. In addition, people scan Web pages and are usually in a hurry. Thus you should use about half the number of words you normally would.

One simple way to cut words is to use headings/labels, lists, tables, and visual aids. The handbook contains specific tips on avoiding wordiness, as well as useful lists of words and phrases to avoid. However, there are a few common causes of wordiness to check for:

- Too many prepositions.
- Weak, stretched-out verbs rather than strong verbs.
- Wordy and redundant phrases.
- Sentences that begin with *There is* and *It is.*
- Vague and intensifying words, such as *very* and *really.*

RELATED HANDBOOK TOPICS

WORDINESS
Active Voice
Clichés
Conciseness
Intensifying Words
Telegraphic Style
Appendix: Useful Lists

▷ Editing Common Web Page Errors

Web readers are concerned about the reliability and credibility of information. You must show objectivity by avoiding marketing/promotional language and exaggeration and by adopting a confident style. A professional tone, page design, Web address, and site name also contribute to your credibility. Link to reliable external sources and support information. Also demonstrate your credibility by using a writing style free of grammar, punctuation, and factual errors.

Errors indicate a lack of quality control and can result in inaccurate information. It is important to edit grammatical, spelling, punctuation, and other mistakes on a Web page because such errors detract from readers' overall impressions of your Web page.

Here are excerpts from home pages for Web design services. There are spelling and other errors, and the authors do not sound confident in their page design. Would you want them to design your site?

☒ **Examples**

-If you web site is not living up to its expectations, we are the ones to turn to.
-Our design team knows how to create an attractive web site and draw the audince you want.
-Amazing, we are so great that we are now offering free money. Well, kind of. We are giving 15% of whatever we make to anyone that refers a customer to us.
-I feel that this site is much more streamlined than the last, and am also hoping that it will help bring me more business. Would you like me to do this kind of professional, streamlining redesign for your site?
-We l;lok forward to helping you build and promote your website!

The handbook does not attempt to duplicate the contents of writing handbooks. It is not an exhaustive handbook of grammar, punctuation, and usage rules. Instead, it lists the most common language errors committed on Web pages and focuses on aspects of writing for the Web that are unique or important.

Print a hard copy of your Web documents. It will help you see document length and notice errors that are difficult to see online. Make sure your message is clear and effective. Also review how pages flow and rearrange the organization if necessary. Have others read and edit your pages, too.

Use a spell checker and grammar checker, which will help you find common mistakes such as agreement, punctuation, common grammatical errors, sentence length, and wordiness. A variety of other useful tools for writing are available on the Web, such as dictionaries, thesauri, encyclopedias, and style guides. The appendix contains a detailed review checklist you can use to check your Web pages.

Related Handbook Topics

Punctuation

Apostrophes
Colons
Commas
Comma Splice
Dashes
Hyphens
Parentheses
Periods
Punctuation Marks
Questions/Question Marks
Quotation Marks
Semi-Colons

Credibility

Accuracy
Clarity
Conciseness
Content
Credibility
Date Stamp
Domain Name
Objectivity
Privacy Policy
Promotional Language
Updated Information

Grammar

Agreement
Dangling Modifier
Misplaced Modifier
Pronoun Errors
Run-On Sentence
Sentence Fragment
Shifts
Tense

Part 2: Handbook

▷ Abbreviations

An abbreviation is a shortened form of a word or phrase.

Tips

- Avoid abbreviations unless you are sure all your readers will understand them. In general, the only acceptable abbreviations are A.M., P.M., bytes, U.S., and some units of measurement.
- Try to avoid brand names and product terminology that may be unfamiliar to readers.
- When you use an abbreviation the first time on your Web page, spell it out; then include the abbreviation in parentheses. If the Web document is long and requires scrolling, consider spelling it out more frequently.
- Don't assume readers have read a definition because they may have not started at the beginning of your site.
- If you use an abbreviation in your site name, spell it out.

> ☑ **Example**
>
> EFF Electronic Frontier Foundation

- To help search engines, avoid abbreviations in TITLE tags without also stating what the abbreviations stand for.
- If you must use abbreviations:
    - Use industry-standard abbreviations.
    - Use consistent abbreviations and spelling.
    - Form plurals by adding an *s* and no apostrophe (unless there are internal periods).
- If you have many abbreviations, use a glossary.
- Avoid using abbreviations as links.
- Try to avoid using abbreviations in navigation bars even when phrases do not fit a narrow column.

> ☒ **Examples**
>
> Int'l Commerce
> VU
> Cust. Care

- Avoid shortening common words by cutting off the beginning or ending.

> ☒ **Examples**
>
> Site has lots of **info** on publicity.
> The HTML **spec** recommends that headers be used in sequential order.

▷ Abbreviations: E-Mail

Abbreviations for phrases are commonly used in discussion groups and e-mail messages.

Tips

- In general, avoid abbreviations because your audience may not be aware of their meaning.
- If you must use abbreviations, use well-known abbreviations. Examples of common e-mail abbreviations are the following:

  BTW: by the way
  FWIW: for what it's worth
  FYI: for your information
  IMO: in my opinion
  OTOH: on the other hand
  TIA: thanks in advance
  WADR: with all due respect

▷ Access to Information

An accessible Web site allows any Web user to read it, regardless of platform and computer capabilities. A good Web page also makes the information easy to find and access quickly. Because readers surfing the Internet want to quickly determine what your Web page is about, they need help accessing information.

Tips

- Decide how readers will use your Web page.
- Decide how readers will determine where they are and how to navigate. Ask yourself the following: if someone comes to my site looking for a specific piece of information, how easy is it to locate?
- Organize information logically.
- If necessary, provide instructions for locating information.
- Provide more than one way to access the information, including shortcuts and a search engine.
- Label links clearly, and make it obvious where they lead.
- Provide user options such as a text-only version.
- Provide a consistent screen design.
- Use platform-independent terminology.
- Use fast-loading graphics so pages load quickly.

▷ Accessibility

An accessible Web site allows any Web user to read it, regardless of disabilities, browser, platform, and computer capabilities. Some readers have difficulty accessing information due to visual, hearing, and mobility limitations.

The "Web Content Accessibility Guidelines" are a W3C (World Wide Web Consortium) specification providing guidance on accessibility of Web sites for people with disabilities. They have been developed by the W3C's Web Accessibility Initiative (WAI).

The WAI develops guidelines for accessibility of Web sites, browsers, and authoring tools. These guidelines benefit people both with and without disabilities and make it easier to use the Web. The specification contains general principles of accessible design. Each guideline is associated with one or more checkpoints describing how to apply that guideline to particular features of Web pages. For complete guidelines, a checklist, and validation tools, visit http://www.w3.org. The National Institute on Aging has produced a checklist on "Making Your Web Site Senior Friendly."

Tips

General

- Consider placing a link to an accessibility statement in the footer.

> ☑ **Example**
> The National Institutes of Health, N I H, is making every effort to ensure that the information available on our website is accessible to all. If you use special adaptive equipment to access the Web and encounter problems when using our site, please let us know . . .

- Provide any necessary instructions.

> ☑ **Example**
> If you are navigating using only the keyboard or using an assistive device and need help, visit our Navigation Instructional page for alternative views and navigation.

Graphics and Multimedia

- Use ALT (alternative) text to describe the function of
  - ☐ Graphics
  - ☐ Image maps
  - ☐ Multimedia
  - ☐ Animations
- Summarize graphs and charts, or use the LONGDESC attribute. LONGDESC specifies a link to a long description.
- Use text for hotspots in image maps.
- Provide captions or transcripts of audio and video files.
- When using special effects such as applets, plug-ins, etc., provide links to plain text versions.
- Avoid blinking and animation.

Tables

- Use column headers.
- Make line-by-line reading of tables sensible, or summarize.

Navigational Elements

- Allow readers to skip navigation and go to main content.
- Write hyperlink text that makes sense when read out of context. For example, avoid "click here."
- For senior adults, use large simple navigational buttons.

Textual Elements

- Place periods at the end of bullet list items and headlines.
- For senior adults, use large simple fonts and increased line spacing.

Page Design

- Do not rely on color to convey important information.
- Provide text-only pages when necessary.

Page Organization

- Use headings and lists.
- Use a consistent organization.
- Make pages independent of Cascading Style Sheets.

Frames

- Use descriptive titles for all frame elements.
- Use the "noframes" element, which provides alternate content for browsers that do not support frames or have frames disabled.

▷ Accuracy

Accuracy is correct, reliable, error-free information. Your writing must be free of errors in not only the facts you provide but also your writing (formatting, spelling, grammar, etc.). The accuracy of information on your Web site is one way readers evaluate your credibility. Errors may make your site frustrating or even useless for users.

Tips

- Provide updated information.
- Give the sources of your information.

☒ **Example**

A Web site on documenting research sources contains a date stamp and the following statement: Please note that these citations have been written using the MLA Citation Format.

There is no information about which version of the MLA Style Guide was used or about the author(s) or sponsors of the site. The page is 27 screens long (one document) with varying sizes of fonts and wasted white space. Thus the reader may not trust the accuracy of the information. A reader may be better off going to the MLA Home Page and clicking the link to the MLA Style Guide: These guidelines on MLA documentation style are the only ones available on the Internet that are authorized by the Modern Language Association of America.

- Define the scope of your material.
- Try to copy rather than retype information to avoid making errors.
- Have several people check crucial information and test it for accuracy.
- Provide feedback links so readers can report errors.
- Link to support information.
- Link to reliable sources.

☑ **Example**

PC Magazine's Product Guides contains links to the manufacturer's Web site. Each review page contains the price, pros, cons, and other detail. It also links to readers' comments, related products, and price checks.

- Edit for spelling, grammar, and punctuation errors.
- If you make factual errors, immediately announce the correction and apologize to anyone negatively affected. For numerous corrections, provide a link to a Corrections page.

☑ **Example**

Electronics corrections
Panasonic DMR-EH50
Correction: We originally indicated that this DVD recorder was incapable of dubbing in high-speed mode from its hard disk to any DVD format aside from DVD-RAM. In fact, the unit can dub in high speed to all of the recordable DVD formats it supports. (6/21/05)

▷ Acronyms

Acronyms are created from initial letters or parts of a phrase. They use all capital letters and contain no periods. A number of acronym dictionaries are available on the Web. They define specialized acronyms, such as those used in the computer industry, government, World Wide Web, and other fields.

Tips

- In general, avoid acronyms. Unfortunately, the computer field and the Internet use many acronyms.

☒ **Example**

This site uses too many acronyms: Whether you're a novice just starting out in Web development, or an accomplished HTML hacker seeking to broaden your range, this guide should have something for you: Internet, HTML, CSS, CGI, graphics, design, Java, JavaScript, DHTML, multimedia, database, XML, etc.

- In general, spell out the *first* occurrence of an acronym on *each Web page*.
- Use the following methods to define acronyms:
    - Clause defining the term
    - Parentheses

☑ **Examples**

-TCP/IP, or Transmission Control Protocol/Internet Protocol, is the protocol used on the Internet.
-World Wide Web Consortium's (W3C) Web Accessibility Initiative (WAI).

- Spell out an acronym again in a long Web page that requires scrolling.
- Explain acronyms that are central to the site or discussion.

☑ **Examples**

-Does the Internet Service Provider (ISP) have a local point of presence (POP)? A POP is a local phone number that can be used for dialing-in to an ISP that is located outside of the local calling zone.
-The Extensible Markup Language (XML) is the universal format for structured documents and data on the Web.

☒ **Example**

This site's name uses an acronym that is not defined on the home page.
PDAStreet.com

- If you use many acronyms and have novice readers, link to definitions. However, realize that numerous links can be distracting.

☑ **Example**

Mobile IP is most often found in wireless WAN environments where users need to carry their mobile devices across multiple LANs with different IP addresses.

- Make acronyms context independent. Don't assume readers have read the definition that appeared on another Web page because they may have not started at the beginning of your site.

☒ **Example**

The acronym CGI *should be spelled out:* Counters require a CGI program to function.

- Avoid acronyms in headings so that readers are not required to read the body text; spell out the phrase.

☑ **Example**

Joint Photographic Experts Group (JPEG)
JPEG or JPG is the second most popular image format on the Web.

- If you must repeat an acronym, try shortening it instead of writing it out or place the phrase in a heading.

☑ **Example**

LAN (Local Area Network) can be shortened to **network**.

- For plurals, add an *s* with no apostrophe (e.g., CPUs). To create verbs, add an apostrophe (e.g., FTP'd).
- Do not spell out acronyms that:
    - ☐ Your audience will know (e.g., FAQ). However, to accommodate other readers, provide a glossary link.
    - ☐ Are commonly used Internet and computer acronyms (e.g., RAM: Random Access Memory).
    - ☐ Have definitions that don't really contribute anything to understanding the term (TIFF: Tagged Image File Format).
    - ☐ Are product names (OS/2).

▷ Action Steps

Actions steps are tasks you want readers to perform. Action steps are especially important in sales/persuasive Web sites. Action steps are also a method of adding interactivity to your site.

Tips

- Make it clear what you want readers to do:
    - ☐ Browse
    - ☐ Call
    - ☐ Check order status
    - ☐ Download
    - ☐ E-mail
    - ☐ Find information
    - ☐ Give feedback and comments
    - ☐ Go to the next document
    - ☐ Interact
    - ☐ Join
    - ☐ Leave a message
    - ☐ Place an order
    - ☐ Play
    - ☐ Purchase a product or service
    - ☐ Read
    - ☐ Register
    - ☐ Request information
    - ☐ Search
    - ☐ Sign a guest book
    - ☐ Sign up
    - ☐ Submit something
    - ☐ Shop
    - ☐ Volunteer
- Make it easy to act by providing the following:

- ☐ Information about credit cards
- ☐ E-mail links
- ☐ Forms
- ☐ Mail order information
- ☐ Contact information: telephone number, address, etc.

- ▶ Clearly state your security and privacy policy for online transactions.
- ▶ Use action-oriented verbs.

☑ **Example**

The Electronic Frontier Foundation contains the following links:
Action Center **Join EFF**

Welcome to the Electronic Frontier Foundation's Action Center! Here you can contact your representatives on impending legislation that will have a direct effect on your civil liberties online. You can also subscribe to our action alerts as an RSS feed, or sign up to receive EFFector, our regular e-mail newsletter.

▷ Active Voice

Active voice emphasizes the person or thing acting. Passive voice emphasizes the action. Passive verbs use a form of "to be" and a past participle. Technical sentences are clearer when you identify the doer. Active voice also helps readers relate more to the text.

☒ **Before**

Frames can be used to keep footers and/or headers visible at all times.
The URL must be entered accurately to locate a site.
All graphics you see were created by me.
To view sites created by us, click here.

☑ **After**

You can use frames to keep footers and/or headers visible at all times.
You must enter a URL accurately to locate a site.
I created all the graphics you see.
See sites we created.

Tips

- ▶ Use active voice to
  - ☐ Make the text more concise.
  - ☐ Make the information more interesting.
  - ☐ Place the main idea at the beginning of the sentence.
  - ☐ Clarify who or what is responsible for the action.
- ▶ Use passive voice only to avoid placing blame or when the subject is unknown.
- ▶ Avoid mixing passive voice with a clause that begins with an -ing word or infinitive.

☒ **Before**

Copyright issues, determining what kind of Web page you want to create, and six steps for developing your Web activity can be found here.

☑ **After**

This site discusses copyright issues, types of Web pages, and development steps. You can find.....on this site.

▷ Adobe Acrobat

Adobe Acrobat is software used for electronic publishing. The file format PDF stands for Portable Document Format. Adobe Acrobat is the authoring program that lets you create and enhance PDF files. Acrobat Reader is a free viewer available on Adobe's site. You can also distribute it on your own site. It allows readers to view and print Acrobat documents.

There are many advantages of providing files in Acrobat format. Acrobat files

- Preserve the look of the original, including fonts, colors, images, and layout.
- Can include text, images, hypertext links, movies, and sound files.
- Are independent of printer and platform.

Tips

- Consider providing a PDF version of your Web documents, including newsletters.
- Tell readers the following:

  - ☐ File name.
  - ☐ File format. Use the Adobe Acrobat icon .

  - ☐ File size.
  - ☐ How to obtain and download Acrobat Reader.
  - ☐ Contents of the file. This information helps them decide whether or not to download it.

The "Citizens Guide to Pest Control and Pesticide Safety" teaches consumers how to control pests in and around the home, alternatives to chemical pesticides, how to choose pesticides, and how to use, store, and dispose of them safely. It also discusses how to reduce exposure when others use pesticides, how to prevent pesticide poisoning and how to handle an emergency, how to choose a pest control company, and what to do if someone is poisoned by a pesticide. (2.5 MB, PDF format)

▷ Advance Organizers

Advance organizers preview what will be discussed. They can be text, links, or graphics and may be located on a main page or a help page. Overviews help readers know what is coming. Forming a mental model of the information helps readers organize details later.

Tips

When to Use Advance Organizers

- Preview the following parts of a Web site or Web page:

□ Topics in the site

 **Example**

The PhotoDisc website is roughly divided into three topical areas. Each area is oriented toward a different information need. The first area outlined on this page is concerned with image research and download. Two sections of our site contain information relevant to both designers and photographers and are below referred to as 'zines'. PhotoDisc company, press and employment information is described in the final section of this page.

This site is organized into three sections: Section 1 contains the introduction (starting with this page) and some basic information about the Web. In Section 2 various ad styles and the survey are presented. Section 3 describes our services and how to contact us.

□ Sections within one topic

Examples

This site previews the two main topics (JPEG & GIF) and what areas will be covered. Then it gives an overview of what specific areas will be covered within each topic.
The World Wide Web supports two types of image file formats, JPEG and GIF. First off, we'll learn a bit about these formats. Then we'll learn which type of format to use based on image content and purpose. For JPEGs, we'll start with a few uncompressed images and subject them to greater and greater compression. We'll see what happens to the images themselves, as well as monitor the effect on file size. We'll look at the data and the images in side-by-side comparisons, and evaluate the results of the image compression.

This example previews the section and contains links to the topics.

Page Layout Guidelines & Assistance

This section contains the guidelines and help for how to layout your page and all the HTML, graphics & other hints needed to do this.

We have divided web page layout into three sections:

- Page Header
- Page Body
- Page Footer

There are also other elements which are further broken down into several areas:

- Page Sizing
- Graphics creation
- Graphic usage
- AIESEC logo usage

□ Page contents and what will be discussed next

Example

Sections of this page include:
Know why you want to run a brainstorming session
Decide how you will run the session and who will take part
Prepare the room and materials
Prepare the participants and issue invites

Previous pages:
How to brainstorm
Rules of brainstorming
Principles of brainstorming
Later pages:
Running a brainstorming session
Title: brainstorming page 3

☒ **Example**

The following is a good overview, but the list does not use parallel construction.

This page will show you -
What is a table?
Why you should use tables
How to construct a simple table
Placing images in a table

The end of the page contains a lead-in to the next topic.

Now you can make a simple table, and display information within the table cells. However, there are a lot of interesting things you can do with tables.

Find out more about tables

□ Headings and subheadings

☑ **Example**

Web Site Location
The location of your Web site will mainly depend upon a couple of factors that will influence your decision:
-Budget
-Server Internet Connection Speed

Budget
There are several different options to consider. The amount of capital available for investment will dictate which option you pursue. So let's investigate some of the possibilities ranging from the least amount of capital investment to . . .

□ Paragraphs

☑ **Example**

This sentence previews the topic of each paragraph that follows.

Effective June 1, 1998, American Airlines and American Eagle announced three new policies that will enhance **convenience**, **safety** and the **airport boarding process**.
Convenience - American Airlines and American Eagle limit the number of carry-on items to two for each passenger in the main cabin of every flight.

The link to this page also previews the three points.

New Policies

To enhance your convenience, safety, and the airport boarding process . . .

- ☐ Lists

☑ **Example**

The International Federation of Library Associations and Institutions provides an advance organizer for their links to Government Information on the Internet.

The following list provides links to the major intergovernmental organizations. Several key locator sites are listed to provide access to agencies not listed individually on this page. There is also a link to a major directory of international organizations, with links to many non-governmental organizations.

- ☐ Examples
- ☐ Graphics
- ☐ Options

☑ **Example**

There are two types of domains that can be added to an existing account, an Additional Domain, or a Parked Domain. Resellers can use this area to add domains to their account and have them up and running within minutes.

What is an Additional Domain

An additional domain is a new domain which is added under a different user name. It comes with its own control panel and package, but is billed to the owner of the main domain.

What is a Parked Domain

A parked domain is a domain name that is pointed to the domain you place below as the master domain. Anyone going to this domain on the Internet will go directly to the master domain below. It is registered with the Internic and allows two different names to point to the same site.

Please choose whether you would like to set up an Additional Domain or a Parked Domain. [Add Additional Domain] [Add Parked Domain]

WRITING ADVANCE ORGANIZERS

- ▶ Use identical wording in the preview and in the text itself.
- ▶ Put information in the same order as the preview.

☒ **Examples**

The overview sentence should say "the graphics and the HTML."

There are two parts to a web page: the HTML and the graphics. The first section of this guide will focus on the graphics, while the second section goes more in-depth with some tips and tricks of HTML.

The following example is confusing because the overview paragraph uses different terms than the listed links. The items are also not in the same order as the list.

Many authors argue against accessibility on the basis of various <u>myths</u>, without realizing the many <u>advantages</u> of accessibility, and without knowing how easy it is to <u>improve the accessibility</u> of a Web page.

Why Write Accessible Web Pages? Why all authors should strive for accessibility in their sites.
Accessibility Tips What you can do to improve the accessibility of your site.
Accessibility Myths Common myths and strawman arguments against accessibility.

FORMATTING ADVANCE ORGANIZERS

- Use any of the following formatting techniques:
  - Paragraph with embedded text links

☑ **Example**

There are several stages that a well-designed World Wide Web site will go through. The first stage is The Creative Stage. Here the author will decide what subjects and materials to include on the web site. Next comes The Planning Stage. The author needs to decide where to publish the site and how to effectively organize it. Following the planning stage comes The Production Stage. The actual code will be written and graphics created. The final stage in the process is Implementation and Testing. The finished web pages are loaded to the World Wide Web and tested for accuracy.

  - List of headings or subheadings with links to detail (layering)

☑ **Example**

A list previews sub-points and contains internal links to each.

Types of Indexing

If you try several search engines when performing a specific search, you will find they often display different results. This difference is directly related to the type of indexing the search engines use. Each search engine indexes text differently. Moreover, even though there are only two main types of indexing, variations exist within each type. Read further for an explanation of the two main types of indexing.

- keword indexing
- concept-based indexing

Keyword Indexing

Most search engines index by keyword. However, if a document does not contain keywords in its meta tags, the search engine must determine them. Usually, repeated words or words at the top of a document are considered important in determining the keywords to be indexed.

This site uses internal links to jump to the topics (listed) and subtopics (at end of paragraph). (Only the beginning of the Elements topic is shown here.)

Discussion Topics

Elements
Aspects

> **Elements**
> The HTML standard determines what markup is supported by the definition of forms. This standard is a living document undergoing continual review and revision, refining features and adding function. The basic form elements are:
> Text Fields Text Areas Buttons Radio Buttons Check Boxes Selection Lists Pop-up Menus

▷ Agreement: Pronoun

A pronoun should agree with its antecedent. (An antecedent is the noun or pronoun it refers to.)

Tips

- Check that each pronoun agrees with the antecedent in
  - ☐ Person
  - ☐ Number
  - ☐ Gender

> ☒ **Examples**
> -Use the top of each page to give the **reader** enough information to decide whether **they** want to stay on that page.
> -No-one likes a badly behaved **plug-in**. But without **them**, the Net is bland.
> -NOW **EVERYONE** CAN BE A WEB DESIGNER!!! But Should **They** Be?
> -Any web **site** can do well in the search engine if **they** are designed and promoted correctly.
> -All web pages are designed with the **client** in mind, allowing **them** to easily update **their** own web content.

▷ Agreement: Subject/Verb

A subject should agree with the verb.

Tips

- Avoid the following types of errors:

SITUATION	SINGULAR OR PLURAL
Compound subjects connected with *and*	Plural
☒ **Examples** -The Internet and the World Wide Web is an electronic media using computers to explore, share information, market, advertise, and sell throughout the world. *(are)* -Using the internet to communicate with clients and using an Intranet to communicate within your organization is becoming more important within the business world. *(are)*	
Collective nouns (groups)	Singular if a group; plural if members act individually

☒ **Examples**

-Web graphics represents an entirely new medium and one that presents many challenges. *(represent)*

-Our design staff of graphic professionals are eager to create the web site you want. *(is)*

-___ is a group of professionals that specialize in providing organizations with a distinctive and successful presence on the Information Super Highway. *(specializes)*

Plural noun comes between subject and verb	Verb agrees with subject

☒ **Examples**

-The content of your WWW pages reside on shared server Internet disk space. *(resides)*

-Whether you're a beginning level programmer or an old pro, this indispensable collection of HTML examples, listings, tips, and explanations make publishing on the Web quick and easy. *(makes)*

Words come between subject and verb	Verb agrees with subject

☒ **Examples**

-Another widely used feature since Netscape v2.0 are frames. *(feature is)*

-Your input with regards to layout, colors, images, logos, music, text, and everything in between, are always invited. *(input is)*

-Our unique designs grab the attention of people who are surfing the net and gets your pages noticed! *(get)*

Relative pronoun *(which, that, who)*	Verb agrees with subject

☒ **Example**

-We can also enhance the aesthetics of your site using many of today's current technologies, which includes Java applets, JavaScript, Active-X, and animated gifs. *(include)*

One, *each* as the subject	Singular

☒ **Examples**

-One of the most important parts of a Web page are navigational aids. *(is)*
-Each of our solutions are unique while conveying your business philosophy and style creatively and professionally. *(is)*

Compound subjects connected with *or/nor, neither/nor*	Verb agrees with closest subject

☒ **Example**

-Information or instructions pertaining to a specific computer platform does not need to be presented except when a user is accessing your website using that specific computer platform. (*do*)

Inverted word order	Verb agrees with subject

☒ **Example**

-Dynamic HTML are what Microsoft and Netscape claims to be a new effective way to make web pages more interactive. *(is, claim). Rearrange sentence.*

▷ Alignment

Alignment is the arrangement of text and graphics. Justification describes the alignment of text on your page. Left-justification is aligned on the left; the text is ragged on the right. Web page text is left aligned by default. However, you can also create centered or right-justified text. Alignment can signal to readers the structure and levels of importance. It shows both content and structure. Left-justified text is easier to read, speeds reading, and draws the eye down the page.

HTML Code: <ALIGN=CENTER> or <ALIGN=RIGHT> control paragraph alignment (centered or aligned right).

Tips

- Use left-justification for text.
- Keep lists of links left-aligned for easier scanning.
- Use consistent alignment.
- Use centered text sparingly.
- Use right-aligned text sparingly. It is difficult to read but can get attention and can be used to fit two-columns on the page efficiently.

> ☒ **Example**
>
> **Tips For Alignment**
> Avoid right-aligned text, especially for large amounts of text.
> Right-aligned text is difficult to read quickly.
> It is hard for the eye to find where to begin.
> It also causes wasted white space on the left margin.

- Indent explanations such as annotations or details.

▷ Alternative (ALT) Text

ALT (alternative) text is text placed in the HTML code. It is used for key graphics on your Web page. If the image does not load, this text is displayed instead. Writing effective ALT text is important for creating accessible Web pages for readers who

- Do not display graphics.
- Use audio-based browser technology.
- Do not want to wait for graphics to load or want to begin clicking links.

ALT text also aids automated indexing programs that use ALT text to determine a page's contents.

HTML Code: Alternative text is used within the <IMG> tag:
<IMG SRC=name of graphic ALT=place alternative text here>

Tips

- Include alternative text (<ALT>) for the following:

- ☐ Animations, including animated GIFs
- ☐ Applets
- ☐ ASCII art
- ☐ Frames
- ☐ Graphical bullets and buttons
- ☐ Graphical representations of text
- ☐ Image map regions
- ☐ Images
- ☐ Scripts
- ☐ Sounds and audio files
- ☐ Symbols
- ☐ Video

- ▶ Write descriptive text.
- ▶ Keep ALT text short, especially inside tables. Many browsers treat ALT text as one long line of text.
- ▶ Use empty ALT tags for purely decorative graphics.

> ☑ **Examples**
>
> ALT="" (empty tag)
> ALT="-------------" (horizontal line)
> ALT="*" (bullet)

- ▶ Use ALT text so readers can click a button rather than wait for the graphic to load. In general, indicate reader action by using verbs.

> ☑ **Examples**
>
> *The following are ALT text comments that appear on real Web pages while the graphics are loading:*
>
> Find out about our network.
> Click here to move to the main index page.
> Links to About ANSI, Standards Info, Conformity Assessment, and Events
> Link to the Past Features section of this website
>
> ☒ **Examples**
>
> *These examples of ALT text are not helpful to readers.*
>
> Please wait . . . Just a moment Click here

- ▶ Place the text in a separate location or beside the appropriate icon.

▷ Analogies

An analogy is a comparison of something familiar to something unfamiliar. Analogies

- ▶ Explain conceptual information or objects.
- ▶ Help readers understand technical information.
- ▶ Elaborate on definitions.

Tips

- ▶ Use analogies for novice audiences.

☑ **Examples**

-You can think of the World Wide Web as a big library on the Internet. Web sites are like the books in the library, and Web pages are like specific pages in the books. A collection of Web pages is known as a Web site.
-The following pages introduce the basic concepts and explain how WWW and the Internet function by using the analogy with the global road system.

- Make sure the analogy is appropriate by comparing items that are similar.

☒ **Example**

Netscape and its kin are clients (which is just a type of computer program). They're like different types of TV sets that show you what's being broadcast. (Not a great analogy, but there you have it).

- Use analogies sparingly.

▷ Animation

Animation includes scrolling and blinking text, animated GIF graphics, flashing banners, and other forms of moving graphics. You can use animation to

- Gain attention
- Entertain
- Show how things work
- Illustrate concepts or techniques

There are several ways to create animation:

TYPE OF ANIMATION	DESCRIPTION	PROS	CONS
Animated GIFs	Series of bitmap graphics shown in sequence.	▶ Simple to use ▶ Can be created in shareware programs ▶ Recognized by most browsers	▶ Large file sizes ▶ Cannot include sound ▶ Difficult to explain complex topics
Dynamic HTML (dHTML)	Uses scripting languages to modify HTML elements and add movement to a Web page.	▶ Recognized by most browsers ▶ Can be created in programs such as Macromedia Dreamweaver	▶ Limited types of animation possible

Java applets	Programs created with Java programming language	▶ Work on all operating systems ▶ Flexible	▶ Require plug-in: a small program that allows a browser to play one type of file
Macromedia Flash and Shockwave animation	Created with Macromedia software for creating animation.	▶ Use streaming so animation can begin playing	▶ Require plug-ins

Tips

- Avoid animation or use it sparingly. It is distracting and creates unnecessary noise and information overload.
- Use animation when explaining a process or concept.

> ☑ **Example**
> *-The* HowStuffWorks *site contains thousands of animations that accompany their explanations about how things work.*

- Keep animations simple.
- Avoid large animation files that will increase download time.
- Provide a text equivalent (ALT text).
- When possible, allow readers to choose whether to view animation files.
- Provide information about file size and type.

> ☑ **Example**
> View Animation 3, "T-rex Running", 11 sec, 1.1 MB; this movie will take about 6 minutes to download on a 56K modem. (Requires QuickTime)

- For instructional animations, consider displaying the animation in a secondary window. Readers can then view the animation separately from the main content.

▷ Annotation

Annotations provide brief summaries of the contents of a link's destination. The destination may be external or a Web page within your own site. By providing information about a site or link destination, you help readers decide whether or not to access the information and thus save them time. Annotation can also be used to warn readers about large file sizes.

Tips

- Annotate both intra-site and external links.
- Use annotations on list pages (pages with lists of links).

- If you have a large number of annotated links, group them under headings. Then provide an introduction to each group of annotated links. This introduction should summarize the types of information the links have in common.

> ☑ **Example**
>
> The **Computers and Information Technology** category includes guides on topics pertaining to all aspects of computers, computer users and computing environments, as well as the technologies that support computer-mediated information storage, manipulation, transfer, exchange or sharing.

- Visit every site you annotate before you suggest readers visit it.

Information to Include

- Include the following types of information:

  - ☐ Site name
  - ☐ Author(s)
  - ☐ Organization or sponsor
  - ☐ Type of information available
  - ☐ Special resources and unique sections available
  - ☐ Number of resources
  - ☐ How site is organized
  - ☐ Search options available
  - ☐ Intended audience
  - ☐ Date of site material
  - ☐ Frequency of updates
  - ☐ Reliability and accuracy of information
  - ☐ Format: E-mail, WWW, FTP program, slideshow, etc.
  - ☐ Links within your annotation to specific pages

> ☑ **Example**
>
> "Tools for Information Architects," including a weblog and other gleanings from the Web (check out the archives, a right-on bibliography, and an interesting collection of articles on information architecture.

  - ☐ Information from the site itself about its purpose and contents

Wording Annotations

- Use parallel construction and consistency (style).

> ☒ **Examples**
>
> *The following annotations do not begin the same way. E.g., each should begin with a verb.*
>
> Contains information on the history and growth of the Web
> A few pointers to resources for people who want to learn about the World Wide Web
> Is a comprehensive meta-site containing links to information about HTML

The following annotations are not consistent with initial capitalization or wording.
Science Learning Network - activities and information for students of science.
Science Matters - This question-and-answer site helps explain scientific theories and facts.

- Avoid simply repeating the page title or being redundant.

☒ **Examples**

-Transparent GIF Tricks: Tricks to make a transparent GIF.
-Introduction to HTML This site is an introduction to HyperText Markup Language (HTML). *(These annotations repeat page titles)*
-Usable Web: Usable Web is a collection of resources and links on user interface issues, and usable design specific to the World Wide Web. *(redundant)*

- Be specific about the type of information on the site.

☒ **Before**

Online Writing Center: Great site for technical writers.

☑ **After**

Online Writing Center: The University of Minnesota Department of Rhetoric virtual writing center for scientific and technical writing. Contains a chat room, skills center with grammar oracle, and other resources.

☒ **Examples**

-In the **Web Resources** section of my page you will find links to all the very best Web page resource pages on the Internet.
-Developer.com**:** Well-organized and up-to-date resources for professional Web developers.

☑ **Examples**

-Helpful, specific:TechWeb: TechTools: Technology tools for users, designers and developers. Stories cover news, reviews, features on browsers, Java, ActiveX, new publishing technologies.
-Specific: This site attempts to list all the HTML tags currently in general use. The HTML tags are listed alphabetically, with examples of each tag's use, acknowledged attributes, and arguments that might modify the tag. One of the nicest features of this site is that they have included charts that show the history of a tag and which browsers support it, including the W3C specs, Netscape, Internet Explorer, Mosaic, Opera, and WebTV.

- Avoid vague generalities.

☒ **Examples**

-There is a lot of information at this site. It's worth the visit.
-On-Line Children's Stories -- Wonderful site

- Try to focus on the site's contents more than the author or purpose of the organization behind it.

> ☒ **Example**
>
> IEEE Professional Communication Society: The primary mission of the IEEE Professional Communication society (PCS) is to help engineers and technical communicators develop skills in written and oral presentation.

- Use a professional, factual tone.
- Use a "you-oriented" style if appropriate.

> ☑ **Example**
>
> Provides you with information and tools related to electronic publishing. If you are interested in learning more about using PDF files in electronic publishing, this site provides you with an excellent PDF Resource Center.

- Provide the same type and amount of information for each site (patterned information).
- When annotating your own site's links, do not apologize for the type of information included (confident style).
- Begin with verbs such as *contains, offers, provides.*

> ☑ **Examples**
>
> -Contains annotated links to bioethics academic institutions, research ethics tutorials, and links to NIH offices and programs associated with bioethics.
> -Employment
> Find your next technical communication job using STC's database of job openings. (Note to employers: this is also the place to post your own job opening information.)

- When possible, use active, task-oriented verbs to describe pages in your site.

> ☑ **Examples**
>
> Use this page to find primary resources, projects, a weekly newsletter, units of study, and a tutorial to help you plan projects and class homepages.

- If the page is unfinished, do not link to it.

> ☒ **Examples**
>
> I don't have a lot at this section right now, but I'm working on it.
> A long page of useful computer information, it hasn't been updated for awhile.

- Be concise. Avoid the following:
    - ☐ Overused adjectives such as *lots of*

- ☐ Beginning with unnecessary words (*this, a, the, Web*) or phrases, such as *this site*
- ☐ Using the phrase *links to*

☑ **Example**

This section of the site provides some introductory practical tips for writing hypertext for the web, and introduces the evolving web-specific debates about hypertext style.

☒ **Examples**

The first two examples begin with wordy phrases.

-Here you can view a long list of six digit color codes and learn how to use them to change your page's backgrounds, text, and link colors.
-These are tutorials to help you once your pages are running.

This excerpt is wordy:

There is a reason why this site has been visited frequently in this past year. The main purpose of this site is to help people build their websites by demystifying, teaching, and demonstrating through examples how the web can be an effective and accessible medium. It promises to explain through tutorials, written in no-geek language, how to effectively build your website as well as pointing visitors to examples of exceptional websites. This is a bookmark site for any website developer!

FORMATTING ANNOTATIONS

- ▶ Use the site name as the link. However, if readers will view your annotated list in print, also include the URL.

☑ **Example**

FoodLines	Experiment and share recipes with others from around the world, learn culinary tips and tricks, read food jokes and stories, and attend food festivals or trade shows. http://www.foodlines.com

- ▶ Make the information easy to scan through use of any of the following layout techniques:
  - ☐ Bulleted statements
  - ☐ Subject headings

☑ **Example**

Compatibility and Accessibility
http://www.pantos.org/atw/access.html
All Things Web index of articles on accessibility.
Technical tips on more accessible pages and designing for accessibility covered.
Destination: All Things Web (11)
Author: Sullivan, Terry (12)
Find more like this

- ☐ Bold site title
- ☐ Indentation. Indent the annotation beneath the site name or beside the title.

☑ **Example**

Copyright Office

Web site of the United States Copyright Office, which operates out of the Library of Congress. For a better understanding of what copyright is and what it protects, read both Copyright Basics and the Copyright Office FAQ (**F**requently **A**sked **Q**uestions).

▷ Announcement of Web Page

An announcement is used to advertise your Web page to publications, newsgroups, mailing lists, etc. An announcement is one way to promote your Web site. Other methods include submitting your site to search engines or advertising in various printed documents.

Tips

- Include the following items:
  - Organization/owner responsible for the site
  - Your e-mail address
  - URL (Web address)
  - Purpose of site
  - One-sentence summary of contents
  - Longer summary of contents (under 75 words)
- Concisely describe what your site is and why it is important. Avoid adjectives and superlatives (best, greatest).

▷ Apostrophes

Apostrophes are used as substitutes for missing letters and to show possession. They are difficult to see online.

Tips

- Try to avoid contractions (don't vs. do not) and possessives.
- Check for the following common errors:
  - Using plural rather than possessive.

☒ **Example**

Formatted to fit *your* companies needs! *(company's)*

- Using possessive rather than plural.

☒ **Examples**

-Custom home page design and internet consulting, for company's and individuals. *(companies)*

-Many times, expectation's are set too high, or clients expect immediate success. *(expectations)*

- Its vs. It's: *Its* is possessive; *it's* stands for *it is*.

 **Example**

It's also very important to check your hyperlinks from time to time.

Example

A non-collaborative Web page is one that is not dependent on others to give data input for it's content.

- Your vs. you're: *Your* is a possessive pronoun; *you're* stands for *you are*.

Examples

Excellent if your stuck for information and don't want to wade through a lot of information to find what you want. *(you're)*

- Whose vs. who's: *Whose* is a possessive pronoun; *who's* stands for *who is*.

Example

Whos Who in the Department of Computer Science. *(Who's Who)*

- Their/they're/there. *Their* is a possessive pronoun; *they're* stands for *they are*.

Examples

-The WebCrawler is operated by America Online, Inc. at there Web Studios in San Francisco. *(their)*
-Even if there written in HTML, documents are NOT always identical on different platforms. *(they're)*
-We have designed this page to help those people who would like to set up there own software. *(their)*

- Contractions.

Examples

-Check out whats New on the net. *(what's)*
-Thats why I have put this tutorial together. *(that's)*

- Possessives.

Examples

-This months HTML Tip is writing the head element of an HTML document. *(month's)*
-It makes the pages download faster due to the browsers cache. *(browser's)*

- Incorrect placement of the apostrophe.

Example

The anchor tag does'nt have to send visitors away from your starting page. *(doesn't)*

▷ Archives

An archive is a collection of documents, records, and other resources, including print or multimedia files. Examples include past issues of a serial document (such as a newsletter, magazine, or journal), press releases, discussion topics, resources on one subject (e.g., *Math Archives*), or a virtual library of resources (e.g., *Internet Archive*). By providing links to archives, you allow readers to obtain resources for a number of applications, including research.

Tips

- Link to the archives from the current content.
- Provide a search engine.
- Provide options of how to view the list, such as by author, title, date, or subject.

> ☑ **Example**
>
> Archives by Author... ▾ Archives by Topic... ▾

- When using dates, arrange items in reverse chronological order. Give titles as well.
- If possible, provide a summary or annotation and specific, descriptive information about each item.

> ☑ **Example**
>
> **PCIN Issue 194** - July 03, 2002
> One billion PCs shipped since the Altair
> **PCIN Issue 193** - June 26, 2002
> Crack Killer combats software piracy
> **PCIN Issue 192** - June 19, 2002
> Lawsuit challenges copy-protected CDs

- Tell readers what format the item is in and software required to view it.
- If necessary, provide instructions on how to download, save, or view the resource.
- Use clickable titles. However, avoid link titles that are several lines long.

> ☒ **Example**
>
> Thursday, July 16, 1998: FastLane lets you load images off the Internet 1-1/2 to 3 times faster-without installing special software.

- Don't use redundant links.

> ☒ **Example**
>
> *This title is confusing because the title and the date link to the same thing.*
>
> New Service Provides Web Health Check And Service Level Metrics
> July 7, 1998

☑ **Example**

2001: Jan Feb Mar Apr May Jun Jul Aug Sep Oct Nov Dec

- Keep lists clean and simple.
- Group items using headings.
- Use horizontal lines to separate items.

▷ Attention

The beginning of each Web page should gain readers' attention. If you do not immediately engage readers—usually within about seven seconds—they may leave your site. Two ways to get attention are the content and page design techniques.

You gain the reader's attention during several stages in the reading process.

STAGE	WHAT IT DOES	HOW TO DO IT
Attract attention	The reader discovers the site.	Title, logo, or graphic; topic reader is interested in; links from other sites
Encourage exploration	The reader decides to read more details about site.	Menu, links, purpose statement
Motivate reading	The reader reads content and follows links.	Worthwhile content, good navigation
Encourage return visits	The reader bookmarks page.	Continual updates of your content

Tips

- Use any of the following techniques to get attention:
  - ☐ Free offer or free trial

☑ **Examples**

Advertise your organization for FREE on INTERNET.ORG!

Free Trials - Download recently updated *free* trial versions of Meridian's award winning CD Net® Software

  - ☐ Catch words (e.g., *free*) and active verbs (e.g., *learn, improve*)

☑ **Example**

learn
Spanish:
a FREE online tutorial

  - ☐ Anything that encourages interactivity
  - ☐ *What's New* information
  - ☐ A *short* statement about what you can offer readers

☒ **Example**

Professional Web page design, Internet consulting, custom graphic design, desktop publishing, scanning, and complete multimedia services, ____ Productions is your full service World Wide Web and publishing solution.

- ☐ A *positive* statement about what you can offer readers

☒ **Examples**

The following are the ***opening*** *sentences on Web page design sites.*

-Productions has no advertising budget and no marketing staff. Our only advertising is this web site and the word of mouth from our growing number of customers.
-These are not just templates for you to pick from. Whatever you can imagine, we will work with you to create your site.
-Why should you advertise on the World Wide Web? Obviously, something brought you to this Web Site...

- ☐ A *brief*, credible quotation about your site from a believable source

☑ **Example**

The #1 Most Incredibly Useful Site on the Web.
--Yahoo! Internet Life

- ☐ Facts and statistics

☑ **Examples**

-The 1871 Great Chicago Fire. The 1906 San Francisco Earthquake. The 1927 Great Mississippi Flood. Natural disasters irrevocably change the course of history. Read how one flood changed America in *Rising Tide*.

-Approximately 148 million people worldwide have access to the World Wide Web. No matter what your business, you can't ignore 148 million people. To be a part of that community and show that you are interested in serving them, you need to be on the Web for them. You know your competitors will.

☒ **Example**

This statistic would be more effective if the source were given.

Did you know that by the year 2000, 1 billion people will be on the Internet?

- ☐ An interesting statement

☑ **Example**

This is the opening of the Neuroscience for Kids site:

The smell of a flower - The memory of a walk in the park - The pain of stepping on a nail. These experiences are made possible by the 3 pounds of tissue in our heads...the **BRAIN!!**

- ☐ Short, simple, informative headlines
- ☐ You-oriented wording

- ☐ Humor (if appropriate)
- ☐ A question

> ☑ **Examples**
>
> -While surfing the web, marveling at all the wonderful sites created by others, have you ever wondered what it would be like to have your very own homepage?
> -Do you have a computer word you want to look up? Search our PC Webopedia with over 4000 terms at your fingertips. New words added daily!
> -Been looking for a law office software application that would make practicing law much easier? Trying to convince the senior partner to computerize the firm with easy-to-use Macintoshes? Then this site is for you!

Tips

- ▶ Place the attention-getting elements in the page's focal point. Avoid making readers scroll to view them.
- ▶ Accompany attention-getting text with the visual techniques of emphasis, such as bold, boxes, color, emphasis, graphics, headings/headlines, large fonts, large initial caps, lines, pull quotes, and white space.
- ▶ Avoid distracting techniques such as animation, blinking, marquees, and background music.

▷ Audience Analysis

One of the most important steps in writing a Web page is identifying your readers and the information they want. Readers include the *target audience* (the primary users) and *secondary audience* (such as Web surfers). Identifying your target audience is a key step in designing a friendly site. The target audience affects your entire site: its purpose, navigation, and design. According to usability studies, most Web readers want information, are searching for something specific, and are in a hurry.

Tips

- ▶ Select a target audience.
- ▶ Determine how this audience will use your site for its intended purpose.
- ▶ Design your site for the audience.

> ☑ **Examples**
>
> *A site for fans of a rock group would use a different design and language than one written for senior citizens, who would need simple navigation, clarity, and simplicity.*
>
> **About NIHSeniorHealth**
> This web site grew out of NIA's research on older adults, cognitive aging, and computer usage. The research showed that, while older adults do experience gradual declines in cognitive abilities as a part of the normal aging process, they can successfully use computers if the online information is provided in an age-appropriate manner.

> Some of the web site's senior-friendly features include large print and short, easy-to-read segments of information. Older users will find it easy to move from one place to another on the web site without feeling "lost" or overwhelmed. Also, the material on the site is presented in a way that increases the likelihood it will be retained in memory.

- Use feedback forms, surveys, interviews, focus groups, bulletin boards, and feedback links to get input and continued suggestions from readers.
- Ask yourself the following questions:

General Information

- Who is my primary audience? My secondary audience?
- Will there be international readers?
- What is their age and gender?
- What are their professions/job titles?
- What is their budget/spending?
- What are their hobbies/pastimes?
- What is their educational level?
- What is their reading level?
- What type of learning style do they prefer?
- What is their environment/workplace?
- In what situations will the Web site be used?
- What are their personal limitations (e.g. vision impairments, learning disabilities)?

Content Background

- Are they novices or experts?
- Do they have prior knowledge & experience with the subject?
- What are their attitudes toward the subject?
- Are they familiar with terms?
- What are their priorities for the information? Which topics are most important to them?

Computer Background

- How much expertise do they have with computers?
- Are they accessing Internet from home/work?
- What are the specifications of the following items they will use?
    - ☐ Platform
    - ☐ Operating system
    - ☐ Browser
    - ☐ Bandwidth
    - ☐ Monitor size & screen resolution

Internet Background

- How much experience do they have using the Internet?

- How familiar are they with using a Web page: hyperlinks, searching, browsing, etc.?

GOALS

- What are their goals and expectations about my Web pages?
- What do they want from my Web pages?
    - Benefits
    - Entertainment
    - Information
    - Solutions to problems
    - Task performance
    - Technical support
- What task are they trying to perform? What problems are they trying to solve?
- How frequently will they use the Web site?

▷ Audience: Writing for Experts & Novices

Although you should write for a target audience, you may still have readers who range from experts to novices. Experts include experienced Internet users and subject-matter experts. New users are unfamiliar with Web terminology, navigating, and configuring their browsers. They may also unfamiliar with the subject. Some Web sites call these readers "newbies" and direct their sites specifically to new Internet users.

Tips

GENERAL

- Consider the following options:
    - If many experts and frequent users will be interested in your Web pages, try to accommodate their needs and interests. Provide shortcuts to information they need.
    - If many new users will be reading your Web pages, consider adapting your site for novices. Provide links to site help, a tour, and FAQs.
    - If both audiences will use your site, use techniques that accommodate both levels.
- Inform readers of the target audience:
    - On the home page.
    - In the title and/or subtitle.

☑ **Examples**

-The Developer's Desk: A Round-Table Place for Web Masters and Fellow Programmers.
-A Beginner's Guide to HTML
-XML.com: A Technical Introduction to XML

- ☐ On a separate page describing the site's purpose.

▸ Consider the following options:

- ☐ Creating separate Web pages for different audiences.

> ☑ **Examples**
>
> For Writers Only
>
> We have prepared a special page to help first-time visitors get the most out of HowStuffWorks. Would you like to view this page?
>
> SELECT YOUR VIEW
>
> - Authors
> - Editors
> - Health practitioners
> - Librarians
> - Careers
> - Advertisers and sponsors
> - Societies
> - Newsroom

- ☐ Putting links to the site guide, site map, and other help information on the home page to accommodate different reader levels.

> ☑ **Examples**
>
> Click here for non-technical, customer service information.
>
> **Just Curious ...**
> I have no specific agenda in viewing this book. Show me something interesting.
> **Media Maven ...**
> I enjoy finding out about the latest innovations in media.
> **Business Person ...**
> My concerns relate to business and industry.
> **Educator ...**
> I'm mainly interested in issues concerning education and learning.
>
> **Check Our Audience Tracks**
> Congress
> News Agencies
> Other Government
> Scientists
> Teachers and Students
>
> **Information For**
> Home Users
> IT Professionals (TechNet)
> Developers (MSDN)
> Microsoft Partners
> Business Professionals
> Educational Institutions
> Journalists
>
> Welcome New Users | buying tips | selling tips | register now

- ☐ Putting information for all audiences on the same Web page, but visually distinguishing the type of information (e.g., putting expert and detailed information in sidebars).

> ☑ **Example**
>
> See the sidebar below for more information on how to measure picosecond time intervals.

- ☐ Layering information (e.g., providing links to explanations and detail).
- ☐ Using color coding to show audience levels.

☑ **Example**

The professionals' part of the site has BLUE BUTTONS on a GREEN BACKGROUND. The learners' part of the site has PURPLE BUTTONS on a YELLOW BACKGROUND.

- Let readers know which sections are appropriate for various audiences.

☑ **Examples**

Distance Learning for Adult Learners: In this section, you will find information on distance learning that is geared towards students in Adult Literacy and Higher Education.

Distance Learning for K-12 Students: In this section, you will find information on distance learning that is geared towards students at the K-12 level.

Beginners – start your Internet marketing journey with our cross-referenced definitions and basic information.
Experts – start your Internet marketing research with our organized collection of only the best sites and articles.

Introduction to Computers
Ever wanted to learn how to use a computer? This course is designed with the absolute beginner in mind.
[details]
Psychologists Students Public

Beginners Builders Masters

A Web page design and authoring page lists target audience:
Targeted Audience:
Marketing and Advertising executives in general
Product Managers
Art Directors and Copywriters
Designers and Web page creators
Professors teaching Marketing

An Internet library site allows readers to
Enter a door for Teachers, Kids, Teens, Parents, Librarians or College Students.

Writing for Experts

- Omit explanations of how to click or use the site.
- Provide links to detailed information in your table of contents.
- Provide shortcuts.
- Direct experts to advanced topics.

☑ **Example**

Had a lot of experience searching the Web? Try the Guide to Power Searching on the Web.

Programmers looking for information, check out the *Software Toolbox* for existing software and code source for MS-DOS, Amiga, Windows, Macintosh, UNIX, Acorn, and others; *All About GIF89a* for a breakdown of the technical structure and links to Compuserve's original specification. Check the *User Guide* for information of what users will see with different browsers.

- Warn readers that the information is for experts.

> ☑ **Example**
> This is a technical article about some new Internet protocols.

- Let readers select appropriate topic areas.

> ☑ **Example**
> **TRACKS**
> Web Authors
> Designers
> Developers
> Strategists

- Provide more interactivity.
- Provide access to detailed graphics, multimedia, downloadable files, etc.

WRITING FOR NOVICES

- Use textual rather than graphical navigational aids.
- Clearly identify any graphics used as buttons/links.
- Consider a sequential/linear navigational structure. Keep navigation simple.
- Use repetition and consistency.
- Avoid a condescending or insulting tone.

> ☒ **Example**
> ... Instructions for Net Newbies and Download Dummies. ... These terminally simple instructions assume you know nada, nothing, njente about computers. ...

- Use an informal style if appropriate.
- Use definitions or provide a glossary.
- Use examples, analogies, and devices to help readers visualize abstract concepts.

> ☑ **Example**
> How Big Is 100 Terabytes?
> Here's how the size of the Archive's collections today — containing material dating from 1996 to the present — compares to some familiar data banks:
> - A copy of your favorite mystery novel 1 megabyte
> - One copy of the Encyclopedia Britannica (2,619 pages per copy) 1 gigabyte
> - A thousand copies of the Encyclopedia Britannica 1 terabyte

- Link to background, FAQs, and conceptual information.
- Create a table of contents that contains sections new users will be interested in.

> ☑ **Example**
>
> Table of Contents
> Welcome to New Users

What is Code Surfing?
Hypertext Conventions Used
Typographical Conventions Used
Terminology Used
Current Limitations
Support Questions? Call Us...

- Provide instructions for using the site (site guide).

☑ **Example**

The HelpWeb: A Guide to Getting Started on the Internet
Just click on the buttons on the left to view a topic index that describes the subjects covered in that section of the HelpWeb. We offer two other options for finding the information you seek. You may scroll through the Site Guide or search our site by Keyword.

*Readers can click a **Panic** button that takes them to a page that provides reassurance and the following options:*

Are You Lost?
If you've lost your way within our site, try the Site Guide for an overview of the topics covered on the Help Web.
If you're looking for help on a particular topic, try searching our Help Web by Keyword.

- Let readers decide what level they need.

☑ **Example**

So in order to get you started, please tell us a little about yourself by clicking on the link below that best describes you:
- I'm new to the Internet.
- I'm pretty familiar with using the Internet, but I've never done web publishing before.
- I've already designed my web pages on my personal computer. What do I need to do to get these online with my WebCom account?
- I'm an experienced web publisher, just give me the details of WebCom's services and capabilities.

- Recommend sections users should read or a specific order for topics.

☑ **Examples**

First Steps-New Users Start Here ▶

We recommend that you begin in the Services section.
It may be beneficial to start with the Basic Definitions page to learn very basic terminology and information on how to obtain web browsers.

Overwhelmed and Overloaded?
If you're new to the Internet, you may be overwhelmed by the amount of technical material about Web marketing you'll find on this site. Why don't you get up to speed with a basic article or two about business marketing on the Web?

Questions Small Businesses Ask About the Internet (Very elementary)
What a Web Site Can Do for Your Business

- Provide shortcuts to appropriate information.

☑ **Examples**

If you only have access to a Macintosh running Mac OS, then you may want to skip the rest of this page and just read about Transparency for the Macintosh.

- Give basic explanations.

☑ **Example**

This excerpt is written for readers new to the Internet:

Your visit to the Web begins, simply enough, with a page—a Web page. Everything on the Web is on a page (including the screen you're looking at right now), but the term page can be misleading. We're used to thinking of pages as being fixed sizes, like book or magazine pages, but Web pages can have any length or width. In fact, since Web pages are really just computer files, they can be infinitely large. The reason they're called pages is because that is what they look like when they are displayed on your computer screen.

▷ Background Color

The background of your Web page can be a color (or you can use a background image). An effective background can help provide contrast with your text for optimal legibility and tie in with your color scheme. Repeating the background color also unifies your site.

HTML Code: The tag for a background color is <BODY BGCOLOR= #rrggbb> (a six-digit hexadecimal color code for the color).

Tips

- Use background colors that help readers find information.
- Select a background that has high contrast with the default text color or colored text, preferably dark text on a light background.
- Avoid bright background colors because they make reading more tiring.
- Avoid a dark or black background. The default blue links do not contrast well, and these colors also make reading more tiring.
- Always specify a background color. Otherwise, the browser-specified background color will be used.

▷ Background Image

You can specify a graphic (photo, texture, design, etc.) as the background of your Web pages (or you can use a background color). You create a small image (usually GIF or JPEG format). Then due to "tiling," a browser repeats this image across and down to fill the browser window. An effective background can tie in with the theme and tone of your Web site. It also gives your site a unity and consistency (layout).

HTML Code: The tag for a background graphic is <BODY BACKGROUND= [name of graphic] </BODY>.

Tips

- Select a background that has high contrast with the default text color or colored text, preferably dark text on a light background.
- Use the background to highlight elements on your page; it should not attract attention.
- Avoid a busy patterned background that conflicts with text legibility. Instead, use simple graphics.
- Use seamless images that look like one background rather than a repeated, titled image.
- When using a repeating logo or watermark, make the graphic light to be less distracting and not interfere with the text.
- Select a background that fits with the design theme or metaphor.
- Use textures when they fit the theme, but avoid textures that make the text difficult to read.
- Avoid using photos because they load slowly and conflict with text.
- To avoid repeating vertical stripes and other distracting patterns, specify a width that will display correctly on monitors with 800 x 600 and 1024 by 768 resolution.
- Keep background images less than 4 KB.
- Always specify a background color. If your background image does not load or loads slowly, your text will still be legible if you specify a color similar to the background image.

▷ Balance

Web page balance is the arrangement of elements on the page. There are several types of balance:

TYPE OF BALANCE	DEFINITION	EFFECTS
Symmetrical	Elements centered horizontally and vertically.	Appears formal, organized.
Asymmetrical	Elements arranged off center.	Can draw eye and emphasize important information.
Radial	Text and graphics spiral out from a central point.	Draws attention to center of page.

Tips

- Use balance to show the grouping/proximity of information.
- Avoid an unbalanced or lopsided appearance.

- Because a computer screen is small, try to use all available page width. Wasting space may keep the important information less prominent and require scrolling to view it.
- Use white space effectively.
- Use a grid created from tables, columns, or frames to help balance text and graphics and to provide "zones" of information.

▷ Banner

A banner appears at the top of a Web site. It usually contains text (site title) and a graphic, often a logo. A banner does the following:

- Identifies the site.
- Identifies the contents.
- Gives visual appeal.
- Catches readers' attention and interests them in your site.
- Establishes an identity and "branding."
- Sets the tone.
- Helps readers identify their location (contextual clues).

Tips

- Use a banner related to your site or product; otherwise, you are luring readers with false advertising.
- Keep the file size of the banner small (under 4KB) so the page loads quickly.
- Repeat the banner on all pages in your Web site. Use part of it or a smaller version at the top and/or bottom (in the footer).
- Avoid using banner ads. If you must use them, place them at the bottom.

▷ Bibliographic Record

A bibliographic record is a formal description of a resource, such as a book, journal article, or other media. This description is longer and more detailed than a bibliography entry. Bibliographic records are found in sites such as online catalogs and reference sites.

Tips

Creating Bibliographic Records

- For each online resource, provide a link, such as Bibliographic Record.

> ☑ **Example**
>
> **Gray's Anatomy of the Human Body**
> **Contents**
> Bibliographic Record
> Preface - 20th Edition
> Illustration Index
> Subject Index

- Include the following types of information:

- ☐ Title
- ☐ Author
- ☐ Editor
- ☐ Publication date
- ☐ Physical details (e.g., number of pages)
- ☐ ISBN
- ☐ Subject headings
- ☐ Abstract
- ☐ Citation
- ☐ Online edition information (e.g., legal information and terms of use)

▸ Provide links to cross references.

> ☑ **Example**
>
> ONLINE ED.: © Copyright 2002 Columbia University Press. Published September 2002 by Bartleby.com. (Terms of Use).

▸ Use a tabular format (fields) so information is easy to skim and search.

USING BIBLIOGRAPHIC RECORDS

▸ Use the bibliographic record to obtain correct citation information.

> ☑ **Example**
>
> "[Entry Title]." *The American Heritage® Dictionary of the English Language,* 4th ed. Boston: Houghton Mifflin, 2000. www.bartleby.com/61/. [Date of Printout].

▷ Biographical Information

A biography page describes people in an organization (officers, faculty, alumni, etc.), Web page creators, team members, etc. This information adds credibility to your content. It is especially important in supporting descriptions of services you provide.

Tips

CONTENTS

▸ Include the following information:

- ☐ Name
- ☐ Position
- ☐ Awards and honors
- ☐ Degrees
- ☐ Work history
- ☐ Memberships
- ☐ Accomplishments

▸ Provide links to support information:

- ☐ Picture
- ☐ Resume
- ☐ Electronic portfolio
- ☐ Work samples
- ☐ Multimedia files
- ☐ Personal page

▸ Provide a link to the directory from the home page (e.g., *About Us*, *Who We Are*).

Format

- Consider a table format providing a list of people, titles, phone numbers, and links to their biographies, home pages, and e-mail.
- For long lists, use alphabetical lookup buttons and letter dividers that mark the beginning of entries for each letter: [A | B | C | D | E | F | G | H | I | J | K | L | M | N | O | P | Q | R | S | T | U | V | W | X | Y | Z]

▷ Blinking Text

Text enclosed between blink tags flashes. Blinking text is used draw attention. However, Web designers almost universally agree that it is an annoying and distracting technique.

HTML Code: <BLINK> </BLINK>

Tips

- Avoid blinking text. It can be distracting and creates unnecessary noise and information overload. Also, users using speech display systems will not be able to recognize it.

▷ Blog

A blog (Web log) is an online journal that provides commentary on a particular subject. It can offer observations, news, commentary, personal thoughts and opinions, and recommended links. Most blogs are text-only; others include graphics, multimedia, and links. Blogs use most of the writing principles discussed in this handbook.

Tips

Writing a Blog

- Pick a focused topic that you are knowledgeable about.
- Write for a specific target audience.
- Use a clear descriptive title.
- State what the focus of your blog is.

> ☑ **Example**
>
> This blog intends to provide interesting and useful info about Technical Communication, FrameMaker, RoboHelp and related issues. Contributors include Vivek Jain, Group Product Manager, Technical Communication products and Aseem Dokania, Product Manager, Adobe FrameMaker.

- Include valuable content.
- Write descriptive, concise, interesting headlines.
- Be concise.
- Provide the writer's name and date.
- Use keywords throughout to help search engines.
- Keep posts short.
- Get to the point quickly.

- Present only factual information. Make it clear when you are offering your own opinion, and support it with details.
- Give credit to sources.
- Entertain readers by providing details, examples, and humor.
- Use a conversational and interesting style.
- Use a consistent style.
- Avoid mentioning personal information to protect privacy.
- Edit for errors.
- Link to related Web sites and other blogs.
- Consider offering appropriate RSS feeds.
- Make it easy for readers to submit comments.
- Provide contact information.
- Use a regular and frequent update schedule.

FORMATTING AND ORGANIZING A BLOG

- Use headings and lists to aid skimming.
- Put entries in reverse chronological order.
- Provide a search engine.
- Distinguish recent posts from archives.
- Consider grouping posts by topics.

☑ **Example**

Blog Topics
Blogging (36)
Podcasting (42)
RoboHelp (6)
STC (8)
Technical Writing (17)
Web 2.0 (29)
Web tools (29)
Wikis (4)
Word (4)
WordPress (14)

▷ Blurb

A blurb is a short summary of a Web document. It is similar to an annotation but is catchier. A blurb accompanies a headline and may appear in a table of contents, "related topics," list of links, or e-newsletters. The goal is to attract ("tease") readers to a page in your site. The description helps readers decide if they want to visit the page, know what to expect, and can save their time.

Tips

- Use a catchy but informative title.

☑ **Example**

Got Cheaters? Ask New Questions
The Web puts answers to most questions—not to mention ready-made term papers—at students' fingertips. One educator says it's time to assign work that truly makes kids think.

- Keep the title close to the blurb.
- In the title and first sentence, let readers know the topic and why they should care about it.

> ☑ **Example**
> **Use Usability to Best Advantage**
> › › › ANALYZING CUSTOMER DATA
> Think usability tests shouldn't concern you? You might be surprised.

- Emphasize reader benefits.

> ☑ **Example**
> **Ten Tips on Writing the Living Web**
> Your information architecture is as smooth, clear, and inviting as a lake. Your design rocks. Your code works. But what keeps readers coming back is compelling writing that's continually fresh and new. Updating daily content can challenge the most dedicated scribe or site owner. Ten tips will help you keep the good words (and readers) coming.

- Summarize the key content and key positions expressed in the article.
- Include the name of the article and author.
- Keep the blurb concise and short—about one or two sentences.
- Use a tone and style that draws in readers.

> ☑ **Example**
> *Verbs such as* Market Yourself, Manage Your Business, Work Your Career *are used to draw readers into these career-oriented articles.*
> Get Oriented
> Whether you're cutting yourself free from your desk job or trying to get started on your own, these articles will point you in the right direction.

- Avoid marketing language.
- Consider using links to parts of the article.

> ☑ **Example**
> Buying Guide: Servers
> Are you ready to buy your first server but have no idea where to start? Don't sweat it. We tell you everything you need to know, including:
> Top 10 Buying Tips • Choices • Budget Considerations • Important Features • Key Terms

▷ Body

The body section of an HTML page appears after the <HEAD> section. The body contains the main content of your Web page. Other HTML tags, such as headings, paragraphs, formatting attributes, and links, occur within the body.

HTML Code: <BODY> </BODY>

Tips

- Place the content of your Web page in the body section.
- Nest all formatting tags within the <BODY> tags.

▷ Bold

Bold text is a heavier and darker type style. It is used for emphasis and to aid skimming.

HTML Code: <B> </B> is the *physical* style. <STRONG> </STRONG> is a *logical* style that also usually results in bold. Logical styles are used when browsers don't display bold.

Tips

- Use bold to emphasize and highlight.
- Use bold to help readers skim your most important points.
- Bold the most important keywords, names, and headlines. Make sure they are meaningful out of context.

> ☑ **Example**
>
> We offer hands-on training classes on **FrameMaker**, **WebWorks Publisher**, and **RoboHelp**.

- Use bold sparingly.

> ☒ **Example**
>
> Provide **valuable, timely information** to the user, not lots of data. Web sites should be **updated regularly**. Stale web sites say been there, done that. For the information to be valuable it should be **well-edited**.

- Use bold for words or phrases. Avoid bolding large blocks of text or paragraphs.

> ☑ **Example**
>
> To see Web pages you use a special program called a **browser**. You are probably using one of the two most popular Web browsers: **Netscape Navigator** or **Microsoft Internet Explorer**.

- Use bolded terms at the beginning of list items rather than scattered in the sentence. Bold is more effective and easily scanned when arranged vertically but looks confusing when scattered.

> ☑ **Example**
>
> - **career growth** opportunities
> - **free education** for you and your immediate family
> - **child care** on site

- Use bold for topic sentences or lead-in sentences.

☑ **Examples**

Using PDF technology, we can help you reduce the amount of paper in your office by digitizing paper documents and forms and archiving them on CD-ROM.
Put simply: we'll take your paper documents and forms, digitize them, and put them on a CD-ROM for permanent storage . . .

How do I make a link that sends e-mail?

An e-mail tag is not very difficult to design. The tag launches the visitor's own e-mail program and fills in the address you specify.

▷ Book: Online

Full-text versions of electronic books and textbooks are available on the Web. There are many initiatives and projects with the goal of making texts (particularly classics and reference material) available online. In addition, many publishers place books or excerpts online as a means of advertising. Online books are timely, easily updated, and allow readers to browse and search. Many use a traditional linear format, while others take advantage of hyperlinks, search engines, and multimedia.

Tips

- Provide various ways to access the books: browse; search engine; by author, subject, title; index; new items.
- Do not attempt to put a copyrighted book online.
- Make the book available in different file sizes and file formats (e.g., divided into parts, one file for downloading, Adobe Acrobat .pdf version, zipped file, etc.).
- Provide the complete publication information.
- Chunk information and take advantage of links. Do not simply dump the book online.
- Provide a table of contents for different divisions of the book.

☑ **Example**

This Web book contains chapter and section links. The titles are used as links.

Chapter Links:

Chapter 1: **Overview: The Mental Landscape** . . .

Sections in Chapter 1: The Fetch-and-Execute Cycle: Machine Language . . .

Virginia Shea's online Netiquette book uses links to page numbers.

Part I Introduction to Netiquette

Chapter 1 When in Cyberspace... 19

Chapter 2 The Many Domains of Cyberspace 25

The Internet 25 . . .

- Provide flexible navigation to various sections of the book.

☑ **Examples**

Skip to chapter [1][2][3][4][5][6][7][8][9]

[First Section | Next Chapter | Previous Chapter | Main Index]

- Provide links to the *next* and *previous* chapter or section.

☑ **Examples**

The end of a chapter introduction ends with an overview of topics covered and links to sections:

This chapter examines the facilities for programming in the small in the Java programming language.
Sections in Chapter 2:
The Basic Java™ Application
Variables and the Primitive Types Loops and Branches

- Remind readers of their location.

☑ **Example**

Online books at Bartleby are divided into separate pages. As shown in this excerpt from Elements of Style*, each page reminds readers of their location and the book's name, and allows them to navigate to the previous and next pages, the table of contents, and the bibliographical information. This navigation is available at both the top and bottom of the page.*

William Strunk, Jr. > Usage > The Elements of Style > III. Elementary Principles of Composition

< PREVIOUS NEXT >
CONTENTS · BIBLIOGRAPHIC RECORD
William Strunk, Jr. (1869–1946). The Elements of Style. 1918.

III. ELEMENTARY PRINCIPLES OF COMPOSITION
Make the paragraph the unit of composition: one paragraph to each topic.

- Provide links to graphics, related Web sites, and other chapters in the book.

☑ **Examples**

Macmillan's ***Special Edition: Using CGI*** *uses a variety of links within each chapter, as well as a table of contents with links to each section.*

Chapter 4, Understanding Basic CGI Elements, deals with each variable in some depth. This section is taken from the NCSA specifications and is the closest thing to standard, as you'll find. In case you've misplaced the URL for the NCSA CGI specification, here it is again: http://www.w3.org/hypertext/WWW/CGI/
Figure 2.8 : Here, Java™ is being used to create an actual interactive spreadsheet.

Sandy Ressler's The Art of Electronic Publishing (Prentice-Hall) contains links to footnotes, cross-references, and external sites. Thumbnails, as shown here, link to larger versions of the figures.

The large capacity of CD-ROMs is an ideal complement to the large space requirements of full text retrieval systems.

Text retrieval is a complex field that is growing in importance as the world gets interconnected ever more tightly with networks.(6) Internet Starting Points used with Web browsers all have one form or another of a text retrieval engine. The possibility of indexing the Web challenges the computer science of text retrieval.
The increased capacity of low cost storage devices like CD-ROMs is also a major factor in text retrieval, because entire databases can be put on-line right at your very own PC. (For more information on text retrieval, see *Section 8 . 5 . 2 Text Retrieval* in *Chapter 8 Document Management*.)

▷ Boxes

Boxes are used to frame selected text to draw attention to the information. For example, they can be used for sidebars, pull quotes, URLs of sites mentioned, or to summarize key points.

HTML Code: You can create boxed areas with tables (e.g., a one-cell table with a fill and border) or graphics.

Tips

- Don't overuse boxes.
- Use boxes to highlight important information, such as navigational menus, headings, summaries, sidebar information, pull quotes, learning objectives, overviews tips, and interactivity.

☑ **Examples**

NEW

2nd Edition
The second edition of Web Style Guide should be available now bookstores in the United States.

HIGHLIGHTS
If content is short, mostly text, and timely, put it all in the e-mail message.

If content is long, includes images and embedded links, and has long shelf life, post it on the Web and provide links to it in the e-newsletter.

Tips on writing e-newsletter blurbs (article summaries)

About the author

see Box, The Esophagus.

- Use a shaded or colored box for more emphasis. However, make sure there is high contrast with the text so it is legible.

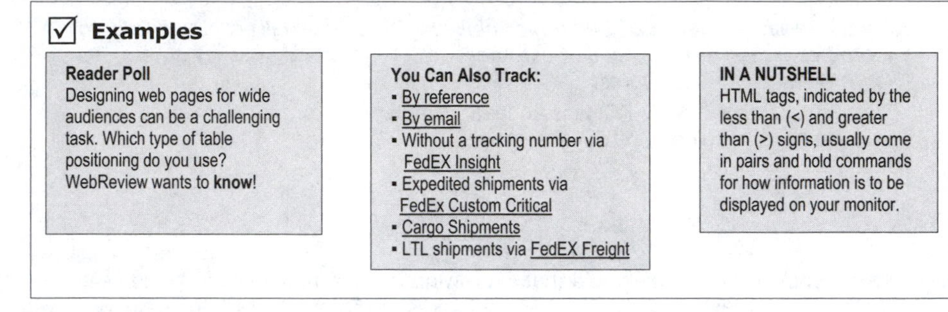

Examples
Reader Poll
Designing web pages for wide audiences can be a challenging task. Which type of table positioning do you use? WebReview wants to know!
You Can Also Track:
- By reference
- By email
- Without a tracking number via FedEX Insight
- Expedited shipments via FedEx Custom Critical
- Cargo Shipments
- LTL shipments via FedEX Freight
IN A NUTSHELL
HTML tags, indicated by the less than (<) and greater than (>) signs, usually come in pairs and hold commands for how information is to be displayed on your monitor.

▷ Brainstorming

Brainstorming is a method of writing lists of words and phrases to generate ideas. One use for brainstorming is to list all possible topics for your Web page or site. You can also use brainstorming for more focused uses, such as generating ideas for a site name, headlines, and changes to your page design.

Two techniques of planning are top-down and bottom-up design. *Top-down* design involves thinking of general ideas and chunking them into topics and more detail. *Bottom-up* design involves beginning with details and organizing them into general categories.

Tips

- Use any of the following techniques:
    - ☐ Enter ideas into a word processor or idea generator.
    - ☐ Use note cards.
    - ☐ Use storyboarding.
- Write as many ideas as you (and colleagues) can think of.
- Remember your site goals and target audience.
- Ask general questions, focused questions, or questions that will lead to descriptive adjectives.

> ☑ **Examples**
>
> "What success stories can we tell?"
> "What promotions can we offer?"
> "How can we increase visits by our target audience?"
> "What image do we want the home page to convey?"

- Write a word that describes your goal or problem. Then write related words. If necessary, select one of these words and generate more words.

> ☑ **Example**
>
> *Problem:* Decide on name/theme for a site on technical writing—general overview of the field. **Technical Writing** will be in the title.

> *Words:* overview, 101, quick reference, introduction, basics, guide, for novices, at a glance, a look at
> *Choose one word or phrase:* at a glance
> *Words:* see, peek, eyes, glasses, magnifying glass
> *Final:* **Technical Writing At a Glance**

- Visit similar sites to get ideas of what to do or not to do.

▷ Branding

Branding is a technique used in marketing to create a distinct image and affect the user's experience. On a Web site, several techniques affect how readers feel about your site:

- Using formatting techniques, such as fonts, colors, icons, graphics, animation, frames, and special effects.
- Using a logo and a unique domain name.
- Creating a positive experience through usability, interactivity, and giving readers what they are looking for.
- Establishing a relationship with readers through tone and communication tools.

Tips

- Use logos and other branding graphics at the top of the page to keep the file size small.
- Use graphics and colors that reflect your company or organization.
- Make your message primary over the layout. Brand through content that attracts readers.
- Make the site easy to use and information easy to find.
- Make help, support, and contact information obvious.
- Encourage reader feedback.
- Attract attention by emphasizing reader benefits and making special offers.
- Establish credibility to gain readers' trust.
- Use a friendly, helpful tone.

> ☑ **Example**
> *Kellogg's uses many of these techniques: their logo and cartoon characters, recipes, fun facts, games, special offers, and nutritional and recall information.*

▷ Browsing Options

Browsing is exploring what is available on a Web page, Web site, or Web resource. It is different from searching for specific information. Browsing allows readers to follow links that interest them. Browsing options organize information, enabling readers to focus on appropriate categories.

Tips

- Using effective labeling and organization.
- Provide a variety of browsing options, such as by the following:

- ☐ Author
- ☐ Category
- ☐ Date
- ☐ Geographical location
- ☐ Title
- ☐ Topic

☑ **Examples**

You can view our latest ad on line three different ways. View by page, merchandise category or brand. No matter which way you choose to browse, you're sure to find exactly what you want with the savings you've come to expect from Kmart.

The informIT site lets you browse a few topics in a left frame. A link to More Topics *displays links to other topics. The number of items is indicated in parentheses.*

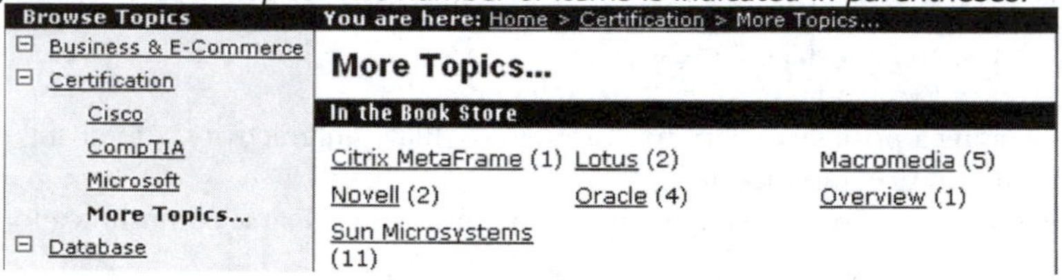

Browse Topics
Business & E-Commerce
Certification
Cisco
CompTIA
Microsoft
More Topics...
Database

You are here: Home > Certification > More Topics...
More Topics...
In the Book Store
Citrix MetaFrame (1) Lotus (2) Macromedia (5)
Novell (2) Oracle (4) Overview (1)
Sun Microsystems (11)

Online Sports provides a variety of browsing and searching options. You can browse by Sport, Item, Team, Suppliers, Players, or Department.

▷ Bulleted (Unordered) Lists

Bulleted lists are called unordered lists. In general, the bullet is a large solid dot, although this may vary depending on the browser. You can also use graphical bullets. Bullet lists can be used to

- ▶ Summarize.
- ▶ Emphasize.
- ▶ Highlight links.
- ▶ Make text easier to skim.
- ▶ Slow readers down.

HTML Code: <UL> </UL> tags mark the beginning/end of the list. Each list item is preceded by <LI>.
You can create round ●, square ■, or hollow bullets ❍ using the TYPE= command:
<LI TYPE=DISC> <LI TYPE=SQUARE> <LI TYPE=CIRCLE>

Tips

Use

- ▶ Use bulleted lists to format a list in which the *order* is not important.
- ▶ Use numbered (ordered) lists when the sequence is important.
- ▶ Use bullet lists for emphasis.

☒ **Example**

Negotiate reciprocal links with other sites, submit your site to web directories, post in appropriate Usenet groups, inform the trade and consumer press.

- Avoid using too many bullet lists.
- Avoid using bullet lists for short words.

☑ **Example**

- facts
- rules
- behavior
- classification

FORMAT

- When using bullet items as links, make both the text and the bullet clickable.
- Insert one or two spaces between the bullet and text.
- Avoid lists of more than seven, plus or minus two.
- If lists become long, break them up with headings.
- Avoid lists with items arranged randomly. Instead, find an appropriate organization.
- Use *nested* lists (indented lists within lists) to show hierarchy or sub-points. However, avoid more than two levels (primary and secondary bullet items).
- To make bullet lists skimmable, bold keywords at the beginning of each list item.
- End each bullet item with a period to help readers using screen readers.

WORDING

- Keep bullet list items short—no longer than one sentence.

☒ **Example**

• Take a Look at the Mural on the wall of old Radio Hall at the University of Wisconsin. It depicts some of the very first distance education taking place via "Telephonics" during the 1930's. And while you are at the UW, be sure to check out their outstanding Distance Education Clearinghouse Web Site. Then, for even more excellent resources drop by the Department of Learning Technology and Distance Education (DoIT for short)

- Check for parallel construction.

☒ **Example**

One way to correct the following list is to use all gerunds or all verbs.

The services we can provide you include
- complete Internet consultation
- projecting your product or image to the world on the internet
- marketing your Web site so that you will get maximum exposure
- create custom logos and eye-catching graphics

▷ Bullets: Graphical

You may use graphical bullets for more visual interest than simple bulleted (unordered) lists. However, readers with non-graphical browsers or disabled graphics will not see them. Graphical bullets add more visual interest to your page and can fit with your design theme. On the other hand, they cause the page to load more slowly.

HTML Code: Graphical bullets are usually small .GIF graphics that use the <IMG> tag.

Tips

- Use bullet graphics that complement your overall design.

> ☑ **Example**
>
> Critical Security Update for IE 12/26/01 (*BugNet: The World's Leading Supplier of Software Bug Fixes)*

- Use bullets appropriate to the tone and image you want to convey.
- Repeat the same bullet rather than varying the bullets. The page will load more quickly because the graphic is cached.

> ☒ **Example**
>
> ◆ Consultation.
> ■ Posting of site to your location.
> ● Maintenance of sites as required for a fresh look.

- Size the bullets consistently.
- Use bullets that are small and do not distract from the text.

> ☒ **Example**
>
> These bullets are large.
>
> These bullets are distracting.

- Use small graphic files (less than 2 KB).
- Use alternative text (ALT text).
- Do not use a bullet when there is no list or when another method of emphasis is used.

> ☒ **Examples**
>
> ☆ Do not use a bullet at the beginning of a paragraph.

▷ Buttons/Navigation Bars

Buttons are small graphics or icons used for hyperlinks and navigational aids. Graphics can be links to an internal or external destination. You can use simple buttons or combine them into *navigation bars* (a type of image map). Individual buttons allow users to see a separate URL (Web address) on the status bar for each link. A navigation bar will also load more slowly. Using graphical navigational aids

- Makes your pages more visually interesting.
- Ties in with an overall design theme.
- Provides a consistent look and feel to your Web site.

Tips

Format

- Make the graphic look like a button or link. If it is not obvious to click, add text.
- Distinguish buttons from icons by using a border.
- Do not use buttons that do not function as buttons.
- Use buttons consistently; do not make some hyperlinks and others not.
- Deactivate the button for the current page.
- Use graphics that represent the type of information they link to and to complement your design theme.
- Keep button design simple so it is clear what they represent.
- Make buttons consistent in design and position.
- If you use graphical text on buttons, make sure it is legible.
- Consider using a *rollover effect.* This technique created with JavaScript can change a button's appearance when the cursor is placed over it. It can also provide additional explanatory text below the button.

Size

- Avoid overly large buttons, which may look like graphics, not links.
- Don't make buttons too small, which are difficult to click.
- Size buttons consistently. However, some sites have two sizes of buttons to distinguish site-wide (main) and local (secondary) navigation.

Position

- Repeat buttons/navigation bars on every page.
- Decide on a vertical or horizontal arrangement.
    - For vertical arrangements, place buttons along the left.
    - For horizontal arrangements, place navigational buttons at both the bottom and top of the page (unless it is only one screen long).
- Group buttons with similar functions.

Wording

- Use text to accompany the button.

> ☑ **Examples**
>
> 
>

- Accompany graphics with alternative text (ALT) for users with non-graphical browsers and for accessibility.
- Write text labels that are clear and easy to understand. Make it obvious what the graphic links to.
- Include all sections of the Web site and a link to Home.
- If necessary, use text to instruct the user what to do.

> ☑ **Example**
>
> *Each button and link provides more information about a business:*
>
> • **Maps & Directions**: Click on the **map and directions** icon or link to receive driving directions and to view the business's location on an area map.

▷ Buzzwords

Buzzwords are impressive-sounding or commonly-used words and phrases, usually related to a particular industry. Avoiding buzzwords keeps your site fresh and free from jargon that readers may not understand. Buzzwords also may make your site unprofessional.

> ☒ **Examples**
>
> -Want your web site to look **cool?** We'll that's what we specialize in - **cool** web sites.
> -Internet Talk Radio is a **neat** use of sound.
> -More and more companies are putting useful information about their products and services online in the form of catalogs and magazines and **lots** of other **stuff.**
> -Remember, if you know of any **cool** tricks or tools - or new **stuff** that I might think is **hot** - just tell me.
> **-JAVA SHAREWARE: Lots and lots** of free Java **stuff**, info, etc. **check it out**
> -You can make your table using a form. **Real neat!!**

Tips

- Avoid using words and phrases that have been overused on the Internet. Examples include the following:

  - ☐ aim your browser at
  - ☐ app
  - ☐ bot
  - ☐ browser-enhanced
  - ☐ check it out!
  - ☐ cookie/cookie monster
  - ☐ cool
  - ☐ cyber-
  - ☐ drill down
  - ☐ for your viewing pleasure
  - ☐ full of resources
  - ☐ functionality
  - ☐ hits
  - ☐ hot

- ☐ hotlink
- ☐ hotlist
- ☐ implementation
- ☐ information superhighway
- ☐ linkrot
- ☐ lots
- ☐ lurk
- ☐ neat
- ☐ net-
- ☐ newbie
- ☐ offline
- ☐ one-stop shop
- ☐ snail mail
- ☐ solutions
- ☐ spam
- ☐ starting point
- ☐ stuff
- ☐ surf/surfer
- ☐ techno-
- ☐ thread
- ☐ tips and tricks
- ☐ ultimate
- ☐ up and running
- ☐ under construction
- ☐ zine

▷ Callouts/Captions

Callouts, or captions, are text used for graphic explanations. Callouts describe graphics and tell the reader what to note. Meaningful captions also make your Web page accessible for readers who cannot see graphics. Captions can be simple titles, descriptions, or explanations.

HTML Code: Use a table and <ALIGN> </ALIGN> the text.

Tips

- Write clear, meaningful captions for all graphics.

> ☑ **Example**
> Figure 1: HP Photosmart 618 digital camera, front view

- Make the caption unique. The caption name should be different from other items on the page.
- Don't state the obvious or repeat information found in headings.
- Explain the graphic.

> ☑ **Example**
> Figure 1. Shows a sample passage of text rendered using 16 point Times New Roman before font smoothing has been activated.

- Link to long captions.

> ☑ **Example**
> Image of Jupiter and Io taken by the Hubble Space Telescope. Caption

- Give the source.

▷ Capitalization

Capitalization is using all uppercase letters or capitalizing the first letter of words. Using all caps emphasizes key words. However, all caps are more difficult to read because the unique letter shapes are lost. A mixture of upper and lower-case letters increases reading speed.

Tips

General Uses

- Use all caps sparingly.
- Avoid large amounts of text in all uppercase because it is difficult to read.
- Use all caps for only a few words, such as labels and warnings.

Special Uses

- **E-mail and Usenet postings:** Avoid all caps because it is considered to be shouting.
- **Electronic addresses:** Follow exact capitalization, because addresses are case sensitive.
- **Code:** Be especially careful when using capitalization for computer documentation and code.
- **Links:**
  - Be consistent when capitalizing link text.

> ☒ **Example**
>
> *The initial caps in this Web site's navigational text are inconsistent.*
>
> New grant competitions | This Week's Chronicle | Chronicle archive |
> Front page | Guide to the site | About The Chronicle | Help

  - Avoid using all caps for navigational items. All caps are difficult to read.

> ☒ **Example**
>
> UNDERSTANDING WEB ADDRESSES

  - Avoid using all small letters for navigational items. Initial caps are easier to scan.

> ☒ **Examples**
>
> table of contents • guidelines • create a web • basics • format • images • tables •
> import files • web resources

- **Headings and titles:**
  - Be consistent when capitalizing headings.

> ☒ **Example**
>
> *The initial caps in this Web site's navigation text are inconsistent.*
>
> Writing your Files
> Browsing your Hard Drive
> Common Errors when editing pages

  - Capitalize the first letter of words of titles or headings. Exceptions are conjunctions, articles, prepositions with less than four letters, and *to* when used as an infinitive.

- ☐ Use capitalization in Web page titles as you would for normal titles.

- ▶ **Web text:** Avoid using all small letters for annotation or other Web text. Some sites use this technique to convey a modern look, but it is difficult to read.

> ☒ **Examples**
>
> ***Web Style Guide***
> general web page style guide from the yale center for a/i media
> ***The Web Developer's Virtual Library***
> assortment of tools and tips for authoring html, java, cgi, and other scripts
>
> using our proprietary web-based software, our customers are able to run and manage their programs online, while eliminating many of the costs associated with traditional offline programs.

- ▶ **Acronyms:** Capitalize acronyms.

> ☒ **Examples**
>
> -New **Html** resources references and tutorials will be added regularly.
> -A **faq** on robots has information on how they work.

- ▶ **Abbreviations:** Capitalize abbreviations.

> ☒ **Example**
>
> 3d special effects animation. *(3-D is short for three-dimensional.)*

- ▶ **Lists:** Capitalize the first letter of items in lists (bulleted and numbered).
- ▶ **Sentences:**

  - ☐ Capitalize the first letter of words at the beginning of sentences. Some Web sites do not begin sentences with a capital letter, but they are difficult to read online.

> ☒ **Example**
>
> you got it. if you haven't gathered it by now, bytes of knowledge is very service oriented. we will walk you through every stage of web site development and train you on the Internet, email and the maintenance of your web site.

  - ☐ Avoid capitalizing the first letter of a word for no reason.

> ☒ **Example**
>
> Very Useful resource.

  - ☐ Edit for capitalization errors.

☒ **Example**

It's easy with an administrative console. We understand your frustration with not being able to edit or change the information on your web site. **we** have the ability to create a site you can maintain by simply using your keyboard and web browser.

- Begin a sentence with a capital letter unless the name is spelled with small letters. If possible, rearrange the sentence to avoid beginning with the word.

☒ **Example**

c|net has developed two proprietary content delivery systems: Prism and Dream.

▷ Cascading Style Sheets

Cascading Style Sheets (CSS) are similar to a template, because they allow Web designers to define styles for HTML elements and apply them to many Web pages.

They contain rules that tell a browser how to display a document or draw an HTML element. Each rule has two parts:

- A selector identifies elements on the page controlled by the rule.
- Properties describe how to draw elements selected by the rule.

Each style sheet is a text file with a .css extension. The style sheet can be embedded in the document HEAD, or the HTML file can link to it (preferred). They're called cascading style sheets because one set of "styles" can override another set of "styles." Externally linked styles, page-level styles and in-line styles form a hierarchy. The browser looks for in-line first, then page-level, then linked ("cascading").
Style sheets have the following advantages:

- Allow you to define the attributes of elements of your Web pages and control the appearance of a Web site more closely, thus maintaining consistency. For example, you can control the fonts, position of text, color, background, and other formatting.
- Simplify formatting because you can specify the format for any element and have it applied automatically throughout one or several documents. They are particularly useful for text formatting.
- Separate layout and style from the content and structure of a Web document.
- Allow content to be accessible from different browsers and types of hardware.

Some older browsers do have difficulty with style sheets.

HTML Code: You create an HTML document and a style sheet, then attach the style sheet to the document. You associate the two using several methods, including embedding the style sheet in the document's <HEAD>, creating a link to the style sheet, or importing the style sheet.

Tips

- Creating and using style sheets goes beyond the bounds of this book. To learn more about using style sheets, see the World Wide Web Consortium site at http://www.w3.org/Style/.

▷ Catalog Page

An online catalog is the equivalent of a print catalog; it allows readers to view and order your products.

Tips

Organization

- Use the following organization:
  - ☐ Main page with introduction
  - ☐ Menu or table of contents
  - ☐ Detail
  - ☐ Links to order form and other action steps

Items to Include

- Include the following items:
  - ☐ Aids for making buying decisions
  - ☐ Availability (in stock)
  - ☐ Benefits
  - ☐ Browse feature
  - ☐ Check order status
  - ☐ Contact information
  - ☐ Delivery information
  - ☐ Demonstration or prototype
  - ☐ Description of features
  - ☐ Downloadable version of the catalog
  - ☐ E-mail order notification and summaries
  - ☐ Guarantees
  - ☐ Identification numbers
  - ☐ Index
  - ☐ Links to bottom-line summaries, specifications, product alerts, accessories, related items
  - ☐ Manufacturer information
  - ☐ Options to print, fax
  - ☐ Options to select product factors (size, color, version)
  - ☐ Ordering information
  - ☐ Payment options
  - ☐ Pictures and thumbnails, with links to product views
  - ☐ Prices
  - ☐ Privacy & security policy for transactions
  - ☐ Product highlights
  - ☐ "Quick" links to new items, best sellers, specials, quick order option
  - ☐ Return policy
  - ☐ Search engine
  - ☐ Shipping options

- ☐ Specials and rebates
- ☐ Support and customer service information
- ☐ Technical specifications
- ☐ Terms and conditions

FORMATTING

- ▶ Use layering so readers can navigate to the information they need and link to details.

> ☑ **Example**
>
> *For each title in the Quick Course books catalog, you can display a summary with thumbnail graphic, then link to an Overview (a brief synopsis of the information presented in each chapter) and a Table of Contents (a detailed look at the topics covered).*

- ▶ Provide contextual clues so readers know their location.
- ▶ Consider using tables to organize the information.
- ▶ Group categories logically.

▷ Centered Text

Centering text is an effective text alignment when used sparingly. Centering can emphasize an element on a Web page. However, it is easier for the eye to track down left-aligned text.

HTML Code: The <CENTER></CENTER> tag can center anything between the tags, including several paragraphs and graphics. The <P ALIGN=CENTER> tag centers one paragraph.

Tips

- ▶ In general, use left alignment. It is easier for the eye to find and scan quickly.

> ☒ **Example**
>
> **Tips For Alignment**
> Avoid centered text, especially for large amounts of text.
> Centered text is difficult to read quickly.
> It is hard for the eye to find where to begin.
> Centered text also causes wasted white space on the right and left margins.

- ▶ Consider using centered text for the following:

  - ☐ Page headings
  - ☐ Headlines in articles
  - ☐ Footers
  - ☐ Pull quotes

- ▶ Use centered text sparingly.
- ▶ Do not use centering for large block of text.
- ▶ Avoid large centered headlines that are more than one line long.

- Avoid centered items in lists of links.

> ☒ **Example**
>
> *Centered links are less easy to scan than left-aligned links.*
>
> Points for hotel stays
> Avis Car Rental
> FTD.com
> GMAC Insurance
> iBank Commercial Finance Center
> International Travel Guide - featuring iGo.com products
> MyPoints.com
> National Car Rental
> Thrifty Car Rental
> USA TODAY®

▷ Charts

Charts include line, bar, pie graphs, and flowcharts. Charts can show trends, changes over time, parts of a whole, and organization.

TYPE OF CHART	WHAT IT SHOWS
Bar graph	Comparison of discrete items
Line graph	Continuous data, changes over time
Pie chart	Parts of a whole
Flowchart	Organization, process

Tips

- Keep file size small to speed loading time.
- Label all graphics clearly.
- Include legends and callouts/captions.
- Provide summary information about your graph/chart
    - In the text near the graphic
    - By linking to a description page by using the LONGDESC attribute or D Link.
- Provide links to larger views so readers can choose whether to view them. Use phrases such as View Chart or Show Graph.

> ☑ **Examples**
>
> View MiniMailer Flowchart
>
> **Adult Asthma Quality-of-Life: Twelve Month Results**
> Results of the adult quality-of-life questionnaire indicated an improvement in all five domains. The level in Total domain of quality of life improved 20 percent. View Graph

- Clearly identify what graph a link leads to.

☒ **Example**

▸ View Chart ▸ View Chart ▸ View Chart ▸ View Chart

- If graphs are detailed, consider showing an overview and linking to parts of the graphic.
- Acknowledge sources.

▷ Children: Writing Web Pages for Kids

Often Web pages are directly aimed at children. Examples include two of the most popular sites for kids, Nickelodeon and Disney. Web pages for kids have a variety of goals, including advertising, public relations, supporting a product or entertainment line, or providing entertainment and activities. Other sites are educational and informational. Kids who review sites look for sites that are "fun" and not boring.

Tips

Formatting

- Use a metaphor, if appropriate.
- Incorporate
  - Informal, eye-catching graphics and icons
  - Informal, large fonts
  - Colorful backgrounds
  - Color coding
  - Multimedia
  - Animation
  - Sound

Navigation and Links

- Provide a site guide, and guide readers through the site. Make these instructions simple to understand.

☑ **Example**

This icon will take you to the periodic table section. If you already know about the periodic table, just click on one of the colored squares above. You can then learn about each individual element.

- Make the site easy to navigate. For example, provide large colorful buttons with clear labels.
- Provide activities that allow interactivity and participation, especially games, puzzles, jokes, surveys, feedback, and free offers. Also provide physical interactivity, such as clickable graphics.

☑ **Example**

This sentence not only encourages action but also explains how to use the site.

Click the launch pads on the left to explore KidsCom.

- Layer information. Let readers read short summaries and link to full stories.

☑ **Example**

Picture a camel, take away a hump (or two), shrink it down in size and place it in South America. What do you end up with? A guanaco.
Click to learn more *(Yahooligans!)*

Writing for Kids

- Target a specific age group.
- Make your privacy policy clear and simple.
- Consider using humor, clever titles, headlines, and links.
- Use an informal style, words, and tone.

☑ **Example**

The Yuckiest Site on the Internet

- Use devices to get attention, especially questions and action-oriented verbs (*see, explore, learn, visit).*

☑ **Example**

The World Kids Network begins with a paragraph that attracts young readers by providing links to the activities available. Be anyone, do anything, or find out almost anything. You can make great friends, join some cool clubs, play games, do your school homework, just hang out in the mall or do a LOT of other cool stuff! You can cruise on in for a visit, or stay around and help out. Every journey into the WKN Galaxy is a unique experience and you are at the controls! Come on in and see for yourself.

- Use short, simple sentences and short paragraphs.

☑ **Example**

Space Shuttle Endeavour will launch soon. The Expedition Four crew will be on the way to the Station, and the Expedition Three crew will be coming home. Endeavour is also carrying supplies and equipment for the Station, and astronauts will go on a spacewalk outside the Station. *(NASA Kids)*

- Use simple definitions.

☑ **Example**

You've asked a good question, but first a definition: Volcanologists use the word magma for molten rock that is still under ground and hasn't yet erupted. Once it has erupted onto the surface, we call it lava.

- Use simple word choices.
- Use "you" and other personal pronouns.
- Incorporate examples.

☑ **Example**

Each page in a Web site, including the home page has its very own address, called a Universal Resource Locator, better know as a **URL**. You have probably begun to see **URLs** in the past year or so on TV and in magazines or maybe you have even heard people talk about them on the radio. Knowing a **URL** for a particular page will always allow you to access that page without searching for it first. You can give a **URL** to your friend across the street or across the ocean in England, and they both should be able to access the same information on that page as you just by knowing a page's **URL**. It is just like a postal address or telephone number.

▷ Chunking

Information chunking involves breaking information into small units or components. Ideally, each unit consists of only one topic, idea, or concept. People effectively handle only seven plus or minus two items of information.

Chunking has the following benefits:

- Allows information to be labeled by function.
- Layers the content.
- Makes information easier to organize.
- Divides text into discrete units or zones of information.
- Breaks large topics into more manageable modules for training.
- Aids memory.
- Makes the page easier to scan, browse, and read.
- Avoids too much information and text on one page or list of links.
- Avoids the need to scroll long documents.
- Makes it easier to create links. Chunks often link to other chunks.
- Lets you reuse blocks of information for different purposes.
- Makes your Web site easier to maintain.

Tips

Creating Chunks

- In general separate long documents into small chunks. However, weigh the pros and cons of the page length.
- Use five to seven chunks per unit of information (seven plus or minus two rule).
- Chunk information by its content, function, and purpose. The following table shows types of information that are considered topics:

TYPE OF INFORMATION	WHAT IT PROVIDES
Background	Material readers must understand before proceeding
Concepts	Explanation of what things are; information needed to understand a subject
Definition	Explanation of terminology
Example	Concrete application/illustration
Facts	Statements, data, statistics
Parts	Components
Principles	Rules and guidelines
Procedures/Instructions	Instructions; steps
Process	How something works; stages or phases
Reference Information	Supplementary information
Types	Categories or groups

- Divide a topic until you can no longer subdivide it. Each topic should focus on one piece of information, idea, or concept.

> ☑ **Example**
>
> How You Can Use Plug-ins
> When to Use Plug-ins
> How Plug-ins Work

- Check that
  - ☐ Each topic does not overlap with any other.
  - ☐ Each chunk is independent of the others (context-independence).
  - ☐ Chunks at equal levels are of equal importance and equal depth.

Planning Chunks with Diagrams

- Use a storyboard method to work with chunks.
  - ☐ Put each chunk on a 3 by 5 card. On the card include the title, description of the content, intended audience, source of information, update frequency, and the contact person. Use lines to show the relationship among chunks.
  - ☐ Indicate the number of times the topic occurs and how often it will change.
- Use a "web" diagram to break a topic into chunks.
  - ☐ Put a topic into a square or circle; then put related topics as branches.
  - ☐ Work from general to specific.
  - ☐ Continue this process until you have broken down the topics as far as you can.
  - ☐ Write a heading for each topic.
  - ☐ Then write a short paragraph for each unique topic.

LAYOUT AND NAVIGATION

- Ideally, fit one chunk on one screen.
- Use the following layout devices to show chunks:
  - ☐ Boxes
  - ☐ Color
  - ☐ Headings/labels
  - ☐ Lines
  - ☐ Lists
  - ☐ Menus
  - ☐ Icons
  - ☐ White space
- Provide a table of contents and index.
- Provide a variety of navigational aids: reading in sequence (*Previous, Next, Topic __ of___*) and jumping to any topic.

WRITING TEXT WITHIN CHUNKS

- Make every topic self-contained. You can never assume readers will read in any particular order.
- Provide any information readers will need (definitions, conceptual information, context/background), or link to this information.
- Label each chunk and the type of information it contains. These labels will help you later create headings and link text.
- Identify several keywords that describe the chunk's central idea.
- Identify the main idea in each chunk with a sentence.
- Write sentences that support the main idea.
- Keep paragraphs short.
- Avoid transitions and spatial references.

▷ Clarity

Clarity is clear wording that avoids ambiguity. Ambiguity results when words have several possible meanings. Vague wording makes readers confused and wastes their time. Because Web readers are in a hurry and are slowed down by reading from a computer screen, clarity is even more important online.

Tips

- Use clear, exact words, especially in headings and text links.

> ☒ **Examples**
> More Information
> Helpful Information

- Avoid vague or ambiguous words.

> ☒ **Examples**
> We offer a very fast web presence. *(Is the Web site quick loading? Do they develop the site quickly?)*

Select Location

This could mean geographical location. It really means location in the site you want to visit (Online Bargains, Latest Products, Instructions, Our Company).

Jobs Careers Career Connection Job Search: *Which of these links to job openings?*

- Explain the context.

☒ Example

Next *Does NEXT mean next page, chapter, topic?*

- Explain everything, including icons and graphics.
- Use active voice to clarify who is doing the action.
- Use positive language.
- Avoid hedging words such as *usually* and *sometimes*.
- Be careful when the word order can change the meaning of a sentence or gives the sentence an ambiguous meaning.

☒ Examples

New Book Sites: *New books or new sites?*
Search Solutions: *Is "search" a noun or verb?*

- Check that punctuation does not change the meaning of a sentence.

☒ Example

You can probably have a pretty effective website gallery.
You can probably have a pretty, effective website gallery

▷ Clichés

A cliché is an over-used, trite expression. Avoiding clichés keeps your writing fresh and concise.

Tips

- Avoid clichés, especially on your home page.

☒ Examples

-Because of the resources we have access to, **the sky's the limit** as far as design creativity goes.
-Most companies are **missing the boat** by only looking at the Internet as a way to sell products.
-This is a real **mixed bag** of browser statistics, simulation and information on the tags.
-Almost **everything you always wanted to know about** HTML **but were afraid to ask**.

-Should you discover this tutorial really isn't your **cup of tea**, try looking -here- for resources.
-The use of color can **make or break** a web page.
-We **go the extra mile** to produce effective, attractive web sites.
-In the beginning, **content was king** on the Internet.
-Bells and whistles have their place, but it's **nuts and bolts** web design that'll make a user appreciate your site.
-If graphics are not your own design, a link should be included on your site to **give credit where credit is due**.

▷ Colons

Colons show a connection between a statement and a phrase or list that follows.

Tips

- Use a colon before a statement that explains or illustrates the beginning of the sentence.

☑ **Example**

Web readers prefer bite-sized chunks of information: fewer words, simpler sentences, more white space, less punctuation.

☒ **Example**

A general color is specified in terms of the three primary colors red, blue and green.
(This is a run-on sentence. Use a colon after colors.*)*

- Use a colon to introduce a list when the introductory sentence is complete.

☑ **Example**

HTML supports three types of lists: ordered lists, unordered lists, definition lists.

- Do not use a colon within a sentence when no formal pause is needed.

☒ **Examples**

-These services include: consultation, World Wide Web site design, custom programming and scripting, search engine registration, and many others.
-Our graphic designers offer a wide range of capabilities including but not limited to: creating original logos or reproducing current ones; creating animated or Flash graphics, image scanning and digital photography.
-Stay away from companies that have no contact sources like: address, phone, e-mail, etc.
(omit the colons)

▷ Color

You can use color for the text, links, background, buttons, boxes, tables, graphics, and elements on graphs. Use color to do the following:

- Draw attention.

- Highlight and emphasize.
- Fit with your design theme or metaphor.
- Show structure and related items by grouping information and color-coding.
- Establish the mood and tone.
- Affect legibility.

HTML Code:
In the <BODY> tag you specify the following colors:
BGCOLOR=background color
TEXT=text color
LINK=hyperlink color
ALINK=active link color
VLINK=visited link color

The standard color model used on the Internet is RGB (red, green, and blue).
Color is described in hexadecimal notation: #RRGGBB
Values range from 0 (absence of a color) to ff (strongest amount of a color); e.g., red is #ff0000.

Tips

- Use color to emphasize important information.

> ☑ **Examples**
> *Use color for headings, key words, required form fields, current location link.*

- Use color sparingly.
- Be consistent with color selection and placement.
- Repeat colors to provide unity and group related information.
- Limit the number of colors used in a Web site to about three or four.
- Reduce the size of graphic files by reducing the number of colors per image. However, avoid less than 16 colors to avoid poorly rendered images. Remember that some monitors only display 16 colors.
- Consider using the 216-color palette (the browser-safe color palette). The remaining 40 colors may vary with some monitors. Some monitors may still display only 256 colors.
- Associate each color with a purpose.
- Consider using color-coding.
    - Make sure it is obvious what the colors represent.
    - Do not rely on color-coding alone; provide another cue to meaning.

> ☑ **Examples**
>
> *Neuroscience for Kids uses color-coding for the four lobes of the brain. The explanation of the four lobes is placed in a table with shaded cells that correspond to the four colors (red, blue, yellow, green) used on the diagram of the brain.*
>
> *The Pantone site uses different colors for each major topic. These colors are indicated by the path shown in the page title, as well as the background color. For example, on the* **PANTONE/Print** *page, "Print" and the background are green. (Pantone is the world-renowned authority on color and provider of color systems.)*

- Use colors that fit your theme, organization, or product.

> ☑ **Examples**
>
> *Use the colors of your school or organization.*

- Use colors that convey the image you want to project, such as cool and calm (blues and greens), lively and warm (reds, oranges, and yellows), and natural (browns and greens).
- Use high contrast between foreground and background color. Dark text on a light background is easier to read than light text on and dark background.
- Avoid colors that may have negative associations, especially for an international audience.
- Avoid colors that the color-blind cannot see (e.g., red/green and blue/yellow), or provide another method of highlighting.

▷ Colored Links

Links are displayed in the color specified by the default browser settings unless the Web page specifies different link colors.
Unvisited links are *blue*.
Visited links are *purple*.

Three types of link states are

- Available (or standard) (not clicked)
- Active (clicked while the page loads)
- Visited (clicked)

Link color makes links more visible and helps readers see which links they have visited. Usability studies show that you should not require that users relearn how to use a Web page. So readers expect that blue underlined text represents a link.

HTML Code:
Available (standard) link: <LINK=#RRGGBB>
Active link: <ALINK=#RRGGBB>
Visited link: <VLINK=#RRGGBB>
Three two-digit numbers are a color code in hexadecimal form that specifies the amount of red, green, and blue.

Tips

- Avoid colored link text. The only exception is special links, such as glossary terms.
- Use the default browser link colors.
- Match link colors with your site's color scheme.
- Plan a color scheme for the following types of text:
    - Headings and headlines
    - Normal text
    - Link text
    - Selected links
    - Visited links
- If you change link colors, make sure that
    - They are consistent throughout your Web site.
    - Readers can distinguish visited and unvisited links.
- If you remove link lines, make sure readers can recognize links (e.g., use blue text).

▷ Colored Text

Colored text in a Web document can be used for normal text, including headings. You can also use colored links. Use colored text to do the following:

- Attract attention.
- Provide emphasis.
- Highlight important information.
- Color code related information.

> ☑ **Example**
>
> *This site uses color coding (shown in parentheses here) to highlight keywords.*
>
> The search for the **exact phrase** *(blue text)*, **Word downloads** *(red text)* in the category, **Full Site Search** *(blue text)*, found no matching records.

- Separate information and show categories (such as glossary items).
- Set off headings.
- Fit the tone or color scheme.
- Provide repetition and unity.

HTML Code: <TEXT=#RRGGBB>
Three two-digit numbers are a color code in hexadecimal form that specifies the amount of red, green, and blue.

With Cascading Style Sheets, you can color text on a page to look like a highlighter marker was used.

Tips

- Always specify text color. This precaution protects your text from users who change the default colors in the browser and may make your text invisible.
- Use high contrast between text and background color.
- Use dark text on a light background rather than light text on a dark background.
- Avoid the following:
  - Blue text (it is difficult for the eye to focus on)
  - Red/green combinations (some readers are color-blind)
  - Red/blue combinations (they focus at opposite parts of the eye)
  - Gray or yellow text on a white background (there is not good contrast)

> ☒ **Example**
>
> Gray text on a white background is used often on Web sites, but it does not provide good color contrast.

- Limit use of colored text and color-coding. Overusing color is distracting, slows reading time, and creates noise and information overload.
- Be consistent.
- Be aware of international and cultural meanings of colors.

▷ Columns

Columns are a way to divide text into vertical blocks. Use columns to do the following:

- Keep line lengths shorter.
- Fit more text on the page, provide balance, and avoid wasted space.
- Organize information.
- Create a grid with "zones" of information.
- Provide visual contrast.

You can also use a narrow column on the left for

- Navigational aids.
- Icons and other small graphics.
- Sidebars, pull quotes, and callouts.

HTML Code: <MULTICOL COLS=#></MULTICOL>

The number (#) specifies the number of columns.
The <BGCOLOR> creates a background color in the column.

Tips

- Use tables or frames to simplify creating columns.
- Use columns primarily when you have large amounts of text.
- Aim at a line length of 60-70 characters per line.
- Use color in columns to provide contrast.
- Use right text alignment in the right column to cut excess white space.

☑ **Example**

The two-column menu with both left/right alignment uses space efficiently.

Auctions	Classifieds
News	Business
Travel	Financial
Career	Education
Computing	Science
Health	Government
Weather	Sports
Local	Living
Explore	Shopping
Hobbies	Kids

- Avoid placing text in columns that must be read like newspapers from top to bottom unless text fits on one screen. This format requires too much scrolling and is difficult to follow online.
- For accessibility, make sure text in columns is read from left to right.
- For links in columns, make it clear which direction to read—especially if the order is important.

☒ **Example**

Often it is unclear if you should read down or across columns of links.

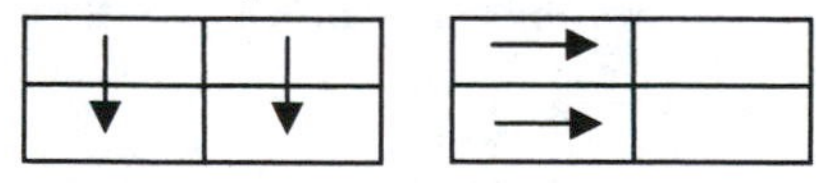

- For sites with lots of text, consider letting readers customize the number of columns.

☑ **Example**

The International Herald Tribune *lets you choose one or three columns for their news stories.*

▷ Comma Splice

A comma splice is a comma between two independent clauses.

Tips

- Use a semi-colon, conjunction, or period between two independent clauses.

☒ **Before**

Some service providers offer free web space when you register with them, this free web space is not efficient enough to be used for business.

☑ **After**

Semi-colon: Some service providers offer free web space when you register with them; this free web space is not efficient enough to be used for business.
Conjunction: Some service providers offer free web space when you register with them, but this free web space is not efficient enough to be used for business.
Period: Some service providers offer free web space when you register with them. This free web space is not efficient enough to be used for business.

☒ **Examples**

-A tag is a code that will not be displayed, instead it will change the appearance of the text.
-Please feel free to explore our site, if you don't see what you are looking for, let us know because your input is important to us!
-Just having a website is not enough, other people will have to find your site too.

▷ Commas

Commas are punctuation marks that help your readers follow your sentence structure and understand your meaning.

Tips

- Avoid over-using commas because they are difficult to see on a computer monitor.
- Edit for the most common comma errors.

The following is a concise summary of rules for punctuating your sentences and examples from real Web pages. *Consult a grammar/punctuation handbook for a more complete list of rules.*

- ☐ Use a semi-colon (or period) between two closely-related independent clauses.

☑ **Example**

You should already know where to store your HTML document(s); if you don't know, check with your Internet access service.

☒ **Examples**

The Web is dynamic, keep your material up-to-date. *(this is a **comma splice**)*
The Web is dynamic keep your material up-to-date. *(this is a **run-on sentence**)*

- ☐ Use a comma between two independent clauses connected by a coordinate conjunction (*and, but, for, nor, or, so, yet*).

☑ **Examples**

-Words and pictures can be a powerful combination, but they must work together.
-Make the form easy to fill out, and your response rate will increase.

☒ **Examples**

-Frames can be useful but there are many reasons not to overdo the use of them. *(add comma before* but*)*
-Each page has a particular topic, and contains hypertext links. *(omit comma)*
-The title of a document is not normally displayed as part of the page, but is often displayed in some sort of special section in most browsers. *(omit comma)*

☐ Use a comma after an introductory dependent clause.

☑ **Example**

If your image is larger than 40 or 50 KB, you should try to reduce it.

☒ **Example**

If you do not update and revise people will not come back. *(add comma after* revise*)*

☐ Use a comma after an introductory word or phrase.

☑ **Example**

Finally, the body of the document should also be marked off with the <body> and </body> commands.

☒ **Examples**

-Today the greatest-looking pages push HTML to its ever-diminishing limits. *(add comma after* Today*)*
-When starting a site you should consider the entry page. *(add comma after* site*)*

☐ Use commas around a word or phrase that interrupts the main clause.

☑ **Example**

Then, after trying out the features, close the window to continue browsing the original document.

☒ **Example**

Statistics prove that more and more people every day, including schools, homes and businesses are using the WWW to provide them with almost all their needs. *(add comma after* businesses*)*

☐ Use commas around appositives.

☑ **Example**

All sites are composed of ***NODES*** or pages, containers for related chunks of information or experience, and ***LINKS,*** which establish the relationships between nodes.

☐ Do not use a comma after a main clause followed by a dependent clause if the clause is essential.

☑ **Example**

Short URLs are better because they're easier to type.

☒ **Example**

Conventional URLs are easier to remember, because there's nothing special or unique about the URL to remember.

▶ Use commas to separate items in a series.

☑ **Example**

You can choose a white, a colored, or a patterned background for your pages.

- ☐ Use commas around a non-restrictive clause (one that is not necessary for the sentence's meaning). Use *which* when the clause is non-restrictive. Use *that* when the clause is restrictive.

☑ **Examples**

-Progressive JPEGs, **which** display progressively better mathematical representations of the image, often look better from the start than interlaced GIFs. *(non-restrictive; use* which *and a comma)*

-A typical example of an image map is a Web site **that** offers national information organized by state. *(restrictive; use* that *and no comma)*

☒ **Example**

This site is enhanced with many writing activities and dynamic web links **which** are a valuable source of supplemental information. *(restrictive; use* that*)*

- ▸ Avoid unnecessary commas.

☒ **Example**

Just creating a web page and getting it online, will not suffice.

▷ Comments

You can include text that is invisible to the user but located in the HTML code itself. Comments are available to people who look at your HTML code. This information identifies you and your Web page. Comments can include dates for updates and other explanations.

HTML Code: Comments appear after the initial <HTML> tag and use a <!-comment-> tag to hide the text. This section goes within the HEAD tags.

Tips

- ▸ Include any of the following information:
  - ☐ Name
  - ☐ E-mail address
  - ☐ Dates: created, modified
  - ☐ Filename
  - ☐ Information about page
  - ☐ Purpose of page
  - ☐ Notes to other authors about the code
- ▸ Keep each comment one line long.
- ▸ Consider using comments to explain links or provide reminders to the Web authors.

☑ **Examples**

<!-- The page layout is via a grid (a table) with three columns. Height and width are set via pixels (absolute spacing). -->

<!--THIS TEMPLATE IS FOR REFERENCE SHEETS -->

▷ Community Features

Community features are elements you provide in your Web site for readers to interact. Providing community features encourages readers to communicate with you and with each other, exchange information, and keep visiting your site.

Tips

- Incorporate any of the following:
  - Bulletin board
  - Chat room
  - Contact information
  - Discussion group/forum
  - E-mail links
  - E-mail list information
  - Feedback mechanism
  - File sharing
  - Message/note board
  - Newsgroup
  - Polls
  - Shared calendar
- List the community features you provide and describe them.

☑ **Example**

Microsoft Technical Communities provide opportunities to interact with Microsoft employees, experts, and your peers in order to share knowledge and news about Microsoft products and related technologies.

- Provide a FAQ that describes the features.
- Provide clear instructions and guidelines for using the features.

☑ **Example**

Microsoft Technical Communities contains a User Guide *with Getting Started information and* Rules of Conduct.

- Create links to the features from a main link such as *Community*.

▷ Company Page

A company page is sponsored by a business enterprise or organization. The URL usually ends with *.com*. The purpose of the page is to establish the company's identity and presence and to establish customer relations.

Tips

Organization

- Use the following organization:
  - ☐ Welcome
  - ☐ Who you are and what you do. This information can be on the home page or a link from it.

> ☑ **Example**
>
> About IBM *links to a separate page that describes the company:*
> At IBM, we strive to lead in the creation, development and manufacture of the industry's most advanced information technologies, including computer systems, software, networking systems, storage devices and microelectronics.

  - ☐ Identification of subject/content
  - ☐ Introductory paragraph with a clear statement of the site's purpose
  - ☐ Highlights section: quick summary (name, address, president) in list format
  - ☐ High-level overview of the organization and links to more in-depth information
  - ☐ Navigational information, menu of links
  - ☐ Quick links to the type of information customers want most. Determine why customers are visiting your site.

> ☑ **Examples**
>
> *Link to downloads, schedules, hours, cost calculators, product information.*
>
> *UPS (United Parcel Service) provides an easy-to find box to* Track Packages & Freight, *as well as* Quick Links *to Create Shipment, My UPS, Open a Shipping Account, Calculate Time and Cost, International Shipping, Find Locations, Order Supplies. There are also multiple ways to access this information, including a drop-down menu and buttons.*
>
> *Hewlett Packard provides an obvious link to* >>Support & Troubleshooting *and* >> Software & Driver Downloads.

Items to Include

- Include the following:
  - ☐ Annual report
  - ☐ Business/service information
  - ☐ Buying or subscription information
  - ☐ Calendar of events, conferences, etc.
  - ☐ Catalog
  - ☐ Contact information (detailed, including phone numbers, addresses)
  - ☐ Corporate information
  - ☐ Dates/currency of information
  - ☐ Description of company, owner
  - ☐ Directory with links to biographical information
  - ☐ Donation information (if applicable)
  - ☐ Downloads
  - ☐ E-mail list information
  - ☐ Employment opportunities

- ☐ Fact sheet
- ☐ FAQs
- ☐ Financial information
- ☐ History, amount of time in business
- ☐ Informational page
- ☐ Investor information
- ☐ Legal and privacy information
- ☐ Locations, local sites, smaller companies
- ☐ Logo and slogan
- ☐ Mission statement and goal
- ☐ Name of company
- ☐ News and news releases
- ☐ Newsletter
- ☐ Office locations
- ☐ Partners
- ☐ Personnel information and organizational structure
- ☐ Philosophy
- ☐ Policies and procedures
- ☐ Press releases
- ☐ Product information
- ☐ Product reviews
- ☐ Prospectus
- ☐ Research
- ☐ RSS feed
- ☐ Resources
- ☐ Questions and answers
- ☐ Services
- ☐ Site map
- ☐ Sites world-wide
- ☐ Source of factual information
- ☐ Special offers or ads
- ☐ Stock quotes
- ☐ Support

FORMATTING

- Make the home page attractive and attention-getting.
- Use a design theme that fits your corporate colors or image.

> ☑ **Examples**
> IBM's home page uses shades of blue.
> Campbell's Soup pages uses shades of red.

- Layer information by linking to details.

> ☑ **Example**
> *The General Motors home page links to the following:*
> THE COMPANY> AUTOMOTIVE> SERVICES>

- Use lists with clear headings.
- Do not require scrolling; put the most important information within the screen size.

▷ Concept Document

A concept document is a concise description of your planned site. It answers questions about the purpose, content, and design. This document is often a "deliverable" to a customer, co-authors, supervisor, or teacher.

Tips

- Write the following types of information:

DESCRIBE THE PROJECT
Title
Type of Web site
Objectives/purpose
Audience
Existing sites
Source material
Specifications (optimal browsers, connections, and platforms) and users' Web access

CONTENTS OF WEB PAGES
Outline/organization
Flowcharts
Topics users will have access to
Links
Home Page Contents
Other Pages
Instructional approach

NAVIGATIONAL PLAN
Method of formatting of links
Overview of types of links planned

DESIGN SCHEME
Description of overall look and feel of site
Description of use of graphics, multimedia, and special technologies
Description of writing techniques to be used, such as definitions, tone, etc.

DEVELOPMENT PLAN
Projected timeline
Potential challenges or issues

> ☑ **Example**
> This document describes some of the e-commerce services that the City can offer through its web site, and how such an "e-commerce enabled" web site might appear.
> *The Concept Document for the City of Grande Prairie (Alberta)*

▷ Conceptual Information

Conceptual information is an explanation of a principle, theory, or process. Conceptual information helps readers understand technical information—usually in educational, informational, or technical support sites.

Tips

- Consider putting conceptual information on the main page or as a link.

> ☑ **Example**
> Children Nodes: Nodes which are parented by grouping nodes and thus are affected by the transformations of all ancestors. See "Concepts - Grouping and Children Nodes" for list of allowable children nodes.

- Consider putting the information in one long document so readers can see the full context.
- Use transitions within paragraphs.
- Relate to known concepts.

> ☑ **Example**
>
> Dynamic HTML is not really a new specification of HTML, but rather a new way of looking at and controlling the standard HTML codes and commands. When thinking of dynamic HTML, you need to remember the qualities of standard HTML-- especially that once a page is loaded from the server, it will not change until another request comes to the server. Dynamic HTML gives you more control over the HTML elements and allows them to change at any time, without returning to the Web server.

- Break up text with headings and internal links.
- Provide overviews.

> ☑ **Example**
>
> There are two main parts to dHTML:
>
> - Document Object Model (DOM)
> - Scripts
>
> **DOM . . . (***this section contains a link to a W3 organization document on DOM)*
> **Scripts. . .**

- Use examples in context and continue these examples throughout (e.g., case studies).
- Provide links to the following:

  - ☐ Expert Internet and FTP sites
  - ☐ FAQs
  - ☐ Newsgroups
  - ☐ Details

> ☑ **Example**
>
> This section provides conceptual information about analyzing a site. For step by step instructions, see analyzing a site (instructions).

  - ☐ Resources
  - ☐ Definitions and glossary
  - ☐ Conceptual diagrams

> ☑ **Example**
>
> Figure 1. Conceptual diagram of a galvanic cell.

▷ Conciseness

Conciseness is using the minimum number of words to convey your message. Conciseness is one of the most important qualities of good Web page style. Reading screens is 25% slower than reading hard copy documents because online documents are more difficult to read. In addition, people scan Web pages and are usually in a hurry. Thus you should use half the number of words you normally would. However, avoid being too brief (telegraphic style) because

- The text becomes open to multiple interpretations.
- Readers must work harder to understand it.
- The tone becomes too impersonal.

See the Reference section of this handbook for a list of wordy phrases to avoid.

Tips

Using Visual Techniques

- Use the following techniques to reduce the number of words on the screen:
  - Use graphics to replace text.
  - Use tables to summarize information.
  - Use lists.
  - Layer information. Link to nice-to-know and background information.

Omitting Unnecessary Words

- Avoid strings of noun modifiers.

☒ **Example**

This page introduces the beginner to the World Wide Web (WWW), the Internet's **distributed hypermedia information** system.

- Avoid strings of adjectives (adjectives used with the following: *up*, *down*, *over*, *together*, *out*, *of*, etc.).

☒ **Examples**

Don't put structural markup **inside of** anchors. *(inside)*
If your business does have a computer, the Internet could be **hooked up** to your system in one day. *(connected)*

- Avoid *that*, *who*, and *which*.

☒ **Examples**

There are many editors **that** you can use **that** will assist with creating the HTML tags. *(Many editors help you create Web pages.)*

- Avoid overusing prepositions.

☒ **Examples**

-The most important part **of** the functionality **of** the site is meeting the marketing goals **of** the site. *(A site's most important function)*
-The main area **in** which we work at ____ is **in** the web page design field. *(Our focus is Web page design)*

-The descriptions used **on** all Web pages **on** this Web site are representations **of** the products or services supplied **by** the distributors or vendors when available. *(This site's Web pages accurately describe)*

- Avoid unnecessary prepositions with verbs.

☒ **Examples**

-I have **separated out** different attributes that can be applied to the same tag onto separate lines. *(separated)*
-If you already have Shockwave installed, **continue on.** *(continue)*

- Avoid weak linking verbs (*am*, *are*, *been*, *being*, *is*, *was*, *were*) and use one strong action verb.

☒ **Examples**

-You **will want** to test your Web pages with several different browsers to see what effects will take place (how your pages are displayed). *(Test your Web pages)*
-One way to make your web site easier to use **is to do testing** with users. *(Test your Web site)*

- Avoid verbs with *to*.

☒ **Examples**

You will **need to** download and install the plug-in. *(Download and install)*
HTML documents **have to be** posted on a server. *(must be posted)*

- Avoid verbs converted to nouns (stretched-out verbs or nominalizations).

☒ **Examples**

-The failure to fill in all required fields *(fail)*
-Take into consideration the cost of maintaining WWW data. *(Consider)*
-Can the **verification of links be carried out** regularly? *(Verify)*
-We offer a full array of services, from custom Web page design and development**, to the hosting, maintenance, and marketing/promoting of** your Web site. *(hosting, maintaining, marketing, and promoting)*

- Avoid redundant adjectives, adverbs, conjunctions, and phrases.

☒ **Examples**

-*Adjective*: close proximity; *Adverb:* rise up; *Conjunction:* and also; *Phrase:* each and every
-The HTML spec recommends that headers be used in **sequential order.** *(sequentially)*
-This **revolutionary new** product lets you add audio to your Web site. *(revolutionary)*

-The Internet is essentially a whole bunch of computers **connected together** by wires. *(connected)*
-Our Web Page Enhancers provide companies with **new and innovative** ways to make their Web sites more effective. *(new)*

- Avoid phrases that repeat the meaning of a word.

☒ **Examples**

-The Writer's Guideline Listing is a service provided **free of charge** to the Internet community. *(free)*
-If your graphics are **too large in size**, they can take forever to load. *(large)*

- Avoid wordy phrases that can be replaced with one or two words.

☒ **Examples**

-For the most part, images do enhance and add to a site when properly used. *(usually)*
-In the near future, the bulk of this site will have this user-friendly, easy-to-navigate design. *(soon)*
-In order to have a popular site, you've **got to** offer something to the user. *(to . . . offer)*
-The main reason is that all graphical browsers support this format. *(because)*

- Avoid unnecessary transitions.

☒ **Examples**

Omit all the bolded words in the following excerpts:

-Well, there is finally a solution to this age-old problem of Netscape browser offsets. **Actually,** there are two solutions.
-What this means is that you will be judged by your web site.

- Avoid intensifying words and vague adjectives, such as *very, really, usually, quite, a bit, as a rule, mainly, many, somewhat, well, actually*, etc.

☒ **Examples**

Omit all the bolded words in the following excerpts:
-GIFs are **very** common and **very** popular.
-There are **many, many** marketing considerations that need to be addressed when designing a web site.

- Avoid phrases that unnecessarily define the noun (such as *city of, in color*).

☒ **Example**

The main title of the **original page** is in a **light green color** that does not stand out well. *(page is light green)*

- Avoid parenthetical expressions.

> ☒ **Example**
>
> *Split the following into several sentences:*
>
> Another problem with using headers incorrectly is that document indexing software will not be able to properly weight the text (that is, text in headers is generally given a higher precedence than other text when you are doing searches--using headers for font control could make it difficult for you or others to search your documents and find the relevant information they are looking for).
>
> A lot of web designers feel the use of splash pages (good looking pages with no info, just a graphic and a "enter" hyperlink) are nice looking.

- Avoid clichés.

> ☒ **Examples**
>
> *Omit all the bolded words in the following excerpts and substitute new text:*
>
> -I have seen **bits and pieces** of information about HTML tips and tricks. I have compiled some of my favorite tutorials and added my **.02 cents** worth. I like them all for different reasons; see if you can figure out why as you go.
> -They aren't all **cut and dry.**

- Avoid phrases that are clues of wordiness, such as *as mentioned above, needless to say, moving right along, the point is that, I don't mean to say.*

> ☒ **Examples**
>
> *Omit all the bolded words in the following excerpts:*
>
> **-It should go without saying** that there are a wide variety of resources for writers online, both on online services and (particularly) on the Web.
> **-What I mean is that** if you just read the text, you shouldn't notice the link unless it was underlined.

- Avoid beginning a sentence with *there is* and *it is.*

> ☒ **Examples**
>
> **-There are** many important features that must be considered when creating a business Web site. *(You must consider several important features)*
> **-It is** our objective to combine all of these skills together to make the best-rounded web site design team on the web! *(Our objective is to)*

- Begin sentences with the important words.

> ☒ **Examples**
>
> **-The reason** the Internet has grown so fast **is because** it is based on the free exchange of information. *(The Internet has grown)*

-What you're doing when you view a web page is telling your browser to fetch a document and display it. *(When you view a Web page)*

- Avoid *is when, is where, is how, is what.*

☒ **Examples**

Inside these two tags **is where** you will put the contents of your web page. *(You will put . . . inside)*
-A browser **is what** you're viewing this page with. *(You are viewing this page with a browser)*
-An affordable web design service **is what you get when** we design your web site. *(You get)*

- Avoid a noun followed by a phrase that defines it. Use adjectives instead.

☒ **Examples**

-The following is a list of **places on the Web** that will assist you in creating Web pages. *(Web sites)*
-Here is a list of Web site services **which are free.** *(free Web site services)*
-This is a comprehensive list of mailing lists **that is searchable.** *(searchable list)*
-These types of graphics can be **of a simple nature.** *(simple graphics)*

- Avoid unnecessary references to yourself or the reader.

☒ **Examples**

In my opinion, the best two HTML editors are HotDog Pro and Microsoft FrontPage. The only problem here is that it's simply invalid HTML and some browsers will terminate it for you, giving the end user the undesired result of loosing some of the markup that **you, the author,** intended. *(omit first phrase, omit "the author," and use shorter sentences.)*

- Combine short, related sentences to avoid repeating words.

☒ **Example**

Index-style sites organize information in some kind of **structure. The structure** is designed by a human being. *(manmade structure)*

- Avoid unnecessary repetition. Use it sparingly for emphasis.

☒ **Examples**

Omit all the repeated bolded words in the following excerpts:

-Did I mention that GIF images are limited to 256 colors? Okay, GIF images are currently limited to 256 colors.
-You have a consultation with a Web site design specialist to over the goals **of the Web site**, the content **of the Web site**, ideas and concepts associated **with the Web site**, and the overall look **of the Web site**.

-In order for your school's site to be effective in communicating its messages, it must be **easy to** find, **easy to** read and **easy to** navigate.
-We go direct to the source and use their guidelines, **their** categories and **their** criteria to your advantage.

☑ **Example**

Repetition is used here for emphasis:

Before publishing anything on the web, proofread, proofread, proofread!

- Avoid hesitant wording because it adds unnecessary words (use a confident style).

☒ **Example**

Omit all the bolded words in the following excerpt.

These pages are **meant to be** a collection of Internet statistics in one place.

- Avoid adding unnecessary endings to nouns (e.g., *-ting, -ation, -ment, -istic, -ology, -ity, -ize*).

☒ **Example**

If you're **architecting** the information in a Web site . . .

- Avoid a telegraphic style.

▷ Confident Style

You convey confidence by your tone and word selection. Many Web sites begin with apologies and words that indicate the author is uncertain of the value or accuracy of the information. Using a confident style helps readers believe your site is accurate and credible. If you begin with apologies, readers have no reason to continue reading.

Tips

- Use a confident style, especially in the introduction.
- Avoid words and phrases such as *hope*, *tries*, *intended to be*, and *meant to be* that indicate you are not confident in your own writing.

☒ **Examples**

-If you get anything at all out of this site, or if you have any comments or suggestions on how to improve and make this site better, please let me know.
-Let me know if there are any confusing sections so I can improve it.
-On this page I am going to try to explain how to write most of the simple html.
-This document tries to tell how PC mouse hardware works and how to read it at the lowest level.
-This page is not simply under construction: it is in fact a warehouse of spare links that I'm planning to test and insert in the appropriate pages over time. It is not *really* meant for public consumption, and I take no responsibility for the consequences... On the other hand, if you feel brave enough to attempt it, here it is...

-I am not an expert, and will never claim to be one. What I have learned has worked for this web site and hundreds of others I and my company have been involved in over the years.

- Avoid apologizing for your Web site.

☒ **Examples**

-This site is incredibly unimaginative and generic. I apologize for this site being very generic, but it will get the job done.
-We apologize for this site being out of date.
-I apologize for this site being incomplete and for typo's.

- If your page is out-of-date, remove it. Use updated information.

☒ **Example**

Due to other projects and work I have not been able to update this site for quite a while. The quality of both the content and the graphic can be done a lot better. However, instead of closing the page until I've got the time to improve it I let it be online instead because there is still plenty of information which is valuable for a lot of people (they say).

▷ Consistency: Layout

A consistent design for your Web site is a method of providing uniformity. You should also use consistency in wording. Consistency gives your Web page a unified and predictable look and feel. Readers also know what to expect, know where to find information, and can quickly learn how to use your site.

Tips

- Be consistent with the following:
  - Format and screen design
    - Overall page grid
    - Background
    - Colors
    - Fonts
    - Graphics
  - Position of elements
  - Labeling, headings, titles, links
  - Organization, sequence, hierarchy
  - Navigational aids (same position, look, sequence). Navigation should also be consistent with standard Web practices.
  - Interaction
- Create a style guide with rules for layout to ensure consistency.
- Consider using Cascading Style Sheets (CSS) to aid formatting consistency.

▷ Consistency: Style

Consistency is using a similar punctuation, style, wording, and tone. Consistency in layout is an important design technique. A consistent writing style

- Helps readers use your Web site and read more quickly.
- Saves time because readers quickly learn how to use each page.
- Reduces information overload.

Tips

- Be consistent in the following areas:
  - Organization and sequencing of information
  - Introductions, overviews, and lead-ins

> ☒ **Example**
>
> This page contains information on:
> Workers' Compensation Investigations
> Liability Investigations
>
> This page discusses the following types of special investigations:
> Background Investigations
> Criminal/Civil Investigations

  - Word choice and terminology

> ☒ **Examples**
>
> Type your name.
> Enter your name.
>
> Select **Next**.
> Click **Next**.

  - Use of business names and proper names
  - Method of defining terms and acronyms
  - Spelling
  - Punctuation
  - Grammar
  - Facts
  - Tone
  - Formality (e.g., use of contractions and personal pronouns)

> ☒ **Examples**
>
> *This excerpt uses informal terms such as* wow, neato, *and* bugs, *along with words such as* utilize *(just say "use") and a technical term,* extensions.
>
> I often come across a web page and say, Wow...How did they do that? Most of these neat-o pages utilize the many extensions (and sometimes bugs) of Netscape Navigator and Microsoft Internet Explorer.

- ☐ Use of numbers, measurements, abbreviations

- ▶ Check for consistent wording of all text, including links, headings, body text, and footer information.
- ▶ When several authors contribute to a Web site, unify their style by creating an *online style guide* and *editorial style sheets*.

> ☒ **Examples**
>
> *Note the way the language in the following is inconsistent:*
>
> A client/server system works something like this: A big hunk of computer (called a server) sits in some office somewhere with a bunch of files that people might want access to. This computer runs a software package (uh...also called a server unfortunately) that listens all day long to requests over the wires.
> The "wires" is possibly a twisted pair network hooked into a local telephone company POP or a cable or fiber optics network hooked up to a corporate WAN or LAN that is also linked up to the national telecommunications/information infrastructure through a local telephone company.

▷ Contact Information

Contact information includes names, addresses, and numbers to help readers

- ▶ Get hold of you or people in your organization.
- ▶ Get any crucial information they need.
- ▶ Ask questions.
- ▶ Provide feedback.
- ▶ Know who is responsible for the site.

Many Web writers do not provide this critical information. Be aware that providing this link may inundate you with messages.

HTML Code: <ADDRESS> </ADDRESS> The *mailto: your-email@address.com* command provides a link to your e-mail address from the Web page. When users click the MAILTO: link, the browser launches the e-mail program. You can provide readers with an electronic mail address so they can contact you.

Tips

What to Include

- ▶ Include any of the following:

  - ☐ All responsible for the site, including the Webmaster
  - ☐ Contact person
  - ☐ E-mail address
  - ☐ Employee and department lists
  - ☐ Fax number
  - ☐ Hours
  - ☐ Link to FTP or other related company/organization sites
  - ☐ Mailing addresses
  - ☐ Names and titles

- ☐ Organization
- ☐ Pages in your Web site they should visit
- ☐ Phone numbers
- ☐ Web addresses

▶ For specialized sites, include contact person/e-mail address for items such as

- ☐ Permissions
- ☐ Support information

LOCATION

▶ Provide contact information in the following locations:

- ☐ On the home page

> ☑ **Example**
>
> U.S. Geological Survey, a bureau of the U.S. Department of the Interior
> This page is brought to you by the Earthquake Hazards Program
> URL: http://earthquake.usgs.gov/

- ☐ At the bottom (footer) or top of each page. This is an abbreviated form of the contact information.
- ☐ On a separate page if it is long
- ☐ As a link (e.g., called Contact or Contact Information)

▶ Be consistent in providing and positioning this information on each page.

FORMAT

▶ Put each item on a new line (using the <BR> tag).
▶ Use headings to make information easy to skim.

▷ Content

A key step in planning a Web page is deciding what you will write about—the type of information you will provide. Successful, award-winning sites provide high-quality content. "Useless" Web sites have no purpose and are a waste of readers' time.

Tips

▶ Determine who your target audience is and the types of topics that are important to them.
▶ Identify sources of content: both internal and external writers and experts.
▶ Provide content that is relevant, useful, interesting, valuable, fresh, and original.
▶ Provide both depth and breadth of content. However, strike a balance between providing too little and too much information.
▶ Provide links to FAQs, archives, background material, relevant sites, interactivity, and other support information.
▶ Provide readers with new content (updated information) each time they visit (information, news, job postings, etc.). Encourage repeat visits.

- Avoid creating a Web page that contains nothing on it but your company logo, name, address, and e-mail link.

> ☒ **Examples**
>
> *These excerpts come from Web pages that contain only this information:*
>
> E-MAIL us here for price quotes and support questions
> Coming soon
> This page is not up yet but will be soon . . . so be sure to come back.

- Even if you provide a simple service, deliver content related to your business.

> ☑ **Examples**
>
> *An eye doctor could provide information about eye care and laser vision correction. A car insurance site could provide information about car ownership, tools for checking status of claims, and information for agents. A hotel could provide a meeting planner. An airline could provide travel tips.*

- If you are publishing content on an organization's site (e.g., a school), check with their policies and guidelines for appropriate content.
- Inform, educate, and entertain rather than write obvious marketing material.
- Don't take your existing content and simply put it online. Rework it using the principles in this handbook, such as chunking, layering information, and making text skimmable.
- Make sure your content is well-organized and well-written.
- Test your site to determine how your target audience feels about your site's content.

▷ Context Independence

Each Web page, topic, and page title should be independent. You never know

- If a reader has entered your site through the home page.
- Which Web pages a reader has visited.
- Which sections on a long scrolling page a reader has read.
- If the page will be printed.

These factors affect your use of the following:

- Contact information and other identification techniques
- Definitions
- Conceptual information
- Pronoun references
- References to sources
- Transitions

Tips

- Make sure each page can stand alone.
- Make sure page titles make sense out of context. For example, page titles appear in reader's bookmarks or shortcuts.

- Try to cover one narrow topic per page.
- Keep pages short.
- Avoid spreading closely-related content over multiple documents.
- Do not require that readers visit another page to understand the content. Repeat the main points readers will need from chunk to chunk so they understand the context.
- Always provide a context for required background information.

> ☑ **Example**
>
> **What we've covered so far in Part 1:**
> The parts of an HTML file (head, title, body)
> Paragraph and line breaks
> Displaying a title
> Font sizes
> Displaying text
> Bolding
> Centering text

- Do not rely on providing critical information through links to other pages or sites.
- Provide links to prerequisite definitions and other background information.
- Provide contextual clues.

▷ Contextual Clues

Contextual clues help readers locate their position in a Web site. These clues include the following:

- Breadcrumbs. Links (separated by >) that indicate the path a readers has taken in a site. Like the breadcrumbs in *Hansel and Gretel*, they help readers find their way back home. You usually can click any of the links to navigate.
- Color-coding.
- Document and chapter names (path showing the location in a large document).
- Headings.
- Identity graphic and logo.
- Menus and navigational aids that show the site content and organization.
- Organization name, site sponsor, and contact information.
- Page number location and total number of pages.
- Titles.

Showing the context helps readers who are browsing a large Web site, coming to your site from a search engine or another Web page, and reading a printout of the page. Web readers also need to know how long the text is and where they are located: beginning, middle, or end.

Tips

Labeling Clues

- Put a title and site name on every page to help readers who do not enter the site at the home page.

- For the page <TITLE>, be specific (e.g., page number/total pages). This technique especially helps readers who find your page through search engines.

> ☑ **Example**
>
> Freeware Guide to Internet Security, Page 4/6

- Consider identifying the type of page somewhere at the top.

> ☑ **Examples**
>
> This is the Start Page
> This is the Introduction Page
>
> **Optics**
> a section of photo.net, maintained by David Jacobson

- Suggest ways readers may have entered and where to go.

> ☑ **Examples**
>
> -If you arrived at this site by traveling through our Ease of Use site, this link will take you back. If you have not visited our Ease of Use website, take the tour.
>
> -If you entered Web66 here, you might want to visit the Web66 Home Page.

Navigational Clues

- Provide navigational clues about the context.

> ☑ **Example**
>
> Home
> **New** | Popular
> Books
> About
> Submit | Feedback

- Show the path the reader has taken to navigate.

> ☑ **Examples**
>
> **Home**: Web Design Services ◀ **YOU'RE HERE**
>
> Home : Business and Economy : Companies : Computers : Hardware : Personal Computers
>
> ▼ **Development** ▶ **Hosting** ▶ **Advertising** ▶ **Maintenance**

- Show the table of contents. Indicate the active page.

☑ **Examples**

Chapters
Table of Contents
Introduction
Basic HTML
Text
Images
Lists
Anchors
Tables
Frames
Forms

III The Boat

1. ★ Terminology
2. Boat Types
3. How to measure length
4. Hull designs and uses
5. Types of hulls

- Use an index and map (site).
- Suggest *next* links at the bottom of the page, if appropriate for the topic.
- Link to related topics.
- Do not link to a graphic or other copyrighted element in its original source. The copyright of this element may be context dependent. For example, an element on a Web site may be covered under the "fair use" law. If you link to this element, you remove the context and thus the copyright protection.

Visual Clues

- Provide visual cues about location.
    - Arrows can show hierarchy.

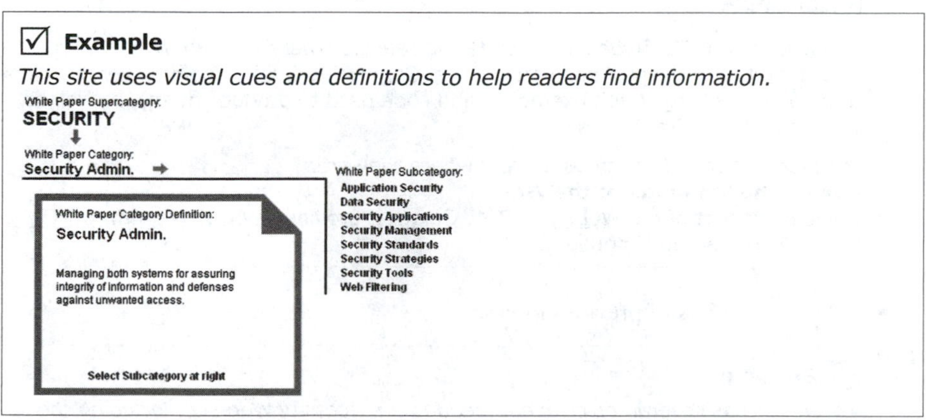

- Tabs at the top and along the side can show location.
- Graphics in a main menu can be repeated in individual topic pages.

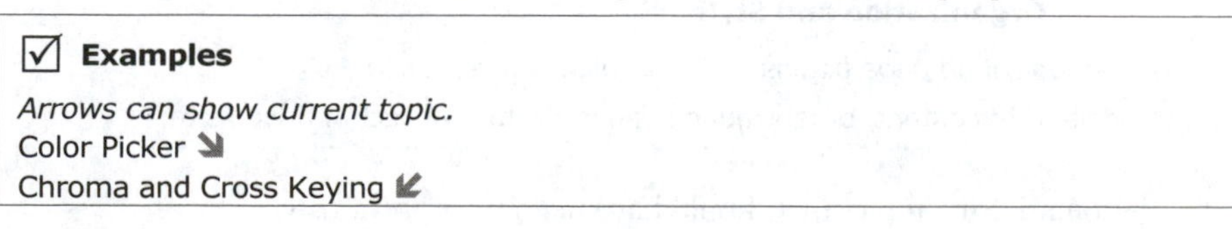

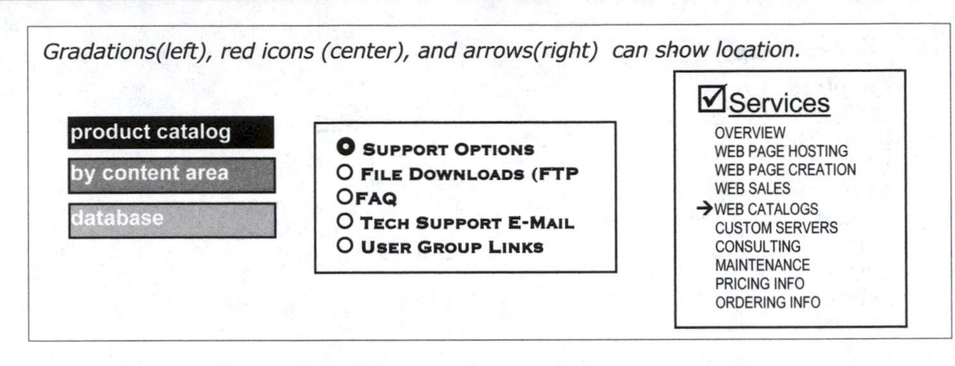

- Make the design of the site distinctive, with a consistent look and feel to all pages.
- Tell readers how large the site or document is by using text or icons.

> ☑ **Examples**
>
> This site currently contains over 5800 pages.
>
> 34 total subcategories. Displaying 1 - 10.
>
> *These icons let readers know they are on page 1 of 7* ■ ▸ ▸ ▸ ▸ ▸ ▸

Verbal Clues

- Provide overviews.

> ☑ **Example**
>
> *The WebCom HTML Guide begins by telling readers where they are in the site:*
> This is the top level entry point for the WebCom HTML Guide. Topics covered include commonly used tags (with examples, and code used to produce them), as well as stylistic and technical issues.
>
> *The first section tells reminds readers where they are:*
> **HTML - the language of the Web**
> This page is part of the WebCom HTML Guide. It provides a brief overview of HTML, (Hyper Text Markup Language).

- Remind readers of previous topics.

> ☑ **Example**
>
> *A Web site has several topics about creating a successful Web site, including the following pages.*
>
> **Deciding on the Content**
> **Organization and Style**
>
> *The Organization page begins with the following sentence:*
>
> In addition to content, organization is an important aspect of a web site.

- Summarize the topics that should have been read previously.

☑ **Examples**

Once you get a grip on the basics (What's HTML? What's a URL?) and master the art of viewing source, you can brush up on all things HTML - from links to font tags to frames.

Once you've got a general understanding of how to organize your whole site, it's time to focus on individual pages. While many aspects of Web page design are similar to those of any other medium, some variables make the Web unique.

Now that you've made the layout and basic design decisions to tie your site together, you're ready to create a **'storyboard'** of your site, a well-planned and easily navigated hierarchy of pages devoted to your message.

- Use keywords in the first sentence.

☑ **Example**

This is the sixth topic (Tools) in a section about Planning your Web site. The phrase in the planning stage reminds readers of the context.

Development Tools and Technology
Use appropriate tools and technologies
In the planning stage, you need to begin thinking about which technologies you will use to build your site.

▷ Contractions

A contraction is a shortened word or phrase in which an apostrophe substitutes for the missing letter. Commonly-used contractions on Web pages include *aren't, can't, don't, here's, I've, I'll, it's, let's, there's, they'll, they're, we're, you'll, you're, you've.*

Tips

- In general, avoid contractions
  - ☐ Because punctuation is difficult to see online.
  - ☐ When you want a professional tone.
- Use contractions when you want
  - ☐ An informal, friendly style.
  - ☐ A more personal tone.
  - ☐ Speedier comprehension. Contractions can make sentences easier to read quickly.
- Check contractions for correct use of apostrophes. Especially check for these common mistakes: it's/its, we're/were, who's/whose, you're/your.

▷ Contrast

Contrast is a design technique that highlights differences among areas of a page. Contrast

- ☐ Adds visual interest
- ☐ Creates emphasis

- Controls focus
- Shows the hierarchy of information
- Aids scanning

Tips

- Make sure there is high contrast between text and the background. Preferably, use dark text on a light background.
- Use contrast to emphasize important information.
- Avoid overusing contrast to avoid visual "noise."
- Use techniques such as differences in the following:
    - Size (large vs. small elements)
    - Color (black and white vs. color; bright/saturated vs. muted colors; warm vs. cool colors)
    - Weight (thin vs. thick or bold elements)
    - Shapes (contrasting or irregular shapes)
    - Direction (vertical vs. horizontal elements)
    - Texture (smooth vs. textured backgrounds)

▷ Copyright Issues

By illegally publishing or distributing someone else's work on your Web page, you may be guilty of violating copyright law. On the other hand, you should create a copyright notice to protect your own material. All original Web pages are protected under copyright laws even if they do not have a copyright notice. Before creating a Web page, you should be aware of copyright issues.

Tips

General Copyright Rules

- Always give credit to sources of information that you use on your Web page by using the following techniques:
    - Linking to a works cited or references section
    - Linking to a bibliography in the table of contents or index
    - Using cross-references to documents on other Web sites
- Do not copy material. This rule even includes Web sites that do not contain a copyright notice.
- Do not use copyrighted material even if you do not charge for it or give credit to the owner.
- Use brief quotations or synopses rather than verbatim copying of text. Then give the source.
- Do not use material from another Web site without permission.
- Obtain permission in writing from the work's owner, creator, or heirs. Remember that the person who posts material on a Web site may not own the copyright.
- Make clear what is your material and what is not.

Site Design and Code

- Avoid copying another site's design too closely.
- Do not copy another site's HTML code, slightly modify it, and then reuse it as your own.

Linking

- Link to other sites, but make it clear that they are not your work.
- Ask permission before using the following types of links:
    - Links that bypass a linked site's home page.
    - Links that result in other sites displaying within frames on your site. This technique makes it appear that their site is part of your site.
    - Links to only images or other elements from the original sources.
    - Links to trademarks.
- Check with the sites you link to for *Terms and Conditions,* a *Link Request* form, citation guidelines, or other instructions.
- Do not use link lists that other people have compiled.

Exceptions

- The exception is fair use: you can copy a brief excerpt of text for study, research, education, teaching, scholarship, reporting, reviews, commentary, criticism, comment, or parody.
  Fair use is determined by the purpose and character of the use, the nature of the copyrighted work, the amount used, and the effect of its use on the market value of the copyrighted work.
- You can copy *public domain* work (work published more than 75 years ago or with lost copyright, most federal government documents, and clearly-donated works). However, most information on the Internet is relatively recent and thus under copyright.

Graphics and Multimedia

- Do not use material from another Web site without permission or unless they are specified as being free. This material includes graphics (drawings, photos, backgrounds, icons, logos or registered trademarks), video clips, and sound files.
- Do not consider that by redrawing a graphic you are free from copyright laws.
- Do not use a person's name, face, image, or voice for commercial benefit without permission.

▷ Copyright Notice

Use a copyright notice at the bottom of each Web page, or provide a link to a separate page (called *Legal, Copyright, Terms and Conditions*, or other descriptive title). You may want to combine the copyright information with a disclaimer. A copyright statement notifies readers that it is your original work. It shows you legally own the contents.

HTML Code: The copyright symbol © is created by using © (this is the form for special characters).

Tips

- Include a copyright notice in the signature for your pages to remind viewers this is your property:

  Copyright © [dates] by [author/owner]
  All rights reserved

- For additional protection:
  - ☐ Register your Web site with the United States Copyright Office.
  - ☐ Use watermarks in your images to mark them with code.
  - ☐ Use meta tags to include your name in the HTML code.
- Consider providing a Permission link to a separate page. On this page provide contact information for requesting permission to use information from your Web site.
- If you would like readers to use your materials, state how you wish to be acknowledged.

> ☑ **Example**
>
> Permission to copy or disseminate all or part of this material is granted provided that the copies are not made or distributed for commercial advantage, the ___copyright and its date appear, and notice is given that copying is by permission of ___. To disseminate otherwise, or to republish, requires written permission. For further information, contact:

- Create a document clearly stating the following *terms and conditions of use*:
  - ☐ Conditions for using materials from your site: documents, graphics, HTML code, databases, name, logo, etc.
  - ☐ Information on using trademarks for names.
  - ☐ Conditions for printing, downloading, publishing, distributing, reproducing, transmitting, reengineering, or using on a Web site.

> ☑ **Example**
>
> No ___ publication is to be copied, duplicated, modified or redistributed in whole or part without the prior written permission of ___

  - ☐ Conditions for linking to your Web site. Some pages provide a *Link Request* form or instructions.

> ☑ **Examples**
>
> Reprint and Linking Permission Information
>
> Instructions and images for linking to desktopPublishing.com can be found here.

- Link to information about using your materials.

> ☑ **Example**
>
> Reuse This Content

- If you would like readers to use your materials, give citation guidelines:

> ☑ **Example**
>
> Cite this Site
>
> David Nagel, "Research: Blended Versus Online Learning," Campus Technology, 3/12/2007, http://www.campustechnology/article.aspx?aid=45404
>
> copy text (above) for proper citation

▷ Course Offering

A course offering page announces a course or seminar. A Web page announcement provides an effective way to distribute and update information and is an effective way for readers to register or request information.

Tips

- Include the following information (if applicable):
  - ☐ Brief description
  - ☐ Certification information
  - ☐ Course benefits
  - ☐ Course objectives
  - ☐ Course scope
  - ☐ Course summary
  - ☐ Course title
  - ☐ Dates
  - ☐ Delivery method
  - ☐ Instructor
  - ☐ Intended audience
  - ☐ Language
  - ☐ Length
  - ☐ Link to other courses, catalog, registration form, sample courses, sample course workbooks
  - ☐ Materials
  - ☐ Meals
  - ☐ Prerequisites
  - ☐ Price
  - ☐ Registration information
  - ☐ Requirements (e.g., computer specifications)
  - ☐ Syllabus
  - ☐ Testimonials
  - ☐ Textbook
  - ☐ Topics covered

- Arrange courses in a logical order; also provide a search feature and ability to browse (alphabetically, by location, by topic groups).

> ☑ **Example**
>
> **Public Speaking**
> Learn how to design, develop, and deliver effective presentations.
>
> > Presentation Strategies for Scientists and Engineers
> > Presentation Strategies for Executives

- Use headings to chunk information: instructor, times, topics, location dates, length, price, contact information (phone, e-mail, address).
- Put the important information (who, what, where, when or a summary/course description) on the first page with links to support information and details.

> ☑ **Example**
>
> **Description:** This course is geared toward Systems Engineers, and Command and Telemetry Database Designers who will be defining and populating the command & telemetry databases; experienced users and operators that might need to modify or read the databases in native format (SML). Knowledge engineers who will be writing scripts and rules that will use that information will also find this class useful.
> **Length:** 1 Day
> **Prerequisites:** Windows 95/NT, C/C++ required for the application building.
> **Price:** $495
> **Target:** Systems Engineers, Command & Telemetry Database Developers
> **Topics:** XML Basics
> XML Extensions
> SML Specification
> SML Tools
> Building an SML Reader Application
> Class Details
> Class Schedule

- Make action steps easy. Provide links to an e-mail address, registration form, and request to be on an e-mail list.

▷ Credibility

Credibility is your believability—how readers trust the information you provide. It is the overall impression readers have of your competence. Web readers are concerned about the reliability of information.

Tips

STYLE

- Use good writing techniques, with no grammatical or punctuation errors.
- Use a confident style.
- Avoid exaggeration.

> ☒ **Example**
>
> We spend countless hours researching the latest technology, tools and techniques to stay on the cutting edge.

- Avoid marketing language, ads, and promotional material.
- Use a professional tone.

Layout and Page Components

- Use a professional page design.
- Use a professional site name/title.

> ☒ **Example**
>
> *Many sites forget to put their site name in the TITLE tag:*
> Web Site Title
> *This type of error may affect your credibility.*

- Use a credible URL.

> ☑ **Example**
>
> *One way people judge the credibility of a Web site is the address. For example, an address with a tilde (~) in the address usually indicates a personal rather than professional Web site. An .edu extension indicates a student or faculty member at an educational institution. An .org extension indicates an organization.*

- Provide a footer with date stamp and e-mail link.
- Use high-quality graphics.
- Avoid using counters.

A Web counter shows how many visitors the site has received: e.g., **1798097**. Many design experts advise against it. Counters are often not accurate and do not really help readers. Also, a low number may make your site appear unpopular. A link to site statistics may be more believable.

Links

- Provide information about the following. This information can be incorporated in the introduction, section such as *About . . .* or links to separate pages or sites.
  - ☐ Author
  - ☐ Contributors

> ☑ **Example**
>
> *The Free Online Dictionary of Computing (FOLDOC) provides a link to a list of over 1960 contributors and the guest editors.*

  - ☐ Your company or organization

- ☐ Group responsible for publishing the information and any sponsors

> ☒ **Example**
>
> Unfortunately I can't devote as much time to it as I would like. If you notice a link has changed, please let me know.

- ☐ Publication information that shows how long the site has existed
- ☐ Number of topics and the depth of coverage (scope)
- ☐ Why the page was written
- ☐ Awards and commendations the site has received

> ☒ **Example**
>
> *Avoid vague accolades such as* Top 5% of All Web Sites. *According to whom?*

- ☐ Testimonials
- ☐ Portfolio
- ☐ Biography
- ☐ Resumes and other credentials
- ☐ E-mail and other contact information

- ▶ Link to respected, reputable external Web resources that verify what you are saying and show you have researched the material.

CONTENT

- ▶ Check facts for accuracy and make it easy to verify information.
- ▶ Use updated information and state how often you keep information current.
- ▶ Use an effective organization.
- ▶ Make the site useful and easy to use.
- ▶ Allow readers to search past content.

▷ Cross-References

Cross-references are links to closely-related information within your own Web site or on other Web sites. Cross-references make related information accessible by providing shortcuts.

Tips

WHEN TO CROSS-REFERENCE

- ▶ Use cross-references for the following:

    - ☐ Sources of statistics.

> ☑ **Example**
>
> According to the Pew Internet and American Life Project, about six million U.S. adults are subscribed to RSS feeds, and in a study by Slashdot, 73 percent of RSS subscribers said they will increase their use of RSS.

- ☐ Closely related topics and other kinds of information about the same subject.

> ☑ **Examples**
>
> **Similar sites:**
> The WDVL | WebDeveloper.com
> A host of other vendors plan to introduce commercial two-way cable modem systems in the next 8 months. See the Cable Modem Vendors page for more information.
>
> The Internet Movie Database provides cross-references to the title, keyword, year, genre, production company location, character name, or cast/crew name.

- ☐ Related pages in your own Web site.

> ☑ **Examples**
>
> Additionally there are libraries of graphics that can be added to Web pages to prevent authors from having to create their own. (See the Links page.)
>
> See 401(k), SEP-IRA, SIMPLE-IRA, and Keogh to learn how other tax-deferred retirement plans offered by businesses differ from a SARSEP-IRA.

- ▶ Avoid excessive cross-references because they disrupt reading, make it difficult to follow idea flow, and make it easy for readers to become disoriented.

> ☒ **Example**
>
> *E.T.* comes home on DVD for the first time. Steven Spielberg's beloved classic arrives in a two-disc package that includes the 2002 remastered version of the film and over eight hours of bonus features. In addition to the behind-the-scenes footage and interviews with the cast and filmmakers (including Spielberg) there's trailers, the evolution and creation of E.T., the music of John Williams, and much, much more.

How to Cross-Reference

- ▶ Let readers know what type of information they will learn from the cross reference.

> ☑ **Example**
>
> **If you need more help**
> - Learn more about buying with eBay's online help.
> - Take a guided audio tour of bidding.
> - Try an introductory tutorial on bidding—at your own pace.
> - Or try placing a practice bid.
> - Learn more about bidding on vehicles at eBay Motors.
>
> **Related Help topics**
> - Getting Started Overview
> - How to Find Items
> - How to Sell

- ▶ Provide annotation for lists of cross-references.
- ▶ Distinguish support/explanatory material from the main material (e.g., distinguish among the title, author, and information about article).

- Be sure to get permission to use the material (copyright issues).
- Make sure cross-references will remain correct even if you update the page (e.g., avoid using section numbers).

WHERE TO CROSS-REFERENCE

- Put cross-references in the following locations:
  - ☐ Embedded in the text.
  - ☐ Before the discussion.
  - ☐ At the end in a *Related Topics* or *See Also* section. Provide titles and a brief explanation.

See also Optimizing Web Graphics for a discussion of Web graphic file formats.

Additional Discussion on These Methods
For more information about these methods, please see the following sections:
Multi-Modal Reading Logs
Panel Discussions
Talking a Draft

CROSS REFERENCES
• Publications and documents
Related topics
Emergency and Humanitarian Action (EHA) • Emergency Nutrition Network (ENN) • ACC/SCN's Refugee Nutrition Information System (RNIS)

Cross-references can appear in boxes:

learn more about...
- How coal was formed
- How coal is mined

SEE ALSO:
Related Resources
Related Books
Related Tutorials

  - ☐ On the page itself with links to additional or less important pages.

☑ **Example**

List more recommended EPA PESTICIDES web pages

- For cross-references to information on other sites, consider the following options:
  - ☐ Link to the information. The advantage is that information in the original may become updated. The disadvantage is that it may change location.
  - ☐ Use the information on your page. The disadvantage is that information will not be updated. You must also check copyright restrictions.
- For cross references to information within your own site, link directly to the information rather than the top of the page or beginning of the topic. Readers become disoriented when a link does not lead to the information they expect.

▷ Dangling Modifier

Dangling modifiers are opening phrases that do not logically modify any word in the sentence. They usually appear at the beginning of a sentence and fail to modify the noun that follows.

To fix a dangling modifier:

- Revise the noun following the phrase.
- Include the noun in the phrase.
- Combine the phrase and the main clause.

☒ **Before**

-By using this filename extension, a Web browser will know to read these text files as HTML.
-When designing a navigation system, it is important to consider the environment the system will exist in.
-When placed with the proper Search Engines, people will be able to access your advertisement page by a click of a button.

☑ **After**

-By using this filename extension, you will let a Web browser know to read these text files as HTML. (or) *Put the correct noun after the phrase.*
-If you use this filename extension, a Web browser will know to read these text files as HTML. (or) *Include the noun in the phrase.*
-Use this filename extension so the Web browser knows to read these text files as HTML. *Combine phrase and clause.*
-When designing a navigation system, you should consider. . . (or)
-When you design a navigation system, you should consider. . .
-When your advertisement page is placed with the proper Search Engines, people . . .

▷ Dashes

Dashes are used to set off phrases that interrupt the sentence, explain a noun, or provide emphasis.

Tips

- Use *em* dashes. They are created by using two hyphens and no white space or the em dash symbol (—).
- Use dashes to surround interruptions or pauses.

☑ **Examples**

Though glitzy sites are attractive—especially to designers—the author argues that too much use of flash technology kills business websites.

- Use dashes to surround a phrase that expands on or emphasizes a noun or noun phrase.

☑ **Examples**

-While there are many HTML tags—the pieces of a Web document that tell the browser how to display a Web page—you don't have to use them all. *(expands on)*
-Publishing a lot of content on a website—and keeping that content fresh—can produce logistical nightmares. *(emphasizes)*

- Use dashes to set off a phrase that contains a series of words separated by commas.

☑ **Examples**

Argues that people need to see themselves—their interests, their questions addressed—when they come to a site, not "about the company" shoved in their face.

▷ Date Stamp

A date usually appears at the beginning or bottom (signature and footer) of a Web page. Date stamps are crucial for the following types of Web pages: announcements, informational and educational sites, and online publications. Dating the material on your Web site helps readers to decide if they want to continue reading and to judge the reliability of your information. Many sites have poor site introductions because they apologize for outdated information.

The date stamp lets readers know

- When the page was last modified.
- How often it is updated.
- When certain sections were updated.

You can use JavaScript to create a date stamp.

Tips

- Provide a date at the bottom or top of each page.
- Let readers know when the information was created, how often it is updated, and the date expected for the next update.

☑ **Examples**

This document will be updated on the 1st of every month.
Procedure updated January 1, 2007.
Revised January 1, 2007
Last updated: 01/01/07 Next update : 01/10/07
This site was last updated June 22, 2007

- Direct readers to new information by using links to *What's New* or *Changes* sections from your main menu.
- Avoid *Under Construction* links.
- Always provide updated information so readers have a reason to return.

▷ Deep Linking

A deep link is a link to a page within a site rather than the home page. This technique lets readers go directly to the page they are interested in. However, there have been lawsuits about copyright infringement.

Tips

- Make each page context independent so readers can enter your site on any page.
- Provide contextual clues so readers know their location in the site.
- To avoid deep links to your site, use a script that checks the referring URL and redirects readers entering from any page other than your own site.
- Avoid deep links when
    - You require that readers read pages in a certain order.
    - You link to copyrighted material.
    - It is not clear that information you are displaying is on another site. This situation often occurs with frames.
    - The site you link to prohibits linking or requires permission that you do not have.

▷ Definitions

Definitions are used to explain technical terms, abbreviations, and acronyms that will be new to the user. Definitions help explain concepts and principles and accommodate novices.

HTML Code: <DL> </DL> definition list, <DT> </DT> definition term
<DD> </DD> data description (the definition)

Tips

Methods of Incorporating Definitions

- Link to definitions. However, avoid too many links because they are distracting.

> ☒ **Example**
> Other extra parts called **peripheral components** or devices include mouse, printers, modems, scanners, digital cameras and cards (sound, color, video).

- Incorporate definitions in any of the following ways:
    - Provide a Glossary link.
    - Use links to small popup windows.
    - Make it clear that the link goes to a definition.

☒ **Examples**

Do these links go to definitions or the Web sites?
Currently there are many Web browsers available. The Microsoft Internet Explorer and the Netscape Navigator are the most popular. A helper app that allows you to download sound files over Web pages in real-time. The player can be downloaded as freeware.

☑ **Example**

This unique address is called your IP address (see our lesson on Terminology for further definition).

- Define terms within the text itself.

☑ **Examples**

Setups that can display 256 colors use a palette, or color map, to determine which 256 colors they can show.

Encoding, which is also known as compression, is the process of turning a video clip on D1 tape into a computer files suitable for sending over the Internet.

Using Windows XP is simple. First, you'll notice the large area on the screen, called the desktop, and the narrow band at the bottom, called the taskbar. Everything you can do on your computer appears inside frames that are called windows.

- Place definitions in a frame.
- Put definitions or links to definitions in a sidebar, box or with other methods of highlighting.

☑ **Example**

Both the compiler and the interpreter search for classes in each directory or ZIP file listed in your class path.

Definition: A *class path* is an ordered list of directories or ZIP files in which to search for class files.

- Put the definitions at the bottom of the page.
- Put links to definitions at the bottom of the page.

☑ **Examples**

Summary | Terms | Exercises | FYIs

Selected Terms Discussed in This Chapter
asynchronous
conferencing
directory
electronic mail (email)

- Separate out definitions with an icon or heading.

☑ **Example**

Learn2.com separates keywords from the main text.

Keywords **Jewel boxes:** the clear plastic boxes in which most CDs are packaged.

- Highlight terms and provide internal links in a sidebar.

☑ **Example**

Access to the Internet is provided by an **Internet service provider (ISP)**, who operates the network hardware to connect a large number of clients to the Internet.

On this page:
» Client
» Domain name
» Home page
» Hypertext Markup Language (HTML)
» Internet service provider (ISP)

- Link to an external dictionary or encyclopedia.

☑ **Example**

Online dictionaries and encyclopedias such as Webopedia allow you to link to their definitions.

- Provide links from the definitions back to the text.

☑ **Example**

The link leads to a definition on the same page. The definition links back to the original context.

He has one of the most common visual disabilities for men: color blindness, which in his case means an inability to distinguish between green and red.

Color blindness (scenario -- "shopper")

Color blindness is a lack of sensitivity to certain colors. Common forms of color blindness include difficulty distinguishing between red and green, or between yellow and blue. Sometimes color blindness results in the inability to perceive any color.

- Explain how to access definitions.

☑ **Examples**

Click a word to see its definition.

For additional term definitions please see the Boating Basics Glossary of Terms.

- Highlight terms by using bold or color (some sites use red).
- Inform readers what special formatting means.

☑ **Example**

NOTE: Words bolded appear in the definition section.

Using Definitions

- Define terms central to your Web site.

☑ **Example**

What's a bot? *is a link on the home page of Botspot.com.*

- Define terms central to your discussion.

☑ **Example**

The main language used to create Web pages is HTML. It stands for HyperText Markup Language. HTML provides the structure for your Web page. Browsers interpret this and provide a layout for your page. A browser is a program like Netscape or Lynx that helps you surf the Web.

- Don't provide a definition link for every occurrence of the term. Instead, define the term the first time you use it on a Web page.

☒ **Example**

This Web document uses links every time the term is used.
A hypertext link is a built-in reference to related information in an HTML document. Following a link in one document can produce text that contains another link or links to other documents or locations. One of the most popular features of the WWW is this built-in linking capability.

- Avoid repeating the term in the definition.

☒ **Example**

Push: Push refers to a technology that sends data to a program without the program's request.

- Avoid circular definitions (using the term to define the term).

☒ **Examples**

FTP (FILE TRANSFER PROTOCOL). An Internet protocol that allows data files to be copied from one connected computer to another.

SEARCH ENGINE: A software program that searches a database and gathers and reports information that contains or is related to specified terms.

- Avoid *is when.*

☒ **Example**

Truncation is used when you want to find all endings of a word.

- Limit use of parenthetical definitions. Parentheses are difficult to see online and interrupt the sentence.

> ☒ **Example**
>
> Hyperlinks are underlined or bordered words and graphics that have web addresses (also know as a URL -- Universal Resource Locator) embedded in them.

- Give the *class* (category) and *differentia* (characteristics that distinguish it from other items in the category).

> ☑ **Example**
>
> A browser is software *(class)* that you use to look at Web pages *(differentia)*.
>
> ☒ **Example**
>
> *(Thing is not a class).*
>
> A network interface is a thing inside a computer that knows how to send information over a wire.

- Define nouns with nouns and verbs with verbs.

> ☑ **Examples**
>
> **Platform**
> A computer operating system such as Unix, Windows, or Macintosh.
> **Download**
> To retrieve a copy of a file from another computer using a modem or computer network.
>
> ☒ **Example**
>
> Sockets - is a name given to the package of subroutines that provide access to TCP/IP on most systems.

- Give the context of a term that has multiple meanings.

> ☑ **Examples**
>
> An **anchor** is a term used in World Wide Web (WWW) publishing to denote the beginning and end of a hypertext link.
>
> **Cookie:** A unique string of letters and numbers that a Web server stores in a file on a user's hard drive.

▷ Disclaimer

A disclaimer is a statement that denies responsibility and liability. A disclaimer warns readers who read your site content and protects you from complaints.

Tips

- Tailor your disclaimer to the type of Web site you have; check legal guidelines that may apply.

- Consider obtaining legal advice to protect your intellectual property and minimize liability.
- Use any of the following types of disclaimers:
  - The author alone is responsible for the contents of the Web pages, not the associated company or organization.
  - The author is not responsible for any links made in the Web site.
  - The links have not necessarily been tested or verified.
  - Presence of a link does not indicate endorsement.
  - There is no guarantee of accuracy.
  - There are multiple contributors to a Web site.
  - Advice given in the Web site could harm someone if misused.
  - Information on the site is for general information and is not comprehensive or a substitute for professional advice.
- Consider including a Linking Disclaimer.

> ☑ **Example**
>
> By providing links to other sites, [name of your website] does not guarantee, approve or endorse the information or products available at these sites, nor does a link indicate any association with or endorsement by the linked site to [name of your website].

- Consider including a detailed disclaimer on a separate *Terms of Use* page.
- Write in clear English rather than legalese so readers understand the disclaimer.

▷ Domain Name

The domain name (hostname) identifies a computer on the Internet and is part of the Web site's URL. Hostnames stand for numeric IP addresses and are easier to remember. Having a domain name makes your business/organization appear more professional. You obtain a domain name ending with extensions such as .com, .net or .org by registering through companies called "registrars" that compete with one another.

Tips

- Decide on the best length:
  - Make the name short if it is simple and meaningful.
  - Make the name longer if it contains your site's keywords. Site names can be up to 67 characters long.
- Choose a name that is easy to remember.
  - Your domain name and site name should be the same.
  - The name should be different from others.

> ☒ **Examples**
>
> | | |
> |---|---|
> | Jobware | TravelJobs.com |
> | JobWarehouse | Traveljobsearch.com |

- Make the name easy to spell.
- If possible, avoid names with hyphens.
- Avoid a name that is too similar to another site.

> ☑ **Example**
> Internic.com *is too similar to* Internic.org.

- Choose a name that is related to your business or organization or that describes your site. A good domain name gives the public an idea of what they will find at your site.

> ☑ **Examples**
> *kidszone.com, autoweb.com, search.com, thesaurus.com, sportingnews.com*

▷ Download Menu

You can create a menu that lists materials readers can download (transfer to their computer). A menu makes it easy to download files, utilities, documents, multimedia, and other materials.

Tips

- Create a separate downloads page or section.
- If the download section is important, provide a shortcut to it on your home page.
- Specify the following:
    - ☐ File size
    - ☐ File type (graphic, video, sound, etc.).
    - ☐ File format
    - ☐ Definitions of the file formats available
    - ☐ Short description of the file
    - ☐ Date file was created
    - ☐ Languages
    - ☐ Operating requirements (hardware, operating system)
    - ☐ Downloading instructions
    - ☐ Installation instructions
    - ☐ Download time

> ☑ **Example**
> **Virtual Desktop Manager**
> Deskman.exe
> 550 KB file
> 3 min @ 28.8 Kbps

- Make text documents available as one long file for printing or in smaller parts for quicker downloading.

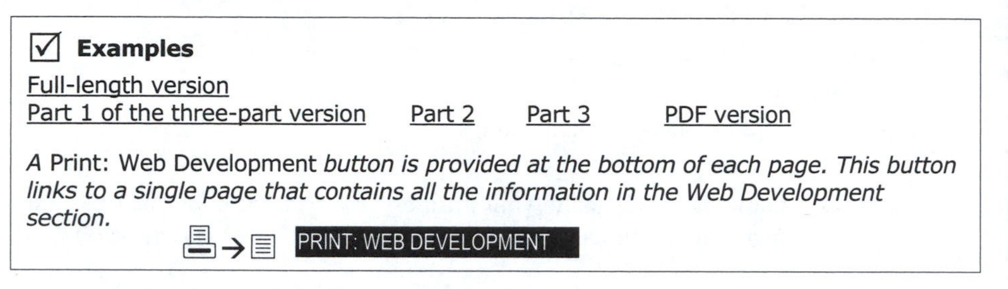

Examples
Full-length version
Part 1 of the three-part version
Part 2
Part 3
PDF version
A Print: Web Development button is provided at the bottom of each page. This button links to a single page that contains all the information in the Web Development section.
PRINT: WEB DEVELOPMENT

- Summarize the type of information available in the download.
- Use icons (with accompanying text) to represent the types of files, such as

  audio file or *Adobe Acrobat PDF file*
- Distinguish between the format (such as .htm) and the access methods (such as ftp).
- Link to a separate page for downloading instructions and help.
- Provide a *Search* feature.
- Group downloads by category. Also provide groups such as *Top*, *New*, and *Featured Downloads*.

☑ **Example**

A *table lets you read across to the type of download you're looking for.*

Documentation		Software		
User Guides	**Warranty**	**Synchronize**	**Handheld Applications**	**Updates & Utilities**
Visor Platinum User Guide Windows Info Mac Info	Warranty	Palm Desktop & HotSync Manager Windows Info Mac Info	Handspring software Get info Third party software Get info	Handspring Updater 3.5.2v1.5 Windows Info Mac Info

▷ Editorial/Submission Guidelines

Editorial and submission guidelines clearly define your policy for reader submissions. Guidelines are used for online publications such as newsletters, magazines/e-zines, journals, e-books, and letters to the editor. (Some sites for book, magazine, and journal publishers also provide submission guidelines for their print versions). These online documents may or may not require hyperlinks. Guidelines let writers know what you expect. They also help you avoid being inundated with incorrectly written and formatted material.

Tips

- Clearly describe the audience, goals, scope, and focus of your publication.

- Develop guidelines for the following:
    - Types of materials (subject matter, types of documents such as reviews, research, commentaries, essays, etc.) you accept
    - Format for submitting material:
        - Hard copy or electronic
        - File type
        - Method of sending (e.g., attachment, ftp, hard copy)
    - Material required with submission (e.g., name, e-mail address, telephone number)
    - Identification (author, institution, etc.)
    - Additional information (e.g., biographical information)
    - Usage and vocabulary
    - Style
    - Format (layout)
    - Required components (keywords, summary, etc.)
    - Citation method and style guide used (e.g., MLA, APA). Provide a link to a sample reference.
    - Links: type of links, number of links, external links, format and wording of links (if applicable)
    - Artwork
- Outline your policy on the following:
    - Initial e-mail query required
    - Editorial process
    - Editorial policy
    - Policy on submission to multiple Web sites
    - Reviewers/referees
    - Deadlines/turnaround time
    - Selection criteria
    - Copyright and ownership of published material
- Provide your editorial calendar.
- Announce any upcoming themes or suggested topics.
- Provide a clearly labeled link (such as *Submissions, Contribute to* , *Information for Authors,* or *Calls for Papers*) to the guidelines from your home page.
- Provide contact information, as well as e-mail links to the appropriate editors.
- Make the guidelines available for download, such as in Adobe Acrobat (.PDF) format.

▷ Educational Information

An educational Web page provides factual information in a specific subject. This information may include the following:

- Articles, white papers, research articles, historical documents, and fact sheets
- Conceptual information
- FAQs

- Multimedia support material, such as graphics, sounds, and video
- Reference material, such as a glossary or encyclopedia
- Resource collections
- Statistics and data
- Training/tutorial

Educational information may be one section of a larger Web site, such as a site for a company, organization, or government agency. It may also be the sole focus of the site (such as school pages, reference sites, special project, museums, or personal home pages). Providing educational information is a method of giving readers a reason to visit your site.

> ☑ **Example**
>
> The Kodak site provides "how to" tips and guides to better pictures, interactive demos, and "educational topics," such as Getting Started with Digital.

Tips

ORGANIZATION

- Provide links to the educational material from the home page.

> ☑ **Example**
>
> *The CDC (Center for Disease Control) Web site provides links to educational material, such as* Health Topics A-Z, Publications, *and* Data and Statistics.

- Provide a summary.

> ☑ **Example**
>
> **summary of resources related to the collection**
>
> *Alexander Graham Bell Family Papers*, 1862-1939, consists of correspondence, scientific notebooks, journals, blueprints, articles, and photographs. The papers document the invention of the telephone, the first telephone company, Bell's family life, interest in the education of the deaf, and aeronautical and other scientific research. The collection includes Bell's experimental notebook with the entry in which he spoke through the first telephone saying, "Mr. Watson—Come here—I want to see you." Bell's various roles as teacher, inventor, celebrity, and family man are covered extensively in his papers.

- Give a preview (advance organizer) of the topics covered.

> ☑ **Examples**
>
> The Reference Book contains tools to help you build upon your understanding of digital imaging.
>
> In this edition of HowStuffWorks we'll take apart a household thermostat and learn how it works. We'll also learn a little about digital thermostats. Let's start by taking a look at the parts.

- Provide choices through menus and navigation bars.

☑ **Example**

Sherwin-Williams Exterior Painting Features
How To's
A step-by-step guide to help you choose the right paint, prepare surfaces, apply the right finish and more.
Surface Preparation
Identify and correct "Common Problems" before painting or staining your home's exterior.

Items to Include

- ☐ Copyright information.
- ☐ Currency of information and revision history (provide a date stamp).
- ☐ Mission statement.
- ☐ Purpose of site and the objective. Is this a public service or advertising?
- ☐ Identity of page owner or sponsor.

☑ **Example**

This Web site contains educational material, but its sponsor is ChemistrySW Software for Windows.
Science Hypermedia, Inc. is developing educational courseware for general chemistry, analytical chemistry, instrumental analysis, optics, and electronics. This Website is supported by sales of our products, and by advertisers and sponsors.

- ☐ Support credits (for funding and sponsors).
- ☐ Project and research behind the site (if applicable).
- ☐ Authors.
- ☐ Qualifications of authors (biographical information).
- ☐ Sources of the factual information. Be aware of copyright issues.
- ☐ Those responsible for accuracy of content.
- ☐ Resource lists with links to useful Web pages.
- ☐ Printed material available and how to obtain it.

Writing Educational Information

- ▸ Write at the appropriate audience level.
- ▸ Use analogies, definitions, and examples.
- ▸ Use advance organizers.

☑ **Example**

This section describes the characteristics of several Kodak films. Knowing the characteristics of these films will help you select the best one for the kind of pictures you want to take. You may want to refer to the complete line of Kodak films.

- ▸ Use graphics and animation.
- ▸ Layer information by providing links to more information.

> ☑ **Example**
>
> A physical examination with a lateral collateral ligament test (varus stress at 25 degrees of flexion) results in unchanged knee joint tightness. This involves bending the knee to 25 degrees and placing pressure on the inside surface of the knee. Other tests may include:
>
> • a knee MRI
> • a knee joint X-ray

- Always provide updated information so readers know that information is current and keep returning.

▷ E-Mail

E-mail is an electronic computer message. It is a method of communicating both informally with friends and formally with organizations and professionals through a computer network. E-mail is quick, cheap, and can be sent to multiple recipients. In general, e-mail is more informal and conversational than traditional letters.

Tips

Organization

Subject Line

- Always include a subject line. It does not have to be a complete sentence.
- Make the subject line clear, descriptive, informative, and concise.
- Describe the message's purpose.
- Avoid vague subject lines, such as "Information" or "Sorry."

TYPE OF E-MAIL	PUT THE FOLLOWING IN THE SUBJECT LINE
Response	Use the same subject, preceded by RE:
Request	REQ or REQUEST
Critical message	URGENT
Non-urgent message	FYI (For Your Information)
Proprietary	PROPRIETARY/CONFIDENTIAL

Introduction

- Include a brief introduction identifying yourself, the purpose, or context.
- Put your message in context.

Body

- Put the main point/most important information first.
- Keep the message brief and to the point.
- Clearly state the action or reply requested.

If you are referencing a previous message:

- *Briefly* quote the original message. Edit the text until it is the minimum necessary to provide the context. Use brackets (<) to indicate a quote.

- If you must include the entire original, place it at the end of the message.

If you are replying to a long request:

- Remind the recipient briefly of the request.
- Do not include the entire message that you are replying to. Cut down the part that you include to the absolute minimum needed to provide the context of your reply.
- Include the most important points or the action requested.

If you are asking a question:

- Begin with the question.
- Give specifics.

If you are sending a long attachment:

- Before sending the attachment, ask the recipient the following:
    - If it is all right to send a large file.
    - The type of file format and encoding preferred.

If you are forwarding a message:

- Include a comment before the message.

If you are subscribing to or removing yourself from a mailing list:

- Make sure you are sending the request to the administrative address rather than the mailing list recipients. This address is a request, LISTSERV, Majordomo, or Listproc address.
- Make the request clear and simple: *Please add (or remove) me to/from the ..... list.*
- Use the correct format to subscribe or unsubscribe. (Most sites give instructions on how to do both.)
    - To subscribe to a LISTSERV, use the form `SUB (LISTSERV name) (your name)`. SUB is short for SUBSCRIBE. To remove yourself, use `SIGNOFF (LISTSERV name)`.
    - To subscribe to a Majordomo, use the form `subscribe (list name)`. To remove yourself, use `unsubscribe (list name)`.

CLOSING

- Do not use traditional closings other than *Thanks* or *Regards.*
- Include a signature (signature file) containing, at minimum, your name and address.

WRITING E-MAIL

- Focus on one topic only.
- Keep sentences short.
- Be concise.
- Consider offering two versions: a long version and short version.
- Watch your tone. Some people may misunderstand or misinterpret your statements. Use a tone appropriate to the audience.
- Respond only after doing research and thinking carefully about your response (and cooling off about emotional subjects).

- Proofread for spelling, grammar, and punctuation errors. If necessary, print it out and read your message several times before sending it.
- When including Web addresses, type the entire URL (http://. . .) because some e-mail programs allow readers to click the link.
- Use your word processor to compose long e-mail messages. You can then copy and paste the text into your e-mail program, or send an attachment. Using a word processor will encourage you to compose your message more carefully and let you use the spell checker.

Avoiding Common E-Mail Practices

- Do not use e-mail for private, confidential messages, or formal messages.
- Do not use REPLY unless you are positive who the recipients will be. Many e-mail users mistakenly send a private message to all members of a mailing list by using the reply option.
- Do not send chain letters.
- Do not use e-mail for anything illegal or unethical.
- Avoid *flaming* (publicly criticizing people in e-mail or discussion groups) and *shouting* (using all caps).
- Avoid *emoticons* (see below) and abbreviations.
- Avoid intentionally misspelling words and omitting pronouns and capital letters.

Formatting E-Mail Messages

Some e-mail programs do not allow you to use special formatting or fonts. Other programs let you send mail as plain text or with HTML formatting. You can use indentation, spacing between paragraphs, lists, and quotation marks. To format your e-mail messages:

- Remember that not all recipients use e-mail programs that can read fancy formatting or special characters.
- Keep messages under 25 lines long.
- Keep line length less than 70 characters, but avoid very short lines.
- Break up the text by using the following:
    - Headings
    - Lists
    - Short sentences
    - Short paragraphs
    - White space
- Use capitalization sparingly; ALL CAPS are considered to be SHOUTING.
- Use *asterisks* for bold and /slashes/ for italics.
- Avoid all lowercase letters.
- Number questions or action steps.
- Avoid using colored fonts or clipart.

Emoticons

Emoticons are punctuation symbols used to show humor or emotions using ASCII characters. Emoticons are substitutes for facial expressions, such as smiles **:-)** frowns **:-(** winks **;-)** An emoticon is also known as a smiley.

Tips

- Many believe you should avoid these symbols because they are overused and can sometimes cause confusion or misinterpretation.

Writing E-Mail Addresses

- When you write an e-mail address within a sentence or paragraph, italicize it.
- Always include all punctuation marks.
- Add a period to the end of the address if it ends a sentence.
- Use the correct case because the addresses are case-sensitive.
- To refer to an e-mail address parenthetically, use parentheses or < >.
- Avoid hyphenating an e-mail address. To break an e-mail address into several lines:
  - Do not add a hyphen (which may look like part of the address).
  - Do not break at the dot (which may look like a period).
  - Try to break after logical units.

Signature File

A signature file (.SIG) is a file attached to e-mail or a message included at the bottom of any message (discussion group) or Web page. Some browsers will automatically append a signature file to your message.

Tips

Items to Include

- Include the following items:
  - Name
  - Address
  - Phone/fax number
  - Web address
  - Advertising message
  - Slogan or quote
  - Business philosophy
  - ASCII art created from text and symbols

▷ E-Mail List Information

Web sites for organizations often give readers information about joining a mailing list, such as a LISTSERV.

Tips

- Identify the sponsor.
- Identify the purpose of the list.

> ☑ **Example**
>
> TECHWR-L is an unmoderated discussion forum for technical communication topics. If you're a technical writer, editor, indexer, teacher, student, or just someone interested in technical communication topics, join the list and benefit from over 4900 subscribers' expertise, education, and experience.

- List and describe any features of your site.
- Explain how to subscribe and unsubscribe.
- Make it easy to subscribe by providing a form or e-mail link.

> ☑ **Example**
>
> **JOIN I-LapTop Warrior**
>
> [ ] JOIN
>
> **Enter your email address, then click JOIN**
>
> To sign up for this discussion list, simply enter your Email address above, and click Join. You will receive a confirmation notice by Email to which you must respond. To respond, simply Reply to the message and send it off. Once you've done this, you'll start receiving the I-LapTop Warrior discussion digest within a week.

- Explain how often readers will receive messages.
- State the policy for submitting messages to the group and the method of posting messages.
- Provide a link to archives.

> ☑ **Example**
>
> **E-Mail Discussion List:** ATTW sponsors the e-mail discussion list ATTW-L. Visitors to this site are invited to subscribe to ATTW-Land review ATTW-L archives. Only official subscribers are permitted to mail messages to ATTW-L.

- Contact information.
- Terms of Service and copyright information.

▷ E-Mail Query

An e-mail query is a message sent to a publisher describing an idea for a book or article. An initial query allows you to find out if an editor is interested in reviewing a proposal before you send it.

Tips

- Check the Web site for the following:
    - ☐ The policy on e-mail queries.
    - ☐ What information should be included.
    - ☐ Whether they accept attachments. Some sites specifically request no attachments, such as a synopsis or samples. If they do accept attachments, determine what file format is required (e.g., resume in ASCII text format).
    - ☐ The e-mail address of the appropriate editor.
- Become familiar with the types of works they publish by researching previous publications. Many publishers have Web sites with online catalogs and submission guidelines.
- Use the following organization:
    - ☐ Include a clear, descriptive Subject line.

> ☑ **Example**
>
> Query: (name of article or book)

- ☐ Begin with a hook, question, and the most important selling feature of the piece.
- ☐ State the name of the piece.
- ☐ Summarize how the document fits with their goals and readers.
- ☐ Include biographical information, including previous publications and experience.
- ☐ Provide links to or the URL of your Web page, online resume, or electronic portfolio.
- ☐ Include contact information, such as your phone number, mailing address, and e-mail address.

- Use a professional tone.
- Check for errors.
- Keep formatting simple because you don't know what e-mail program the editors will be using.

▷ Emphasis

Emphasis is a technique used to visually distinguish types of information. You can emphasize important items by a variety of formatting techniques. Techniques of emphasis draw attention and increase the scannability of your page.

HTML Code: <EM> </EM> (usually appears as italic) or <STRONG> </STRONG> (usually appears as bold) are used for emphasis. These are logical tags (in contrast to physical tags such as bold <B> or italics <I>.

Tips

- Create emphasis using a variety of formatting techniques:

- ☐ Animation
- ☐ Blinking
- ☐ Bold
- ☐ Boxes
- ☐ Capitalization
- ☐ Color
- ☐ Fonts
- ☐ Italics
- ☐ Labeling
- ☐ Lines (rules)
- ☐ Position
- ☐ Pull quotes
- ☐ Sidebars
- ☐ Size
- ☐ White space

- Highlight the most important information on your Web page, including the introduction and headings. Don't bury your message.
- Use formatting, such as font size, bold and color, to help emphasize one word, idea, or phrase over another.
- Use subheadings such as *What's New, Tip, Note,* and *Important* to highlight the information that is to follow.
- Use emphasis sparingly. If you overuse emphasis, nothing will stand out. Your document will look confused and cluttered and create noise and information overload.

▷ Examples

Using examples is a method of making your writing concrete. Examples can include specific situations and applications, case studies, and graphics. Examples can be used for a variety of purposes, such as to help explain abstract conceptual information or improve productivity.

Tips

When to Use Examples

- Use examples to do the following:
  - ☐ Explain definitions and other background information.
  - ☐ Make instructions and procedures more concrete.

> ☑ **Example**
>
> **Instructions**: Type specific words, phrases, or Names in the boxes below to personalize your news. Click the **submit all** button when you are done.
> **Examples:**
> Put phrases in quotes, like **life on Mars.**
> Separate terms using commas, like **Steve Jobs**, **Bill Gates**, **Microsoft**, **operating system**, **monitor**

  - ☐ Provide real situations and applications.

> ☑ **Examples**
>
> Please note the filename are CASE sensitive so make sure that you are using the correct filename. For example, `MyPicture.gif` is a different file than `mypicture.gif`.
>
> **A quick example:** To demonstrate this feature and hopefully give a better understanding of its use, let's step through a small example. We are going to create a new type of button that highlights whenever the mouse cursor passes over it, and then use this new button several times within the same Quest frame . . .

- ☐ Provide scenarios and applications of principles.

Example

See "Testing a Comment: Sample Code" for an extended example.

- ☐ Accompany theory.

Example

Many Web sites about HMTL and programming language use examples of actual code.

- ☐ Help readers visualize concepts.

Example

You can have directories within directories as well. In the example shown below, there is a subdirectory of the example directory. The subdirectory is named new:

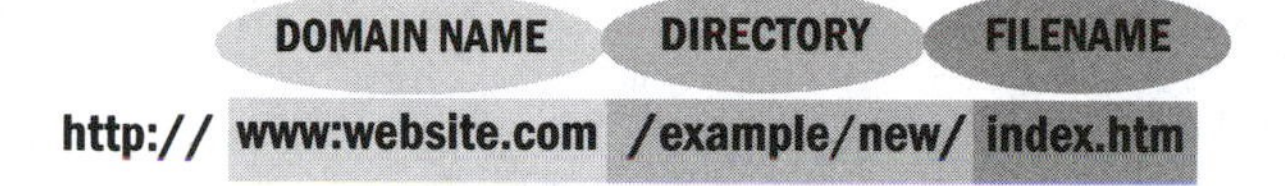

QUALITIES OF GOOD EXAMPLES

- Use examples that are relevant and easy to understand.

Examples

Web forms let a reader return information to a Web server for some action. For example, suppose you collect names and email addresses so you can email some information to people who request it. For each person who enters his or her name and address, you need some information to be sent and the respondent's particulars added to a database.

A protocol is simply a method to follow in order to get something done. When a protocol is established, everyone knows what to do and more importantly what to expect. For instance, when you go to the BMV/DMV to get your driver's license renewed, you take a number, sit and wait (and wait, and...). You know you had to do this and it's what is expected of you. That's the protocol for getting your license renewed.

- Use examples that help readers visualize concepts.

Example

If a packet size is smaller than it should be, transfer speed will suffer. This would be roughly analogous to moving a pile of dirt across your yard with a 5 gallon bucket rather than a wheelbarrow. The sheer number of trips with the bucket slows down the overall time it takes to transfer the dirt. On the other hand, too large of a packet size will also dramatically slow down transfer. This is because the server must then fragment the packet in order to send it on. Using the wheelbarrow example, this would be like overloading the wheelbarrow, stopping halfway to your destination to relieve some of the load, bringing the rest to your destination then returning to pick up the stuff you offloaded.

FORMATTING EXAMPLES

- Set off the examples by using a different format, such as a sidebar, box, indentation, bulleted lists, icons, frames, etc.

> ☑ **Examples**
>
> **Input devices** are the hardware components you use to talk to a computer. You use them to place requests, send messages (to the computer or to other people), move around in virtual worlds, or even shoot at enemies in some computer games.
>
> A few examples of commonly used input devices are:
>
> - Computer keyboard - Joystick - Microphone - Mouse
>
> *In frame:*
> Example 11
> Audio:
> [Modem] [Broadband]
>
> . . . the pattern can be regularized through the application of the Metrical Contraction rule (Example 11).

- Use a subheading or bold to label an example.

> ☑ **Example**
>
> **Example:**

- Layer information by linking to examples from your explanation.

> ☑ **Examples**
>
> Take a look at the following pages for some samples of the variety of effects that can be created:
> Fancy Photos Text Effects Web Buttons
>
> Macromedia Flash2 [EXAMPLE]
>
> **Sending Cover Letters Electronically:**
> Example Cover Letter
>
> **Traffic Reports**
> Summary Explanation See Example

- Provide a context for examples.

> ☑ **Example**
>
> The following examples highlight use of some accessibility solutions:
> Online shopper with color blindness (user control of style sheets)
> Reporter with repetitive stress injury (keyboard equivalents for mouse-driven commands; access-key)

- For many examples, provide a links list or samples index.

☑ **Examples**

The Official Netscape JavaScript 1.2 Book (Netscape Press) lists examples by chapters.
Chapter 3
Example 3.1: Two Simple Scripts
Example 3.2: A Quick Look at Functions
Example 3.3: Calling the Function

Each example page provides flexibility in viewing examples.
Previous Chapter's Examples-|-Next Example-|-Return to Chapter Listing
ASP Examples
Basic
Write text using ASP
Format text with HTML tags

▷ FAQ

A FAQ (Frequently Asked Questions) is a listing of questions and answers. FAQs are common in newsgroups, discussion lists, and customer support sites. They answer the most common questions new readers ask.

Tips

Organizing a FAQ

- Use a specific title.
- Begin with an introduction. If the FAQ is extensive, the introduction should contain the following types of information:
    - History
    - Contributors
    - Mirror sites or other methods of availability
    - Versions, changes, new information
    - Organization
- Include a table of contents that lists all the topics in order. (Provide internal links from these topics to the answers and back to the top).

☑ **Example**

Name of FAQ
Introduction
Question 1
Question 2
Question 3

Question 1 Top
Answer
Question 2 Top
Answer
Question 3 Top
Answer

- Put the questions in a logical order: e.g., from simplest to most complex, most frequently asked to least frequently asked questions.

- Make the organization obvious.
- Use levels: an overview paragraph followed by details (general to specific).
- Break into small subtopics or information categories for quicker reading and information retrieval. If the FAQ is extensive, list topic groups alphabetically.

☑ **Examples**

The NOAA site organizes questions into eight categories.

1. Regarding El Niño in general:
2. El Niño over the ages:
3. On the weather effects of El Niño:
4. El Niño and marine life:
5. Oddities and potpourri about El Niño:
6. Regarding the regional effects of El Niño:
7. Regarding our prediction capabilities:
8. Web links to other El Niño FAQs:

FAQ Topics
Active Directory (84)
BackOffice Server (23)
Backup (27)
Batch Files (23)

☒ **Example**

This list of FAQs includes 50 questions. Even this short excerpt from the list is difficult to read quickly.

What is everyone using to write HTML?
Where can I get a hit counter?
Where can I find a list of all the current HTML tags?
Where can I put my newly created web?
How can I include comments in HTML?
Where can I announce my site?
How can I check for errors?
How do I use forms?

Formatting a FAQ

- Bold the questions.
- Consider using the following formats:
    - ☐ Numbering the questions
    - ☐ Using an outline format
    - ☐ Using a tabular arrangement

☑ **Examples**

[1] General DVD
[1.1] What is DVD?
[1.2] What are the features of DVD-Video?
[1.3] What's the quality of DVD-Video?

Licensing

Q	A
What will be the result of Sony's licensing agreements with other companies?	We expect to see a great diversity of compatible products from these licensee companies before too long. Sony is not, however, in a position to comment on what these products might be.

- Use color coding for different topics, if appropriate.
- Use indentation for the answers.
- Use short paragraphs.

- Keep the line length under 75 characters.
- Keep backgrounds simple to keep text legible.

WRITING A FAQ

- At the top of the page, clearly identify the name of the FAQ topic, product, or service.
- Date the FAQ. If necessary, date each question/answer.
- Provide a version number.
- Identify the FAQ authors and their credentials.
- Write clearly and concisely.
- Use first person (e.g., How can I . . .?).
- Check for accuracy.
- Avoid jargon and abbreviations.
- Provide specific information, such as troubleshooting.
- If necessary, provide a disclaimer.

USING LINKS

- Use internal links from the list of questions in the table of contents to the question-answer.

> ☑ **Example**
> *Make each question the link, or consider a method like the following:*
> Q1 How do I set the file permissions of my server and document roots?

- Put the list of questions in any of several locations:
  - ☐ At the top of the page.
  - ☐ On an introductory page.
  - ☐ In a frame.
- Provide links from each answer or group of topics to the top of the page.
- If the FAQ is long, provide an introduction page with links to separate pages for each category.

> ☑ **Example**
> *The XML FAQ provides links to separate pages:*
> The FAQ is divided into four sections: General, User, Author, and Developer.

- Consider the following page size options:
  - ☐ One long scrolling page.
  - ☐ One general topic category per page.
  - ☐ One question and answer per page.
- When appropriate, include links to:
  - ☐ Cross references
  - ☐ A *What's New* section

- ☐ Glossary
- ☐ Companies or sites mentioned
- ☐ Contact information
- ☐ Reference information
- ☐ Support information

▷ Feedback

Feedback is a way for readers to respond and comment about your Web site. Feedback provides a way to add interactivity and community features. It also provides you with a source of corrections and other ideas for improvement.

Tips

- Encourage readers to provide site-specific feedback such as reporting the following:
  - ☐ Stale links and suggested links.
  - ☐ Incorrect information.
  - ☐ Technical difficulties.
  - ☐ Comments, suggestions, messages, questions.

> ☑ **Examples**
>
> Tell Us What You Think
>
> Was the information on this page helpful?
> ○ Very helpful ○ Helpful ○ Not helpful
>
> ?! Click on any Comment icon to browse/contribute to discussion on this article

- Provide a brief introduction.

> ☑ **Examples**
>
> Your feedback is important to us. We carefully review all customer feedback and often make improvements to our site and support services based on customer feedback.
>
> **Site Feedback**
> Thank you for taking the time to send us feedback on Adobe.com. Your help—by reporting a bug or by making an enhancement suggestion—helps us to focus our improvement efforts and build a site that meets the needs of our customers. We read all feedback we receive and take what you have to report seriously. We truly value your input.

- Make it easy to provide feedback by providing a form and e-mail link.

> ☑ **Examples**
>
> feedback **Did you find this page useful? Yes or No**
>
> To send feedback to Boston.com, please complete the form below.
>
> *A feedback button at the bottom of the page can link to a page to e-mail a comment to the page owner or a form/questionnaire.*

- Consider placing a feedback link at the bottom of each page.
- If appropriate, group feedback topics into sections.

> ☑ **Examples**
>
> *CNN has readers select the CNN network they would like to contact. For example, CNN.com contains the following options, with links to specific topics:*
>
> COMMENT ON THE TOP STORIES OF THE DAY?
> GIVE US YOUR FEEDBACK
> SUBMIT WEB SITE ERRORS (Editorial, Grammar, Audio, Technical, Video)
> COMMENT BY SECTION (e.g. World, U.S., Weather, etc.)

- Be sure to respond to readers who provide feedback. If you do not respond, inform readers that they will not receive responses.

> ☑ **Examples**
>
> Comments are acknowledged by our auto-reply, which is your confirmation it has been received, and provides information to you on how your comment will be used. If your email requires an answer from our research department, please return to the main Feedback page and follow the instructions to submit a question.
>
> Please note: Due to the high volume of responses we receive each day, we are unable to respond to each feedback individually.

▷ File Management

You should carefully plan the names and organization of your Web pages. Using meaningful and logical filenames and directories helps you (and team members) maintain and update your site. It is also a way of assuring that your links will work and graphics will display correctly when you move folders.

Tips

Organizing Files

- Organize files into logical directories/folders to create a directory "tree." You can use one folder if your site is small. This structure allows you to make relative links among files in the site. A relative path is a location defined in relation to the page from which the link originates. By using relative paths, you can easily move your files without breaking links.
    - Make the Web site the main folder.
    - Put your home page in the main folder (the "root" of the tree).
    - Put folders containing related files within the main folder.
    - Group graphics files into a separate directory.
- Link to pages within your own site using relative paths.
- Create a map that lists all the filenames in your Web site.
- Keep an inventory of all Web page filenames, titles, and ID numbers.

Naming Files

HTML documents have an .html (UNIX) or .htm (Windows) extension. UNIX filenames are also case-sensitive.

- To avoid link errors, keep filenames to eight characters or less.

- Name the home page *home.htm, index.htm*, or *default.htm.*
- Use meaningful filenames.
    - ☐ Relate the names to a section of the site.

> ☑ **Example**
>
> *personnel.htm* is better than *page5.htm*

- ☐ Use keywords in filenames so search engines find your site.
- ☐ Use descriptive names for graphic files.
- ☐ Use descriptive names for navigational buttons. When readers place their cursor over the button, the filename will appear at the bottom of many browsers. They can then click the button without waiting for it to load.

> ☑ **Example**
>
> *Buttons can include the word "button;" navigational button bars can contain the word "navigbar," etc. A navigational button can also contain the name of the page it links to (e.g., services.htm) so readers can click without viewing the button graphic.*

▷ Flowchart

A flowchart illustrates steps in a process. It uses special shapes to represent different types of actions or steps.

A box represents a step.

A diamond represents a decision or branch.

A terminator symbol marks the beginning or end.

Lines and arrows show the sequence of steps and their relationships.

Tips

- Use flowcharts for the following purposes:
    - ☐ To provide a map or blueprint of your site.

      A flowchart can help you plan the overall structure of the topics in your Web site, including the outline, relationship among topics, organization, and links. This diagram can later become your site map.

    - ☐ To accompany explanations of a process or procedure.
    - ☐ For troubleshooting instructions.
- Use a simple flowchart; then link to more details.

☑ **Example**

Steps in the flowchart (partially shown) that require the student to complete a form or access information are hotlinked to the corresponding form or information needed to complete that step.

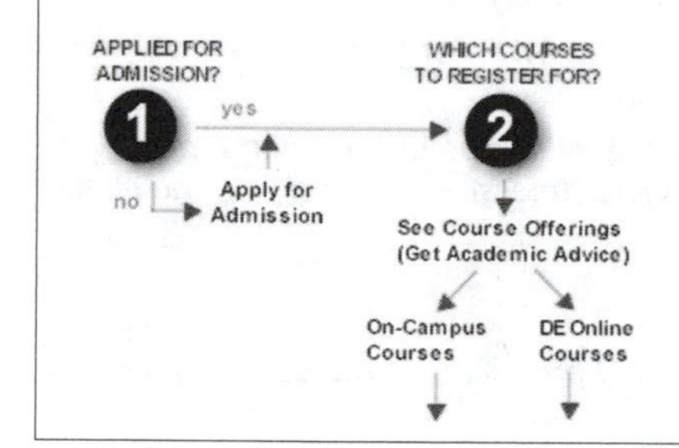

- Use color coding to help distinguish types of information on the chart.
- If the flowchart is large, provide a thumbnail or link to it.

☑ **Example**

Compensation Flowchart #1: Additional Compensation

▷ Focal Point

The focal point is the dominant eye catcher, the location on your Web page where the reader's eye is drawn. On a Web page with no focal point, everything looks the same and has equal attention. Thus readers are uncertain where to begin and where to go. They may also believe that they must work to read the page.

Each page should have a focal point that

- Draws attention
- Provides an entry point
- Directs the eye
- Provides a "hook"

The type of focal point and its position depends on the audience, culture, and type of site. For example, a Web site for children would use animation, bright colors, and large, unusual fonts. In Western cultures, the primary locations for the focal point are the top center and upper left of the visible part of the screen.

Tips

- Make sure your page has a focal point.
- For each Web page, make it clear where to begin. This is especially important on pages with numerous links.
- Use large text and well-worded headings and headlines.
- Use the following techniques to achieve a focal point:

- ☐ Alignment
- ☐ Animation
- ☐ Color
- ☐ Contrast
- ☐ Grouping
- ☐ Isolation
- ☐ Placement
- ☐ Size

▶ Place the most important information in the focal point.

☑ **Examples**

Delta Airlines places three colorful boxes along the left side of their home page. These allow readers to quickly Log In, Book a Trip, view Itineraries & Check-in, and check Flight Status.

On Amazon and Ebay, most readers come to find products, so the search boxes are prominent.

▶ Do not require users to scroll to see the focal point.
▶ Place navigational aids at the top, bottom, or sides.
▶ Use the focal point to capture attention, create emphasis, and direct eye movement.

☑ **Example**

▶ Avoid filling the focal point with a large graphic. Many home pages contain a large graphic surrounded by white space. Readers must then scroll to find text and links.
▶ Avoid overwhelming readers with too many devices to attract attention.

▷ Font

A font is the complete set of characters in one typeface in one size and style. Variations in the font style

▶ Affect legibility.
▶ Convey your meaning and tone.
▶ Help readers see how information is structured.
▶ Can serve as design elements.

With HTML, you have little control over fonts. Pixels on a monitor can't display the curves of type. You also do not know a user's system, browser, or configuration. Methods of specifying fonts include the following:

▶ Cascading Style Sheets
▶ Embedded fonts
▶ "Generic" fonts such as serif or sans serif so the browser uses the closest available face.

- Transparent GIF graphics to display special fonts. This method assures than the font will be displayed in any browser. It also allows you to include text using special effects, such as 3-D, embossing, and shadows. Negatives of this technique are speed issues, accessibility issues, and inability of readers to select text.

HTML Code:
<FONT> </FONT> changes the font. The latest versions of HTML recommend not using the <FONT> tag.
SIZE= changes the font size on a scale from 1 to 7.
COLOR=creates colored text.
The FACE= attribute changes the font name. However, if the font cannot be found on the user's system, the default font will be displayed. Also, few browsers support this attribute.

Tips

Locations of Fonts

- Define the various places you will use text:
  - Body text
  - Captions
  - Footers
  - Levels of heading
  - Links
- Use serif fonts for body text. (Serif fonts have feet on the ends of the letters.) Sans serif fonts are best for headings.
- Use font styles with consistency to avoid clutter and confusion and to provide unity.
- Use fonts to show the document's organization and structure.

Font Styles

- Use common fonts (Arial and Times Roman) that most users will have on their computers.
- Avoid specifying fonts. If you do use specific fonts, list alternative fonts that can be substituted.
- Use large, solid, uniform fonts. Avoid narrow, script, shadow, and embossed fonts that display poorly on monitors.
- Use high contrast between text and the background.
  - Use black text on a white background for most legibility.
  - Avoid busy backgrounds.
  - Use colored text sparingly.
- Use no more than three different fonts in one document.
- Use legible fonts.
- Use font styles appropriate to the tone and topic.

☒ **Example**

Online Tutoring Services

▷ Font Size

The font size is the size of the typeface. Factors that affect the size of text include the following:

- Monitor resolution. A 12-point font will look small in 1024x768 mode.
- Default font size set in the Web browser. Browsers let readers increase or decrease the font size of pages viewed.
- HTML <FONT SIZE> tag

HTML Code:
You can change font sizes using the following:

- Headline tags <H1> - <H6>

  Use the <HEADING> tag only for headings, not to change the font size.

- <SMALL> and <BIG> tags
- <FONT SIZE=n>

  The font sizes are relative, ranging from 1 to 7 (largest). The default is 3, which results in a point size of 12 (small). You should avoid using the <FONT> tag.

- Cascading Style Sheets, the preferred method of changing font size.

Readers can change the font size using browser options.

Tips

- Vary the font size and position to help readers see how information is structured.
- Avoid too many font sizes.
- Size your text consistently.
- Keep fonts 10-12 points or larger and no smaller than 10 points. For older adults, use a larger font.
- Avoid small text because screens are more grainy and difficult to read.
- For maximum accessibility, use relative rather than absolute font sizes.
- For readers with vision problems, allow them to change the font size.

☑ **Example**

Font Size [-] [+]

▷ Footer

A footer is crucial information at the bottom of each Web page. A footer is important for the following reasons:

- Identifies the author, affiliation, credentials, and date. Many readers use this information to judge the reliability of the information.
- Provides information for readers about how they can contact you and give feedback.

- Provides consistency and continuity to pages on your site.
- Indicates to readers that they have reached the bottom of the Web page.

HTML Code: <ADDRESS> </ADDRESS>

Tips

- Include any of the following in the footer (or links to separate pages for detailed information):
  - Address
  - Author's name
  - Copyright notice, legal status
  - Credits
  - Date created, updated
  - Disclaimer
  - E-mail address
  - Frequency of updates
  - Institution, organization, or company sponsoring site
  - Link to credentials
  - Link to the home page
  - Phone number
  - Site owner's name and address
  - Small logo, seal, trademark
  - Software used to create the site
  - Title
  - Version/revision information
  - Web address (URL) for readers who print the page
- Put a footer on every Web page. All footers should be consistent.
- Consider separating the footer from the body text with line or shading.
- Use space efficiently by balancing information.

☑ **Example**

▷ Footnotes

Footnotes on Web pages have the same function as those in print documents (usually scholarly articles). However, they can be in the form of links–hypertext footnotes. Footnotes give credit to your sources, such as Web pages (using external links) or print documents. They can also provide explanatory material and annotations.

Tips

Formatting Footnotes

- Use any of the following methods to format footnotes:
  - Use the familiar superscript numbers placed at the end of a sentence. However, realize that small links are difficult to click and see.

☑ **Example**

. . . psychological principles drawn mainly from the work of the Gestalt school.[(53)]

- ☐ Use numbers in parentheses or brackets.

> ☑ **Examples**
> -Stick with HTML 3.2 (1) and your pages will look good on all browsers that support it.
> -In the first round, copyright, through the courts, seemed to defeat legions of Napster users [5].

- ☐ Use marks such as ^ to indicate a footnote. The theory of this method is that hypertext footnotes do not need to be numbered because the links go to the appropriate note. This is not a good technique if readers will print the page.

LOCATIONS FOR FOOTNOTES

▶ Use any of the following options for placing the references:

- ☐ Put footnotes at the end of each paragraph. They are usually set apart from the text by blank space and lines.
- ☐ Put footnotes at the bottom of each page. Use internal links, with the citations at the bottom of the document. Remember that you cannot control where the end of the page appears.

> ☑ **Example**
> *This article contains a link (Jakob Nielsen) to the citation at the bottom. The citation has a link Go to citation.*
> Organizing information into an expanded article format (as in the article you are reading now) is similar to an idea expressed by Jakob Nielsen; he describes **clusters of nodes arranged in mini-networks**. This would mean grouping a number of closely related nodes, or chunks, together in a way that facilitates easy movement among them, with probably a more deliberate effort required for the user to exit from one cluster and go on to another.
>
> ---
> **Cited**
>
> Nielsen, Jakob. *Hypertext and Hypermedia*. San Diego: Academic Press, 1990.
> Go to citation

- ☐ Put footnotes on a separate page. Cite the reference; then create a list of references. At the bottom of the page, put a References Cited link. Realize that if you make your footnotes links, the Web page is replaced with another. Readers must wait for the page to load, then return to the document.

> ☑ **Example**
> **References**
> HTML 3.2 - W3C's specification for HTML, replaces the expired HTML 3.0 draft. Includes popular tags such as tables, body bgcolor and text colors, applets, text flow around images, and superscripts and subscripts. Also see the newest HTML draft.

- ☐ Use a table column to place the footnotes next to the referring text.

Managing Footnotes

- Provide instructions, if necessary.

> ☑ **Example**
> This document contains hypertext footnotes. The underlined text in the body of the article is linked to a footnote at the end of the article. Clicking on the title of the footnote will return you to your place in the text.

- Provide a way to return from the footnote to the text.
- Avoid overwhelming readers with footnotes.

▷ Form

A form contains spaces for entering information or selecting options. A form lets you obtain information, get feedback or reader information, take orders, conduct surveys, and receive registration information or reservations. It is also a method of making action steps easy for readers. A poorly designed form is hard for readers to understand and increases the chance of frustration and errors.

HTML Code: Forms are created using an e-mail link (mailto:) or CGI scripting.
The tabs used are <FORM> </FORM>
ACTION= tells the server what to do with the data.
METHOD= tells the server the method (GET or POST) for returning data to you.

Tips

Organization

- Inform readers what the goal of the form is and how information will be used.
- If appropriate, begin by thanking readers for participating.
- Put information in a logical order.
- Arrange items in order from top to bottom, left to right.
- Clearly explain your privacy policy.

Items to Include

- Include the following types of information:
  - ☐ Name
  - ☐ Address (business and home)
  - ☐ Phone/Fax
  - ☐ E-mail address
  - ☐ Business/profession
  - ☐ Payment
  - ☐ Areas of interest
  - ☐ Product/service information
  - ☐ Comments

Formatting

- Keep the form simple and consistent.
- Keep the form short.
- Don't cramp the information.
- Group related items.

- Align elements with a consistent and clear grid.
- Make type legible.
- Avoid colored and textured backgrounds.
- Use familiar units.
- Request specific types of information only once.
- Distinguish field labels from data fields.
- Use clear field labels that indicate what data should be entered.
- Add explanatory text to help readers fill out the form. Give readers hints about what format is required.

> ☑ **Example**
>
> Contact phone Format: (xxx)xxx-xxxx

- Indicate required fields.

> ☑ **Examples**
>
> Name: [ ] **(Required)**
>
> **Items with a red asterisk are required.**
>
> * Your name: [ ]

- Use appropriate interface elements (controls):

USE THIS ELEMENT	WHAT IT LOOKS LIKE	TO HAVE THE READER
Radio button	○	Select one item from a list
Selection list	option 1 ▾	Select one or more items from a list
Check box	☑	Select multiple items from a list
Scrolling menu		Select multiple items from a list
Text area		Enter free-form text
Text field		Enter small amount of text
Submit button	Submit Query	Click button to submit the form

- Use free-form elements (such as text areas) when possible because readers have more control.
- Consider reader experience and frequency of use when deciding which types of elements to use (e.g., drop-down lists or fields).

▷ Frames

A frame is a smaller window (or pane) within the browser window. Each frame

- Contains a separate HTML document.
- Is independent of others.
- Can scroll.
- Can have a background. Each frame contains a separate HTML document.

- Can have a size and location.

☑ **Examples**

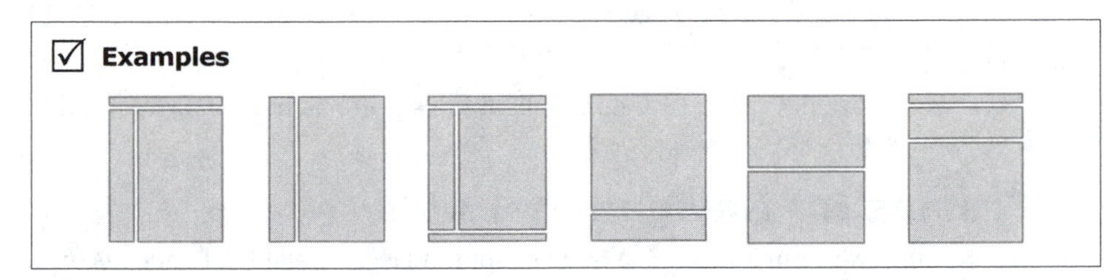

Advantages of Frames

Frames can be used to:

- Contain navigation, such as the table of contents.
- Keep the site name and other global features always visible.
- Simultaneously display information.
- Provide a window for examples or pop-up text.
- Call up external sites. However, these sites lose their context, which can often be deceptive.
- Make updating easier.

Disadvantages of Frames

There are negatives of using frames:

- The URL (Web address) is the frameset, not the page.
- Pages in frames are difficult to bookmark.
- Search engines may have difficulty indexing framed pages.
- Frames disable many browser features, such as going back.
- It is difficult to print the framed page, only individual frames.
- Pages that appear within frames often disorient and confuse readers.
- Pages load more slowly.
- According to usability studies, a majority of readers dislike frames.
- Screen real estate is reduced.
- Although most browsers now support frames, devices such as PDAs and palmtops have difficulty with them.

HTML Code: The frameset document tells the browser the frame layout, links and targets.
<FRAMESET> </FRAMESET>
<FRAME> identifies each frame.
<NOFRAMES> tag is used to offer the page without frames.

Tips

- Limit the number of frames to two. Otherwise, the windows will be too small.
- Use each frame for a specific function.
- Be sure to put the following on every frame page:
    - ☐ Identification of the site name and topic.

- Contact information.
- A link to the home page.

- Offer a non-frames version.
- Use vertical frames rather than horizontal frames. Vertical frames can be used to create narrow text columns.

▷ Frames for Navigational Aids

A frame is a window within a window. You may opt to place your table of contents or main menu in a frame. The frame keeps the table of contents visible at all times. This technique can also reduce overall file size because the menu does not have to be included on every page. However, many Web experts advise against using this technique.

HTML Code: <FRAMESET> </FRAMESET>

Tips

- Weigh the pros and cons of using frames.
- When using a frame for navigation, place the frame on the left side of the Web page.
- Use frames for navigation rather than have links that scroll off the page.

▷ Glossary

A glossary is a list of technical terms, abbreviations, acronyms, and jargon and their definitions. The glossary is usually a separate page or section within a Web site. There are many glossaries available on the Web for a variety of disciplines. When you use definitions many times throughout your Web site, a glossary allows you to define each term once. Readers can then access the definitions from anywhere in the site. A glossary is also useful because you do not know what Web pages readers have read first. However, placing definitions in the glossary requires that readers follow a glossary link, then return to the Web page.

HTML Code: A glossary is sometimes called a definition list because HTML definition tags are used to format the entries. Tags used include the following:
<DL> </DL> Definition list (beginning and end of list)
<DT> </DT> Definition term
<DD> </DD> Definition text (definition data)
Use anchors to provide links directly to terms using the < A HREF> tag.

Tips

CONTENTS OF THE GLOSSARY

- Explain technical terms, abbreviations, acronyms, product names, and proprietary terms.
- Don't assume all readers know Internet terminology. Consider using a glossary of Internet terms and terms related to your subject.

Formatting a Glossary

Location

- Use any of the following locations for the glossary:
  - Separate page
  - Several pages (one definition per page, one group of letters per page)
  - Bottom of the page
- Recognize that if you use one or more glossary pages, readers must wait for each page to load.

Pages

- Avoid putting the glossary online as one long page with no links. Use the following methods:
  - Menu at top with terms and definitions on one page.
  - Main menu page with links to sections of the alphabet (e.g., *Glossary A-B*).
  - Main menu page with links to one term per page.

Links to Terms

- Use internal links to jump directly to the term and definition.
- Use the following methods to link to terms and definitions:
  - Alphabetical lookup buttons and letter dividers that mark beginning of entries for each letter.

Example

A B C D E F G H I J K L M N O P Q R S T U V W X Y Z

  - A frame with terms on the left and definitions on the right.

Example

Access Provider Active X ADSL Agent Anchor Archie ARPANET ASCII Attribute	**Active X** Most web pages are static documents with little interactivity. Microsoft has created a programming language, called Active X, to remedy this situation. Active X "controls" promise to make the web surfing experience comparable to that of highly produced CD-ROMs, where you can listed to music, watch animation and video clips and interact with the program.

- Provide links to the top of the page and to the main glossary menu.

Linking to the Glossary

- Put a glossary link in the navigation bar.
- Use a glossary link the first time a term appears on one Web page.
- Don't use too many glossary links, or the reader will become distracted.

☒ **Example**

A document is a conforming HTML document if Its document character set includes [ISO-8859-1] and agrees with [ISO-10646]; that is, each code position listed in section The HTML Coded Character Set is included, and each code position in the document character set is mapped to the same character as [ISO-10646] designates for that code position.

- Make it obvious that a highlighted term links to the glossary.

☑ **Example**

The Learn the Net site links terms to the glossary.
A modem translates the digital signals from your computer into analog signals that can travel over a standard phone line. Modems come in different speeds and can be installed inside your computer (internal), or connected to your computer's serial port (external).
Related articles within the site are separated at the bottom of the page.

Related Articles:
About Modems
Monitor Settings

External links are indicated by the context. However, some readers may confuse these links for glossary links.
The World Wide Web (WWW) was originally developed in 1990 at CERN, the European Laboratory for Particle Physics. It is now managed by The World Wide Web Consortium, also known as the **World Wide Web Initiative**.

- Provide cross references to the glossary.

☑ **Example**

Look up more definitions in our Virus Glossary.

- Add links back to topics that discuss defined terms or links to WWW documents.
- If your site focuses on the glossary itself, allow a variety of ways to find terms, such as browsing and searching.

INTRODUCING THE GLOSSARY

- Explain how to use the glossary. For certain audiences, explain in a site guide how to use the glossary:

☑ **Examples**

-If you're not familiar with a word, try clicking on the glossary link. This interactive list of definitions includes hypertext links to other sites, so you can get more information about a new term.
-Entries will take you to a subject page for each term. On each subject page you will find a definition and explanation of the term as it applies to Web design and links to expanded information or illustrations.
-ALL TERMS IN BOLD ARE DEFINED IN THE GLOSSARY.

- Date the glossary and tell how often it is updated.
- Identify the source and scope of the definitions.

☑ **Examples**

-These are my own definitions and have been written in layman's terms.
-Definitions given here are a compilation of general terms used in British Columbia, Canada. The glossary does not include all forestry terms used in other countries... Definitions have been based on a variety of resource material documented in the bibliography.

Writing the Glossary

- Make each page is independent of others (context independence).
- Define nouns with nouns and verbs with verbs.

☑ **Example**

Backup
Noun: A duplicate copy of data made for archiving purposes or for protecting against damage or loss.
Verb: Create duplicate data.

- Check for consistency.
- If definitions are long, use headings.

☑ **Example**

About.com uses headings such as
Definition, Also Known As, Common Misspellings, Related Resources, Elsewhere on the Web.

- Use cross-reference links within the glossary. Put cross-references within the definition or separated out.

☑ **Examples**

This site includes cross-references in the definition.
External storage is all addressable data storage that is not currently in the computer's main storage or memory.

This site separates out cross references to other glossary terms.
See also: HTML, Java

- Use layering to link to more detail.

☑ **Example**

The World Wide Web Consortium uses a More link to let readers jump to more information. The detail page contains cross-references, as well as external links.
Hypertext
Text that is not constrained to be linear. (More...)

▷ Graphics

Graphics can include charts, clipart, scanned images, photographs, and diagrams. You need to plan which and how many graphics to include on your Web page. Sources of graphics include a digital camera, clipart (on CD-ROM), scanner, Web sites, and drawing your own. Paint programs, such as Paint Shop Pro, allow you to create buttons and other graphics, edit images, and convert them to different file formats.

TERMS

The most popular graphic formats are GIF, JPEG, and PNG.
GIF (Graphics Interchange Format) is best for line drawings, screen captures, and logos.
JPEG (Joint Photographic Experts Group) is best for photographs and continuous-tone images.
PNG (Portable Network Graphics) can replace GIF and TIFF images, is compressed, patent and license free, and can support 48-bit color.
SVG (Scalable Vector Graphics) is a new vector graphics language written in XML. There are many benefits to SVG: high-quality, small file size, resolution independence, ability to incorporate real-time data, ability to search for text within images.

Graphics that appear on the Web page are called *inline* images. *Transparent* images do not display their own background color but rather the one displayed in the browser. Graphical buttons can also be anchors for links.

USES

Do not use a graphic unless it relates to your purpose and message.
Do use a graphic to

- Add visual interest to the page.
- Affect the tone/mood.
- Attract attention.
- Break up the page.
- Complement text.
- Create a focal point and direct eye movement.
- Create a hotspot link to an action, document, or image.
- Entertain.
- Maintain a special font.
- Present a logo.
- Provide backgrounds and watermarks.
- Provide contextual clues.
- Provide examples.
- Replace words.
- Represent content areas.
- Serve as navigational buttons.
- Show comparisons, relationships, and trends.
- Show processes and flow.
- Show spatial relationships.

- Show what a product/object looks like.
- Summarize information.
- Visualize concepts and abstract information.

HTML Code:
<IMG SRC=URL of graphic> inserts the image.
ALT=[text] specifies text to display.
ALIGN=position (ALIGN=alignment left, right, top, middle, bottom, right, or left). For example, <IMG SRC=image name ALIGN=TOP>
The BORDER attribute is used to add a border.
The VSPACE and HSPACE attributes control space around the image.
The HEIGHT= and WIDTH= attributes tell the browser what size to reserve for the graphic so the page can begin loading.

Tips

Using Graphics

- Use graphics only if they are necessary and if they contribute to the overall design theme. Graphics slow download time.
- Use graphics sparingly; use no more than 2-3 per page.
- Avoid graphics that look like banner ads.
- Use callouts and, if appropriate, figure numbers.
- Use <ALT> text to provide alternative text so your site is accessible.
- Check if there are copyright issues.

Increasing Speed

- To keep the *file size* and *loading time* of graphics small:
    - Reduce the number of colors. Use an 8-bitcolor depth and the "safe" 216-color palette supported by Macs and PCs.
    - Keep image resolution under 72 dpi (dots per inch).
    - Reduce the amount of detail and number of layers.
    - Crop unwanted portions.
    - Use *aliased* (jagged) rather than anti-aliased graphics, which contain more file information.
    - Use a compressed file format, such as GIF.
    - Repeat graphics to speed downloading time; the browser keeps a graphic in memory (cache).
    - Specify height and width for images. The browser reserves a space for the graphic and continues to load the page and alternative (<ALT>) text so readers can continue to work.
    - Use *interlaced* graphics; they give the impression they are loading faster because they are drawn in stages.
    - Avoid graphics referenced from other sites (which take time to load).
    - Use thumbnails (small versions) that link to larger images so readers can see a preview and determine whether to follow the link.

- ☐ Link to graphics so readers can choose whether to view them. Warn the audience about the file size and format. Inform them that they can click on graphics.

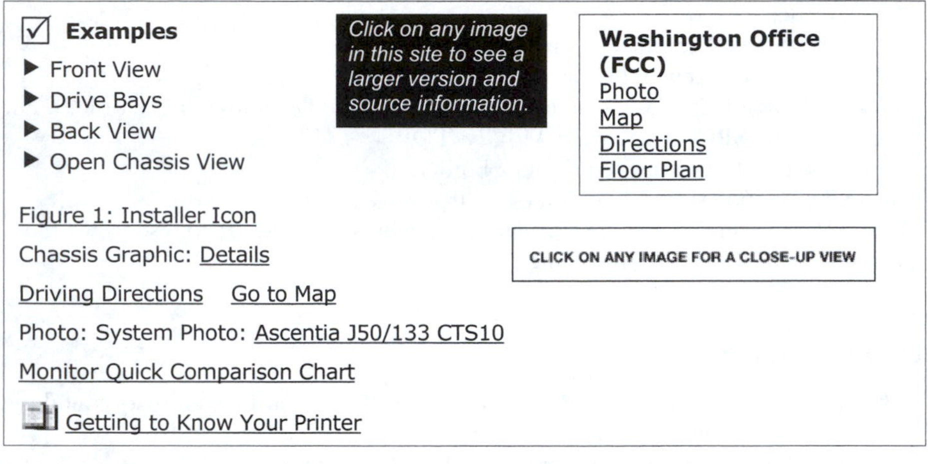

- Avoid using graphics larger than one screen because they require scrolling. For large or detailed graphics, use layering: begin with a simple graphic and link to details, or allow readers to zoom.

Examples

zoom Enlarge image

POSITIONING GRAPHICS

- Refer to the graphic in the text.
- Put the graphic as close as possible to the referring text. Place the graphic
  - ☐ *Before* or to the *left* of the text when the graphic is more important.
  - ☐ To the *right* of the text if the text is more important.
- Use the ALIGN attribute to control how text is aligned with an inline image. Text aligned with the bottom of the image is preferable because it will not leave blank space.
- Use white space to separate graphics.
- Avoid spatial references to the graphic (above, below). Instead, use *preceding* or *following*.

▷ Grid

A grid divides and organizes your Web page into functional areas or zones of the screen. Each area contains text and/or graphics and is devoted to a specific purpose.

A grid has the following uses:

- Provides an underlying pattern or design you can use to plan page arrangement. It determines where blocks of text, graphics, and components such as the header/footer, titles, and navigation will be placed.
- Divides the screen into categories and levels of information, creating a visual hierarchy.
- Shows the chunking and grouping of information.
- Provides consistency.
- Provides predictability. Lets readers learn
    - Where to look for different kinds of information on each page in your Web site.
    - Where they are located.
    - Where to go next.
- Determines the focal point and affects eye movement.
- Conveys planning and organization to readers.
- Creates a modular structure. You can provide updated information and leave other material (such as navigation) consistent.

HTML Code: You create a grid by using tables, frames, columns, lines, and white space.

Tips

- Create a horizontal grid so that your page is designed for the shape of a 680 by 480 pixel monitor.
- Divide screen into three to five information zones, such as
    - Title
    - Navigational controls
    - Content
    - Footer

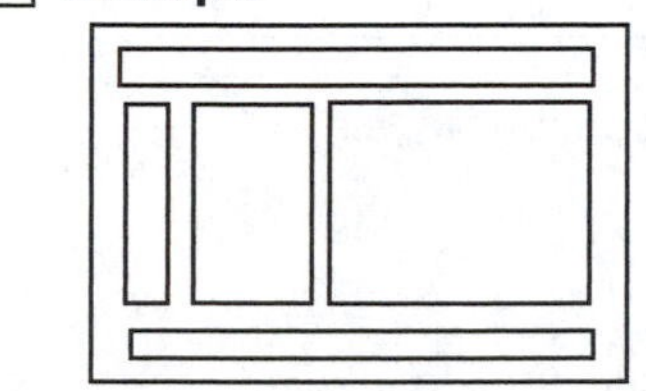

- Use no more than seven zones.
- Arrange information from left to right.
- Put important information in the focal point.
- Be consistent: use a repeating pattern, and put the same information in the same locations from page to page.
- Position elements to provide balance.

▷ Grouping

Grouping is chunking related information into categories. Grouping helps avoid long lists and menus and shows hierarchical relationships.

Tips

- Break up long lists and menus into categories and sub-categories.
- Group related text, graphics, and navigational items.
- Group items *visually* using the following methods:
  - Boxes
  - Bullets
  - Color and shading
  - Columns
  - Graphics
  - Grid
  - Indentation
  - Lines
  - Tables
  - Varying fonts styles, sizes, and attributes (e.g., bold)
  - White space

☑ Example

The idocs Guide to HTML *groups information about HTML tags: tag name, code, a table summarizing usage recommendations and browser test, links to attributes, and shaded samples.*

<ADDRESS>

Usage Recommendation
use it

<ADDRESS> denotes contact information for the author or organization of the web site.

this code	produces this
Our address by snail mail:<P> <ADDRESS> 404 Sandia Court, Blacksburg, VA 24060 </ADDRESS>	Our address by snail mail: *404 Sandia Court, Blacksburg, VA 24060*

- Group information *verbally* using the following methods:
  - Headings
  - Labeling
- Group information that changes regularly to make updates easier.

☑ **Example**

Jakob Nielsen's site places Permanent Content in a yellow column and News in a blue column.

Permanent Content	**News**
Alertbox Jakob's column on Web usability 10 High-Profit Redesign Priorities (March 12) Several usability findings lead directly to higher sales and increased customer loyalty. These design tactics should be your first priority when updating your website.	Usability Week 2007 Conference Washington, D.C., April 22-27 London, May 6-11 San Francisco, June 18-23 3-day camp: usability in practice 3-day immersion: interaction design

▷ Handheld Devices: Designing for

Some Web sites are displayed on PDAs (Personal Digital Assistants), cell phones, and other handheld devices. It is important to remember that their screens are small (150 by 150 pixels or less), have less resolution, and may have limited or no color.

Tips

- Keep pages small and simple.
- Use simple, clear navigation. Provide tables of contents and indexes.
- Use layering (links) rather than requiring scrolling.
- Use basic HTML tags.
- Avoid special techniques:
    - Frames
    - Image maps
    - JavaScript
    - Java
    - Cascading Style Sheets
- Avoid graphics or include only essential graphics that have been redesigned for small, low-resolution screens.
- Use black and white or contrasting, simple colors.
- Use fonts designed for handheld screens (e.g., Verdana).
- Keep text short and concise.
- Avoid requiring users to input information.
- Choose only essential content for your target audience.
- Use emulators to test your pages.
- Provide PDA versions of resources on your site.

☑ **Example**

Download to PDA

▷ Head

The HEAD is the HTML tag at the beginning of each document. Each HTML page should have a <HEAD> and matching </HEAD> tag directly following the initial <HTML> tag. The HEAD defines an HTML header and contains document information, which is not visible within the browser. This section can identify file information, graphics and tables used for the top of the page; tell others about the file's contents; and contain other tags. Comments in the HEAD section are also used when a site has multiple authors.

HTML Code: <HEAD> </HEAD>

Comments are created using <!-comment->

Tips

- Use the HEAD block for information such as the following:

- ☐ Author's name
- ☐ Background color
- ☐ Comments
- ☐ Content description
- ☐ Date modified
- ☐ Date of creation
- ☐ Document title
- ☐ E-mail address
- ☐ Filename
- ☐ Purpose of page
- ☐ Software used
- ☐ Style sheet link

☑ **Example**

```
<html>
<head>
<title>ATTW -- Association of Teachers of Technical Writing</title>
<meta name="description"
content="ATTW is an interactive site dynamically developed by ATTW members and
site visitors. The site provides members with access to a dynamic collection of
resources, including syllabi, calls for papers, job listings, information regarding ATTW
publications, and links to professional journals and organizations.">
<meta name="keywords" content="ATTW, Association of Teachers of Technical
Writing, Technical Writing, Professional Writing, Technical Communication, Writing,
Technical, English, proposals, reports, memos, resumes, conferences, jobs,
bibliographies, calls for papers, teaching, teachers, teach, publications, ATTW Bulletin,
TCQ, Technical Communication Quarterly, writing awards">
</head>
```

- ☐ Tags: BASE, LINK, META, TITLE, STYLE, AND SCRIPT

▷ Header

The header is similar to a masthead in a newsletter. It can be a logo, banner, text, or combination of these. Sometimes navigation is considered part of the header.

The header is used to do the following:

- ☐ Grab attention: it is the first thing readers see.
- ☐ Provide unity and continuity to the site.
- ☐ Identify the site name.
- ☐ Identify the page topic.
- ☐ Identify the person/organization responsible for the page or site.

Tips

- ▸ Include any of the following in the header:
  - ☐ Web site name
  - ☐ Web page name
  - ☐ Logo

- ☐ Navigational links

- Identify the person or organization behind the site.
- Put a smaller version of the same information in the footer on each page.
- Use a logo image with a small file size to speed download time.

▷ Headings

Headings are text used to divide the page into zones of information. Along with links, they are a method of labeling information. Headings are one of the most important parts of a Web page because readers can scan them quickly. Also, many search engines use this information to index documents.

HTML Code: There are six levels of heading tags (e.g., level one is <H1> </H1>). These sizes are relative. H1, the largest heading, is generally used for the page title.

Tips

Using Headings

- Use headings to do the following:

  - ☐ Aid in skimming.
  - ☐ Attract readers. Usability studies show that over three-fourths of readers scan a page before reading.
  - ☐ Chunk and categorize information.
  - ☐ Help orient readers to their location (contextual clues).
  - ☐ Help readers decide if they want to read the section.
  - ☐ Identify organization, hierarchy, and levels of importance.
  - ☐ Preview information by serving as advance organizers.

> ☑ **Examples**
>
> Internet Traffic Can Slow Your Surfing Speed
> Modem Speed Makes a Difference
> Large Files Will Take Longer to Download
>
> ☒ **Example**
>
> A Few Techniques

  - ☐ Summarize and describe the section or paragraph.

> ☑ **Example**
>
> Dynamic HTML: You're in Control

Writing Headings

- Use keywords and phrases. Emphasize the most important information that summarizes the section.
- Avoid clever or phrases, film titles, references to current events, and other creative headings that are unrelated to the topic. This technique will only confuse search engines and make it more difficult to distinguish your site from others.

☒ **Example**
A search for "The Way We Were" results in hundreds of Web sites.

- Use exactly the same headings that appear in the main menu/table of contents and arrange them in the same order listed.
- Use headings that focus on only one point.
- Write headings that are informative, unambiguous, and clear.
- Make headings context independent.
- Avoid humor unless it is clear what content follows.
- Keep headings relatively short.

☒ **Example**
Listing of Graphics and Multimedia Files, and Which Programs You Need to View Them, and Where They May be Downloaded

- Use action verbs (gerund or imperative) and dynamic wording.

☑ **Example**
E-Mail vs. Using E-Mail Effectively (*better*)

- Check headings for parallel construction. To check your headings for consistency, list them all in outline form.

☒ **Example**
Search Information
Searchable Zones
Boolean and Proximity Operators
Wildcard Characters
How to Write Verity Search Queries *(not parallel)*
Verity Search Query Language Reference

- Avoid abbreviations and acronyms.
- Use terms your readers will understand.

☒ **Examples**
-Listing of ISO Codes used on URL Extensions and the Countries They Represent
-The Neophytes Guide to Effective E-Mail

Formatting Headings

- Show levels of importance, but use no more than three levels.
- When subdividing, try to use at least two subheadings at each level.
- Use the heading tags (<H1> to <H6>) only for headings, not to create large or small fonts.
- Never skip a level: use the heading tags in order (<H1>, <H2>, etc.).

- Avoid capitalization of all letters; all caps are difficult to read and look dense. Capitalize the first letter of words except for articles, conjunctions, "to" in infinitives, and prepositions less than four characters long.
- Use end punctuation for complete sentences or questions.
- Check for consistency (layout): use of punctuation, initial caps, size, etc.
- Use left alignment.
- Put headings on a separate line, or make them lead-ins to the paragraph.
- Consider highlighting headings by placing them in colored table cells or using colored text.

LINKING TO HEADINGS

- Provide overviews of headings and subheadings.
- Use links so readers can jump to the appropriate heading.

> ☑ **Example**
>
> This chapter contains the following section:
> Downloading PDM Software
>
> **Downloading PDM Software**
> This section includes the following topics:
>
> Download the Latest Software from the Web
> Download the Latest Software with FTP

- Indicate whether the headings are on the same page (internal links) or other pages (intra-site links).

▷ Headlines

Headlines are headings used in Web sites that use news/newspaper, newsletter, or magazine formats. They are also used in sales and marketing. The types of headlines used depend on the type of site and audience.

Tips

USING HEADLINES

- Use headlines to do the following:
  - ☐ Attract attention.
  - ☐ Let readers know what they will read about.

> ☑ **Example**
>
> Disney Launches New Web Site for Kids

  - ☐ Sell or persuade.
  - ☐ Set the tone.
  - ☐ Summarize the key information in the story or its conclusions.

☑ **Examples**

IBM to acquire Rational Software
The Not-So-Golden Pages

Headline Techniques

▶ Use the following techniques when appropriate:

TECHNIQUE	EXAMPLES
Alliteration	Banishing Bandwidth Blues
Colon *This technique identifies the source or captures attention.*	Bush: Economy under control Study: Internet Users Climb to 92M Deadly Feast: Can Venison Kill You?
Command	See why you should kiss 56K goodbye
How/why	Why Tai Chi Is the Perfect Exercise
How to	How to protect your network
Humor	What to do when an ATM eats your card
News headline	California house explodes; 10 injured
Numbers	10 job skills bosses want
Opinion	Why you shouldn't buy gold
Parallelism	All Right, If You Insist Technical tips for those determined to force formatting Push Me, Pull Me Using client pull tags as a navigational aid
Question	Is a Tablet PC For You? Microsoft's grant has strings attached?
Sentence	Thousands flee fatal Europe floods
Solution to problem	Improve your PC's performance with the Memory Configurator
Statistics	More than 1/3 of the Contiguous States in Drought; U.S. and Global Temperatures Warmer than 100-Year Average in June
Teaser	The uninvited: ghosts or just the ice machine?
Testimonial	How I Saved Money When My Children Were Small
Title format	The Science of Wildfires
Warning	New strain of virus hits computer e-mail
Word play *In general, avoid using puns.*	Hair fiasco: Do or dye? (pun) The Storm Before the COM (play on word *calm)* Pulling the Plug-Ins (play on words) Ready, Aim, Hire (play on a saying) Windows on the World (double meaning of windows)

Writing Headlines

- Use informative headlines that include the keywords and phrases from the section and clearly convey the content. If you cannot summarize the text in a few words, consider focusing and chunking the article.

> ☑ **Example**
> *Headline:* Researcher cracks Net encryption software
> *Excerpt from article:* researcher . . . cracked the standard encryption software used in most commerce on the Internet . . .

- Place keywords first.
- Write headlines that make sense out of context. Because they must appear in bookmarks, search engine lists, and lists of links, they must be independent of the article itself.

> ☒ **Example**
> Extreme Chips *(This article is really about new semiconductor technology)*

- Omit words like "a" and "the" at the beginning of the headline so they appear in lists with keywords first.
- Avoid puns, teasers, and clichés.

> ☒ **Examples**
> Apple Harvest *(this article is about Apple Computer "software harvest.")*
> What's Holding You Back? *(This article is about seat belts)*

- Be specific.

> ☒ **Example**
> It's Everywhere *(this article is about toxic mold).*

- Use positive wording.

> ☒ **Examples**
> -Why cyberterrorists don't care about your PC...
> -No End To The Turbulence For Airlines

- Use simple wording for a global audience.
- Use action verbs to emphasize reader benefits.

> ☑ **Example**
> Gain better control of the flow of information on your intranet and Web sites!

- Avoid abbreviations and acronyms if your readers will not understand them.

☒ **Examples**
-Bandwidth Face-off: 56K, ISDN, and xDSL
-Impossible Data Whse Situations
-Airport WLANs Lack Safeguards

- Link the headlines to the full story.
  - ☐ Use headlines as links.

☑ **Example**
Dow Closes Down 206; Nasdaq Falls 38

  - ☐ Combine headlines with a brief annotation or summary and link to the full article.

☑ **Examples**
the whole story >>

Fed Leaves Interest Rates Unchanged
With the economy advancing only in fits and starts, the Federal Reserve held short-term interest rates steady on Tuesday but left the door open to future reductions.
Full Coverage

International Herald Tribune lets you click a Clippings icon , which saves articles you are interested in reading.

Formatting Headlines

- Group headlines under headings.

☑ **Example**
Latest Headlines
Categories: Nation, World, Technology, Sports, Travel, Science

- For extensive headlines, allow readers to sort and search.

☑ **Example**
1stHeadlines lets you sort headlines by source (as shown below) or time.
ComputerWorld - More Headlines
Review: Apple TV just plain works
Fri Mar 23, 2007 2:43 PM EDT

- Use left alignment.
- Match headlines with any visuals.
- Use colored text to attract attention.
- Make headlines easy to read by using legible fonts.
- Avoid techniques such as ALL CAPS and extra punctuation!!!!

▷ Home Page

A home page is the first page of a Web site–the page people start with and return to when navigating your site. It is the most important page on your site because it serves as the following:

- An introduction that indicates the content, purpose, and scope of the site.
- The page that sets the tone and creates a first impression.
- The location of the main menu/table of contents, where readers can find information.
- The location of any information or aids that help readers accomplish their goals.
- A home base to which readers can return when they are lost.

The home page uses many of the components and techniques described in this handbook. For more details about these topics, refer to the specific handbook entries.

Tips

Naming

- Put *Home Page* in the page name and title.
- Name the file *home.htm*, *index.htm,* or *default.htm.*

Contents

- Include the following:
    - Header and footer.
    - Tagline: a description of what you do.
    - Hook to get readers' interest and attention.
    - Benefit or value of the site.
    - Introduction to the site (including the purpose and description of content).
    - Overview (advance organizer) of the site's organization.
    - Identification of sponsor.
    - News.
    - Information about system and browser requirements for viewing site.

Formatting

- Keep the home page simple.
- Keep the home page one screen long, or put the important information within the focal point.
- Use a grid so readers can find information quickly.
- Use small, necessary graphics to keep download time small.
- Emphasize what your audience wants from your site.

Links

- Include a menu or table of contents. One technique is to include both a menu and the same menu with annotations.

- ☐ The menu is repeated in the same position on every page.
- ☐ The annotated menu informs readers of the contents.

- Include a link to the home page from every page in your site. If you use graphical links, include text links as well.
- Use layering: link to details.
- Link to special sections, such as *What's New*, a site guide, mission statement, information about your organization, and a search engine.

WRITING

- Use direct, simple sentences.
- Provide a clear focus on your site's purpose.

▷ HTML Editing Software

To create a Web page, you need software that will create the HTML tags. Although you can create a Web page using any text editor, HTML software allows you to create the HTML tags without knowing any code. In addition, many HTML editors work with related technologies such as Cascading Style Sheets (CSS), Extensible Markup Language (XML), and JavaScript. You can buy software that specializes in creating and updating Web pages, sites, and blogs. Many HTML editors are also available as shareware. Most simplify applying HTML code through buttons and menus.

Tips

- Select from the following types of programs available. They vary in ease of use and price (some are free). Programs also differ in support for special features to your Web page, such as frames, forms, link checks, site management, multimedia, and built-in templates and clipart.

EDITORS

- ☐ *Text editors* allow you to work with the HTML code. Any text editor (such as Windows Notepad) will let you type the text and tags.
- ☐ *Text-based HTML editors* are text software customized for Web authoring. You work in text mode, but they provide tools to facilitate markup and verify syntax.
- ☐ *WYSIWYG (What You See is What You Get) editors* allow you to work on a Web page while software generates the code in the background. You can opt to view a preview of the page or the code.
- ☐ *Word processors and desktop publishing programs:* Many word processors and desktop publishing programs (such as Microsoft Publisher) also include HTML features. You can save the file in HTML format.
- ☐ *Web authoring programs* (such as Microsoft FrontPage and Macromedia/Adobe Dreamweaver) are full-featured programs for creating a Web page. They provide features to add enhancements (such as JavaScript and Cascading Style Sheets) and to upload your site to a server. Advanced programs contain tools for developing high-end Web sites.

Graphics Programs

- *Graphics programs* (such as Paint Shop Pro) allow you to create Web graphics, convert images to the file formats used on the Web (such as GIF and JPEG), and to edit images by reducing the size and number of colors.

Other Tools

- *HTML validation tools* allow you to test your links.
- A *browser* is software that interprets the HTML tags and displays the Web document. You need to view your Web pages in several browsers because each displays various elements differently. Two of the most well known browsers are Microsoft Internet Explorer and Netscape. You should also view your Web pages in a text-based browser such as Lynx.
- A *wiki* is a site (or server software) used for collaborative authoring. Visitors can create content, add links, and edit content.

▷ Humor

Humor is appropriate on some Web sites and for certain audiences. Usability studies indicate that Web readers prefer some humor, such as word play or cynicism. Humor can take many forms: parody, jokes, cartoons, etc. It can be the sole purpose of the Web site or used only occasionally. Humor can help to break up a long dry passage of technical information and give readers a break. It is especially appropriate in writing for children.

Tips

- Analyze your intended audience before using humor.
- Use humor to make your site less dull.
- Use humor cautiously because it can be misinterpreted over a computer.
- Avoid the following:
  - Puns and humor that may be taken wrong by an international audience.
  - Localized humor.
  - Plays on words (*site* vs. *sight).*
- If your page or site is humorous:
  - Use a humorous Web address and site name.

☑ Examples

ISBW (IShouldBeWorking.com) is dedicated to slackers, goof-offs, procrastinators, loafers, "long lunchers," and web addicted employees worldwide.

-Modern Humorist
-My Site Stinks

- Use humor in your title so readers expect humor.

☑ Examples

Wacky HTML obviously links to a humorous Web site. The site's slogan is **The HTML Your Mother Never Told You About**.

Shop Talk 101: A Humorous Look at the Retail Industry

- ☐ Explain the purpose of your site in the introduction.

☑ **Example**

HTML For the Conceptually Challenged
Want to learn HTML fast? If you mostly watch television, have an attention span measured in microseconds, and think reading is a waste of your valuable time, this page is for you.

- ☐ Explain the humor in an annotation.

☑ **Example**

If Elvis Were Alive ...
A humorous look at what 'The King' might be up to today

- ▶ If necessary, provide information somewhere on the site that explains that the site is satirical.

☑ **Example**

The Onion is a satirical weekly publication published 47 times a year on Thursdays.

- ▶ Use humorous headings or headlines if you provide annotation or it is clear what the story is about. However, this technique may make it difficult for search engines to index your topics.
- ▶ Use humorous graphics and cartoons if appropriate and if they fit the content.
- ▶ Use humorous link text if appropriate, but make sure it is clear where the links lead.

☑ **Example**

This humorous link, a play on the words "design/divine," is annotated with specific topics contained on the destination page:
Accept Design Intervention.
Learn the more subtle aspects to web design.
Keeping graphics small. Designing for technology AND the user.

- ▶ Use informal writing and a personal tone.

☑ **Example**

Jeffrey Glover's site on Web page design uses a variety of techniques to provide humor: humorous headings, questions, conversational language, contractions, and examples.

Does your site suffer from **Link Lunacy**?
Here's an example of what I mean by Link Lunacy:

Does this look familiar? Notice how it's full of links **everywhere**! Words seem to be randomly linked throughout the paragraphs. So you might be saying to yourself, Well, hey, Jeff... Isn't that the whole power of this net-thing! The links are what makes it cool! Well, you're right... Sorta... Putting lots links in your paragraphs make them very difficult to read! It breaks your concentration while trying to read the paragraphs and in some cases people won't even finish the paragraph if they find a link they like.

- Avoid placing humor on a Web page that is accessed often as a path to other pages. It may be read too often and become irritating to readers.
- Be careful using subtle humor in e-mail because it may be misunderstood.

▷ Hyphens

The hyphen is punctuation used to link and separate words.

Tips

- Use a hyphen between compound modifiers (two or three words that modify a noun and serve as adjectives) if they express one thought.

☑ **Examples**

-What makes for a well-designed Web site?
-Web-based training

☒ **Example**

We offer high quality, well thought out, custom designed web sites at extremely competitive rates.

- Do not use a hyphen when compound modifiers come after a noun.

☑ **Example**

Your site should be well organized.

- Do not use a hyphen if any one of the words can modify the noun alone.

☑ **Examples**

-Netscape Composer is an easy-to-use tool that makes creating Web-based documents as easy as writing a memo with a word processor.
-All-In-One Listserver Discussion Group Guide.
-Internet Relay Chats (real-time group discussions)
-User-centered design
-Easy-to-use Web site

☒ **Examples**

-A step by step guide to building your Web site. *(step-by-step guide)*
-Make sure links are the same color throughout the whole site to avoid confused web-users. (*web or Web users)*
-Complete Hosting Services with the most up to date features. *(up-to-date)*
-Word-processors seem to have taken over the job of hyphenating broken words at the right-hand end of our lines. *(word processors)*

- Use a hyphen after prefixes in the following situations:
  - After *ex-*, *self-*, and *all-*.
  - Before a capitalized word.

☑ **Example**

pre-Internet days

  - When double letters or confusion result. (Consult a dictionary because there are many exceptions.)

☑ **Examples**

-You can use the item tags in un-numbered (bulleted) or descriptive lists.
-Sabres Re-Sign McKee, Brown

- Use a hyphen with compound numbers.

☑ **Example**

Fifty-nine

- Use a hyphen with figures or letters.

☑ **Examples**

Figure 5-1
Table 1-A

- Use a hyphen after a letter or number that modifies and noun (e.g., 32-bit color).
- Use a hyphen to separate key combinations (e.g., Control-Shift-3).
- For "cyber words," such as e-mail and e-commerce, there are still no standard rules for hyphenation. Some suggest using a hyphen for nouns and modifiers and no hyphens for verbs.
- Use hyphens carefully with compound words that form modifiers, verbs, and nouns. Examples include the following:
  - add on (verb)/add-on (noun, modifier)
  - command line (noun)/command-line (modifier)
  - dial up (verb)/dial-up (modifier)
  - direct access (noun)/direct-access (modifier)
  - double click (noun)/double-click (verb)
  - drag and drop (noun)/drag-and-drop (modifier)

- end user (noun)/end-user (modifier)
- entry level (noun)/entry-level (modifier)
- file sharing (noun)/file-sharing (modifier)
- follow up (verb)/follow-up (modifier)
- hard copy (noun)/hard-copy (modifier)
- high level (noun)/high-level (modifier)
- high resolution (noun)/high-resolution (modifier)
- look up (verb)/look-up (modifier)
- low end (noun)/low-end (modifier)
- low resolution (noun)/low-resolution (modifier)
- off line (adverb)/off-line (modifier)
- source code (noun)/source-code (modifier)
- turnkey (modifier) (not turn-key)

▷ Icons

An icon is a small graphical symbol that usually represents an object or abstract concept.

The most well known and recognizable icons are *home* and *search*. However, it is difficult to create icons that all users will be able to understand. Most icons are small GIF files that load quickly.

Tips

- Use icons for the following:
    - Buttons.
    - To coordinate with your design theme.
    - To identify categories of information and help readers quickly scan.

> ☑ **Example**
>
> **Tip:** Text in a link uses the color specified in the link attribute of the <window> tag.
>
> **For More Information**: SMIL files can also define hypertext links that may override the link you set here. For more information, see Chapter 15.
>
> **View it now!** (requirements for viewing this sample)

- To help readers identify where they are located.

> ☑ **Example**
>
> *This site uses icons in the navigation. Larger versions of each icon appear on each page.*
>
> City Government

- To identify types of file formats.

☑ **Examples**

This icon represents downloading to a PDA.

This Web document uses icons to classify chunks of information: description, graphics, and related topics.

5.2 Pointer

Description

The pointer is a pointing device (usually an arrow) displayed on the screen. The position of the pointer can be controlled by moving the finger on the track pad (of course there are other, additional pointing tools to control the pointer, however, the principle remains the same). The pointer can be moved across the entire screen. Depending on the situation, the pointer appearance can vary.

Actions such as pointing, clicking and pressing can be carried out with the pointer, the track pad and the track-pad key, resulting in an interaction between the user and the control.

Fig. 16 Examples of Different Pointer Layouts

Related Topics

- Remote Control Device
- Focus

- Use both a text label and the icon.
- Make it clear what icons represent.

☒ **Example**

It is not clear that this icon represents a cross reference.

- Keep icon design simple; omit detail.
- Limit the number of icons.
- Make the icons large enough to be distinct.
- Make sure icons are legible at different resolutions.
- Use icons with consistency: use the same icon, size, shape, and labels throughout the Web site. Use icons with the same "look."

☑ **Example**

▷ Image Map

An image map, or clickable image, is a graphic (usually a transparent GIF) with hotspots that execute an action. The graphic is divided (or mapped) into regions that link to other pages; you associate a URL with each region. One large image is used instead of several small buttons or icons. Four parts of an image map are the image, hyperlinks, map file containing coordinates of the clickable areas, and the script to process it in the HTML file. Image maps are primarily used for navigation or educational/informational graphics.

There are two types of image maps:
Client-side: This type of image map is preferable because it resides in the HTML document and uses the browser to interpret the hotspots as links.

Server-side: This type of image map depends on the server to translate the script files that reside on the server. This process creates network traffic.

There are negatives of image maps:

- Some browsers don't handle them well.
- They download slowly.
- It is often not clear where to click unless you move the cursor over a hotspot.

HTML Code: <MAP> </MAP>

Tips

- Use an image map to
  - Tie in with the metaphor you've chosen for your page.
  - Provide a map or graphical representation of your site.
  - Contain navigation links.
  - Provide an informational graphic with links to definitions, explanations, and detail.

> ☑ **Examples**
>
> *An image map would let you create hyperlinks on a geographical map, highlight features of a product, or explain parts of an object.*

- Use ALT tags.
- Use text links to provide alternative text.
- Make it obvious which regions are hotspots. If possible, make them look like buttons, or use a rollover effect.
- Make clickable areas large.
- Create hotspots in a logical order, or use the TABINDEX attribute to specify the tab order. This method allows readers to tab through an image map to find links.

▷ Important Information

The important information is the message or main point of your Web page or site—what you want people to see first and remember. The most important information should appear within the first few lines on the page. This text is important because

- You let readers immediately know the page's contents and message.
- It provides contextual clues for readers who have come from other Web sites.
- Search engines often use the first 200 characters of a Web page when they display search results.

Tips

Determining What is Important

- Analyze your audience and determine what is important to them—why they are visiting your site.

> ☑ **Example**
>
> *The most important information will vary depending on the type of Web site. For example, a news site would emphasize the most important headlines. A used book site would emphasize a search engine. A site that sells a product would emphasize a special sale. An airline site would emphasize schedules and a fare finder.*

- Prioritize components on your page: header, navigation, search engine, etc.
- Prioritize information categories, such as news.
- Arrange topics in a logical order.

Writing Important Information

- Write a strong introduction to your Web site (introduction to site) and an introduction at the top of each Web page.
- Put high-level summaries of important information first and use links to details (layer information).
- Link to lengthy information.

> ☑ **Examples**
>
> *The American Heritage® Dictionary of the English Language, Fourth Edition.*
> Copyright © 2000 by Houghton Mifflin Company.
> Published by the Houghton Mifflin Company. All rights reserved.
> Other Important Information
>
> **Important Information**
> ● Important Addresses ● Important Phone Numbers

- Use the inverted pyramid organization.
- Separate need to know (relevant) from nice to know (supportive) information.
- To reach your entire audience, make important content text rather than graphics or hyperlinks. If an image conveys important information, use both ALT text and an extended description.

Formatting Important Information

- Make the most important information prominent at the top of the screen and within the focal point.
- Put the important information at the beginning and top left of each Web page and within the top third of the screen.
- Do not require scrolling to read what is important.
- Draw readers' eyes to what is important by
    - Using devices of emphasis (size, color, bold, bullets, graphics, and white space).
    - The arrangement on the page (e.g., left to right, top to bottom).

> ☑ **Example**
>
> *FedEx places navigation to* PACKAGE/ENVELOPE services (e.g. SHIP, TRACK) *at the top of the page. Most visitors to this site are probably seeking this information.*

- Show levels of importance by using headings, boxes/sidebars, and indentation. Showing a hierarchy of information makes information easier to find.
- Emphasize important information through repetition.
- Highlight key announcements and notices.
- Avoid any techniques that make important information look like advertising.

▷ Indentation

HTML does not provide for paragraph indents. However, you can create indentation by using the following:

- Cascading Style Sheets (CSS) (The preferred method to indent a page, section of a page, paragraph, or first line of a paragraph.)
- A small blank graphic
- A graphic that is the same color as your background
- Non-breaking spaces
- Block quotes
- Tables

Indentation can be used to show the hierarchy of information. For example, indenting the body of a paragraph beneath the topic sentence highlights it. Indentation is also used with bulleted, numbered, and nested lists.

HTML Code: –non-breaking space.
<BLOCKQUOTE> </BLOCKQUOTE> indents text on both sides. It should be used for quotes.

Tips

- Indent text to
    - ☐ Direct eye movement.
    - ☐ Prioritize and show the hierarchy of information.
    - ☐ Emphasize headings.
    - ☐ Emphasize summaries and topic sentences. Indent explanations and detail.
    - ☐ Separate paragraphs.

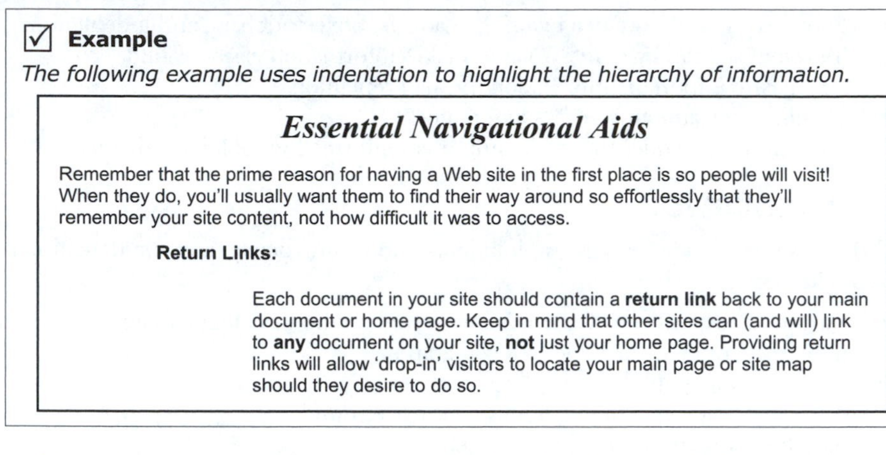

☑ **Example**

The following example uses indentation to highlight the hierarchy of information.

Essential Navigational Aids

Remember that the prime reason for having a Web site in the first place is so people will visit! When they do, you'll usually want them to find their way around so effortlessly that they'll remember your site content, not how difficult it was to access.

Return Links:

Each document in your site should contain a **return link** back to your main document or home page. Keep in mind that other sites can (and will) link to **any** document on your site, **not** just your home page. Providing return links will allow 'drop-in' visitors to locate your main page or site map should they desire to do so.

- Use indentation to indicate the start of paragraphs when you decide not to insert blank lines. This technique is best for large amounts of text.

▷ Index

An index (sometimes called an alphabetical guide) is a long alphabetical list of keywords and phrases the writer has assigned to each page, section, or topic. Each entry contains jumps to a page or part of a page. It is usually a separate page on the site accessible from the navigation bar or menu.

An index is most useful for large Web sites. You can create an index manually or use an indexing program to build the index for you. (The latter is beyond the scope of this book and usually appropriate for sites with large documents.)

An index has the following advantages:

- Familiar to readers because it is just like a book's index.
- A simple way to let readers find topics by browsing.
- Useful for readers who are not sure what terms to look up.
- Visually scannable.

Tips

Writing an Index

- Determine what information your readers consider important.
- List each topic, concept, and task in your site.
- Group related entries.
- For each topic, page, or section of a document, think of alternative terms or phrases a user would use to look up that topic. Include synonyms, related words, and grammatical variations (i.e., stemming).

☑ **Example**

Mailing List Netiquette
Netiquette | Mailing Lists

- Use words readers will recognize.
- Provide annotations if necessary.

> ☑ **Example**
>
> Offices (OSHA Office Directory)
> Earthquake (Factsheet)

- Begin each entry with a noun or verb.
- Avoid broad terms such as "using."
- Check for consistency (style) and punctuation.

FORMATTING AN INDEX

- Alphabetize entries using the first keyword.
- Consider chunking the entries by category.

> ☑ **Example**
>
> *This site contains a link to the site index, which contains links to the major sections of the index (partially shown).*

Site Index
Articles
Forum
Headlines
How-to

Authoring	Collections * Design * Graphics * Languages (HTML, XML) Programming * Site Management * Style * Tutorials
E-Commerce	Articles * Collections * Commerce server providers * Commerce server sofware * Digital Certificates * News * Payment systems * Tutorials

- For a large index, use navigation letters and letter dividers marking the beginning of entries for each letter.

> ☑ **Example**
>
> **Site Index: A B C D E F G H I J K L M N O P Q R S T U V W X Y Z**

- If the index is long, consider using separate pages.
- Left-justify first-level items. Indent second-level items and intersperse with the first-level entries.

> ☑ **Example**
>
> **About FedEx**
>   FedEx Overview
>   Family of Companies
>   Worldwide
>   Technology
>   Social Responsibility

▷ Informal Writing

Informal writing is a conversational style created by your word choice and sentence style. An informal writing style has several advantages:

- Many Web readers prefer informal, conversational, down-to-earth writing to formal writing.
- A conversational style lets you interact with and engage readers.
- Formal writing is more difficult to understand and slows reading.

Because the Web is considered a more informal medium, an informal style is often appropriate. However, you should use a style appropriate to your audience, content, and purpose.

Tips

- Use the following techniques for an informal style:
  - Contractions (although avoid too many of these for international readers).
  - "I," "You," and other personal pronouns.
  - Short paragraphs.
  - Short simple sentences.
  - Personal tone.
  - Simple, colloquial, or lay terminology.
  - Humor.
  - Questions and answers.

> ☑ **Examples**
>
> Need a Map? Got lots of 'em right here.
>
> *All articles on howstuff works.com use an informal style. Potential authors are asked to match the site's tone and style and avoid boring, technical explanations of how things work.*

- Use a conversational, informal style only if it is appropriate for your content and the image you want to project.

> ☑ **Example**
>
> PC Plain Talk located in the Gateway Computer technical support site is a helpful but cool PC guide: we decided to make a fun-to-read guide to PCs written in regular English.

- Avoid being too informal when it affects your credibility or is inappropriate in context.

> ☒ **Examples**
>
> There are some funky ways around this limitation.
>
> But since the introduction of tables, everything is just hunkey dorey in HTML land.

- Avoid using informal language in excess.

> ☒ **Example**
>
> You'll find gobs of groovy graphics, *gratis.*

▷ Information Overload

Information overload is the anxiety readers feel when overwhelmed with data. Web readers feel information overload when they must make too many decisions, cannot find the information they need, or do not understand it. Instead, readers of Web pages are in a hurry and want to find information quickly.

Tips

- Limit memory load to help readers scan your Web pages quickly.
- Avoid overwhelming readers on the home page. Use simplicity.
- Avoid the techniques shown in the first column of the following table and use the techniques in the second column.

AVOID	USE
Organization	
Too many decisions to make	- Seven plus or minus two rule - Layering
Too many links	- Menus and other navigational aids - Grouping
Disorganized, uncategorized information	Grouping
Language	
Negative language	Positive statements
Promotional language	Facts and statistics
Technical terms, abbreviations, acronyms	Glossary
Passive voice	Active voice
Unannotated lists	Annotation
Layout	
Too much information on a page	Layering
Large blocks of text and scrolling	- Chunking - Lists - Separate pages
Inconsistency	- Consistency - Grid
Overuse of emphasis, animation, blinking, color, fonts	Simplicity and moderation

▷ Informational Page: Business or Service

An informational page provides factual information and details about a business or service. Creating a Web page allows you to

- Establish a presence.
- Promote your product or service.
- Establish good customer relations.
- Conduct business.
- Answer reader questions.

Tips

Types of Information to Include

- Provide a tagline that clearly describes what you do.
- Answer the questions *who, what, where, when,* and *how.*
- Include the following types of information:
    - "About us"/background
    - Address
    - Contact information
    - Directions/map
    - Fees/costs
    - Guarantees
    - Hours of operation
    - Mission
    - Phone number/Fax number
    - Rules/regulations
    - Services
    - What's new/announcements
- Make the important information easy to find on the home page.

> ☑ **Examples**
>
> *American Airlines emphasizes making reservations on their home page because that is one of the main reasons readers visit the site.*
>
> *Hewlett Packard's Home/Office site emphasizes shopping, learning, support & troubleshooting, and software & driver downloads.*

- Consider providing useful content related to your business.

> ☑ **Example**
>
> *Hewlett Packard has a Learning Center:*
> The Learning Center is the place to learn more about your HP and Compaq personal computers, related products and technologies. Get tips and tricks, sign up for free online classes and more.

Links to Provide

- If information is detailed, provide a clearly labeled link from the home page.
- Provide links to other information that will be of interest to readers, such as an online catalog, publications, support information, elements that provide interactivity, a map for directions, and action steps.

Formatting

- Organize detailed information.
- Provide a variety of ways to access the information.

- Use formatting techniques such as tables, headings, bold, and lists to make the information easy to read.

▷ Instructions

Instructions explain how to perform tasks. Instructions on a Web site may be used for a number of reasons, such as in the site guide or as part of support (technical) section.

Tips

Writing Instructions

- Begin with an introduction, overview, and preview of steps.
- List conditions, prerequisites, and equipment separately.

> ☑ **Examples**
>
> **The Steps**
> **Intro:** Before you begin **Step 1:** Treat strained muscles **Step 2:** Avoid strained muscles
>
> **Time**
> **Mild strain:** 30 minutes a day for 3 days
> **Severe strain:** 60 minutes a day for 7 days
>
> **What You'll Need**
> A combination ice and heatpack, or tray of ice cubes and a towel, or a paper cup filled with water and frozen, or a bag of frozen peas.
> A towel or washcloth.
>
> ---
>
> **How to Get Software**
> This page provides detailed instructions on how to download Netscape's browser suites and servers.
>
> ---
>
> **Requirements and Prerequisites**
> This guide assumes that you have a working knowledge of common desktop deployment technologies and networking components.

- Use present tense.
- Make steps short and simple.
- Combine short related steps.
- Omit obvious actions.
- Put one action per step.

> ☒ **Example**
>
> *Step one is actually a combination of several steps:*
> 1. Save the image above to your hard disk (click and hold down your mouse button if you're using a Mac and right click if you're using a PC, which will bring up a pop up menu that allows you to save the image to your hard disk).

- Put steps in chronological order.
- Use simple terms and known abbreviations.
- Keep terminology consistent.

- Use parallel construction.
- Begin with imperative verbs.

☒ **Example**

This excerpt explains how to use a search engine. The steps are not parallel because steps 1 and 5 do not begin with imperative verbs.

1. Searching on more than one word produces much better results.
2. Use more specific terms.
3. Be explicit.
4. Insert spaces between words.
5. Spelling matters.

- Use task-oriented verbs.

☒ **Example**

The following are links to instructions. The headings are not parallel or task-oriented, and they use caps inconsistently.

What is Excite Search? Advanced search View by Web Site
What Can I search? Search results More Like This Search tips Relevance rating Browser errors

- Insert *Danger, Warning, Caution*, and *Note* labels as appropriate:
    - DANGER: Injury or death will occur if the instructions are not followed.
    - WARNING: Injury or death is possible if the instructions are not followed.
    - CAUTION: Damage to equipment is possible.
    - NOTES: Notes are added to give more information, usually in a procedure.

☑ **Example**

Warning Before you work on any equipment, be aware of the hazards involved with electrical circuitry and be familiar with standard practices for preventing accidents.

! Caution To prevent ESD damage to electronic components, always use an ESD wrist strap when handling modules or coming into contact with internal components.

- Include all important information and leave no gaps in readers' understanding of a step.
- Present the most common and simple way to do something rather than every possible method.

Linking

- Use layering: provide simple steps with links to details, examples, definitions, and *next* links.

☑ **Examples**

These basic steps explaining how to use a search engine site link to more detail.

1. Pick your search site
2. Learn to use the site
3. Choose your words carefully
4. Vary your spelling

This step links to definitions.

Use truncation and wildcards to include word variations.

▶ Provide internal links to headings and steps.

☑ **Examples**

Here is an overview of the steps:

Select the pictures to print

Locate and align the infrared (IR) ports

Transfer pictures to the printer

step 1: select the pictures to print

Your digital camera should have a mechanism (usually on the back of the camera) for selecting the pictures stored in it.

1. Find the pictures in your camera that you want to print.
2. Select the ones you want to print.

Cisco documentation uses a frame on the left for general navigation. Links to the previous and next links appear at the top and bottom of each page. Internal links are used for sub-steps.

Contents:
Quick Installation Guide - Catalyst 5000 & Catalyst 5505
1 Prepare to Install
2 Rack-Mount
3 Connect to the Network
4 Connect the Power
5 Start the System

Additional Procedures:
A Inserting/Removing Modules
B Inserting/Removing Power Supplies

[Back to "Prepare for Installation"] [Forward to "Connect the Switch to the Network"]

2. Rack-Mount the Switch

To rack-mount the Catalyst 5000 or Catalyst 5505 switch:

1 Attach the L brackets and cable guide to the chassis

2 Install the switch in the rack

▶ Link to related steps.

☑ **Example**

■ If there is a "Setup.exe," you will need to install the Make Student Disk program and then run it as needed to obtain data files. Go to Step 5.

▶ Use simple graphics.

▶ Link to detailed or large graphics or multimedia files.

☑ **Examples**

6. Once the strip is in place, trim the excess at the ceiling and the baseboard using broad knife as a guide for razor knife. The blade must be sharp to get a clean cut, so change blades often.

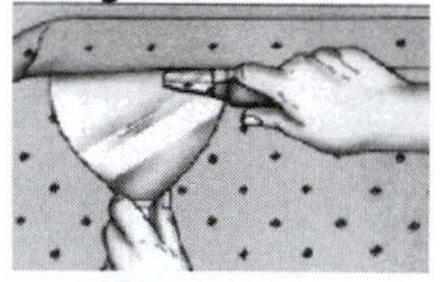

■ If the WinZip self-extractor dialog box opens, click Unzip to extract your file (see Figure 6).

view how-to video

- For numerous sets of instructions, provide an annotated table of contents and link to separate pages.

☑ **Example**

Windows Vista migration step-by-step guide

Find out the steps to use when upgrading a computer from Windows XP to Windows Vista, including the migration of existing files and settings.

FORMATTING INSTRUCTIONS

- Number the steps.
- Use tables or a vertical list.

☑ **Example**

This site uses a table format to link to details for each step.

Creating Your Product Catalog	
Step	**See**
Create categories to organize your products.	Creating Product Categories

- Separate comments and feedback from the steps.
- Do not make options steps; instead, separate them in a bullet list or table.

☒ **Example**

Step1: See if Dial-Up Networking is already installed. Go to My Computer and see if you find the Dial-Up Networking folder. If you find the Dial-Up networking folder in My Computer, then Dial-Up Networking is already installed. Proceed to the section on Installing TCP/IP. If you don't find the Dial-Up Networking folder then proceed to the next step...

- Break up the list by using headings or outline form; use no more than about seven steps per group (seven plus or minus two rule).
- Use layout devices, such as white space, color, bulleted lists, and boxes.
- Consider using screen captures when explaining how to use software.

☑ **Examples**

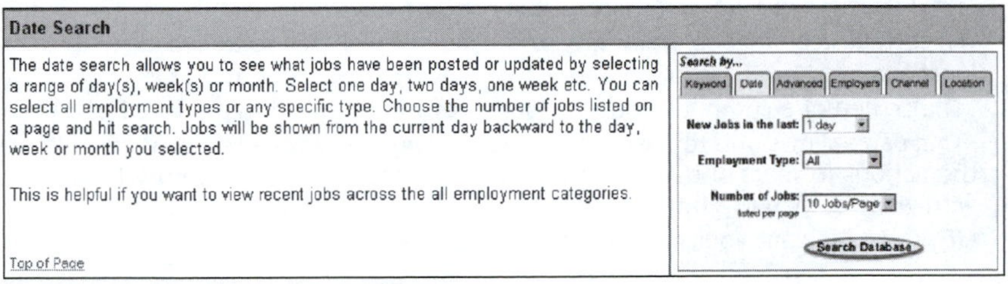

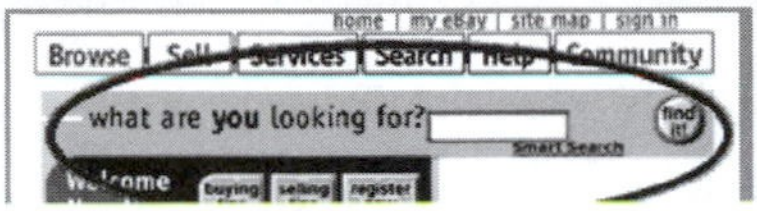

Date Search

The date search allows you to see what jobs have been posted or updated by selecting a range of day(s), week(s) or month. Select one day, two days, one week etc. You can select all employment types or any specific type. Choose the number of jobs listed on a page and hit search. Jobs will be shown from the current day backward to the day, week or month you selected.

This is helpful if you want to view recent jobs across the all employment categories.

Top of Page

Search by...
Keyword | Date | Advanced | Employers | Channel | Location
New Jobs in the last: 1 day
Employment Type: All
Number of Jobs: 10 Jobs/Page
listed per page
Search Database

home | my eBay | site map | sign in
Browse | Sell | Services | Search | Help | Community
what are you looking for?
find it!
Smart Search
Welcome buying selling register

Searching If you know exactly what you're looking for, just type a few words into the search box on the Home page and you'll get a list of relevant items.

This site limits the number of screen by putting one or two steps per page and linking to Next and Previous steps or pages.

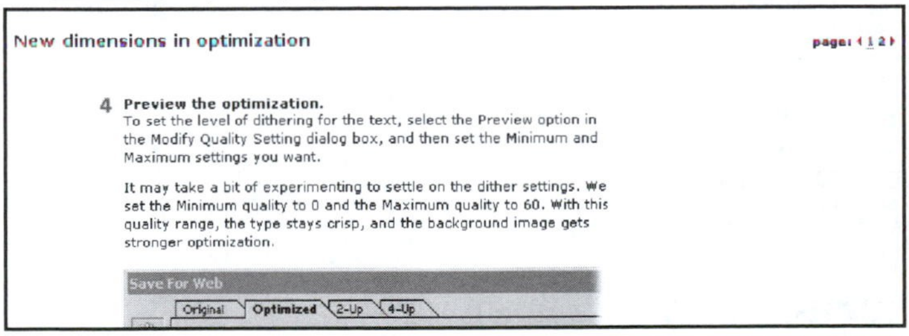

New dimensions in optimization

page: ◂ 1 2 ▸

4 **Preview the optimization.**
To set the level of dithering for the text, select the Preview option in the Modify Quality Setting dialog box, and then set the Minimum and Maximum settings you want.

It may take a bit of experimenting to settle on the dither settings. We set the Minimum quality to 0 and the Maximum quality to 60. With this quality range, the type stays crisp, and the background image gets stronger optimization.

Save For Web
Original | Optimized | 2-Up | 4-Up

- Give examples.

☑ **Example**

3. Complete Form FS# 354 (see Example #5)
Provide downloadable instructions in other file formats, such as Adobe Acrobat PDF.

- Let readers know where they are.

☑ **Example**

A progress bar can remind readers of the steps.

CHECKOUT PROGRESS MONITOR
Current location = order details

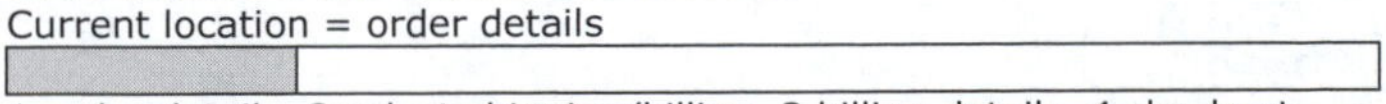

1 order details 2 select shipping/billing 3 billing details 4 checkout

▷ Intensifying Words

Intensifying words are used for emphasis. They add little to a sentence and contribute to wordiness.

☒ **Examples**

-That's all the basic HTML you need to know to write a **really** impressive home page. It may seem like a lot, but most of it is **really** rather intuitive.

-Hypertext itself means, **basically**, the ability of words in text files to jump to other locations.

-The screen of a computer monitor looks **very** different from a printed page.

-This page allows you to ensure your Web navigator is correctly configured to handle the various formats of text and multimedia objects you will **certainly** have to deal with while surfing the Internet.

-If you **truly** want your new web site to succeed . . .

Tips

- Avoid the following words:
  - □ Absolutely
  - □ Actually
  - □ Any
  - □ Basically
  - □ Certainly
  - □ Completely
  - □ Definitely
  - □ Extremely
  - □ Of course
  - □ Particularly
  - □ Pretty
  - □ Quite
  - □ Rather
  - □ Really
  - □ Significantly
  - □ Simply
  - □ So
  - □ Somewhat
  - □ Specifically
  - □ Truly
  - □ Utterly
  - □ Very

▷ Interactivity

Interactivity includes ways for readers to participate in your Web site. Interactive Web pages attract readers and give them something to do.

Tips

- Make it clear what you want readers to do.
- Engage readers and let them interact with your site by incorporating the following:
  - □ Chat room
  - □ Community features
  - □ Contact link
  - □ Contest
  - □ Discussion group
  - □ E-mail list information
  - □ Exploration
  - □ Feedback link
  - □ Forums
  - □ Freebies
  - □ Games and puzzles
  - □ Guest book
  - □ Message board
  - □ Multimedia
  - □ Order information
  - □ Questionnaire
  - □ Questions to experts
  - □ Quiz
  - □ Registration
  - □ Shared pictures & stories
  - □ Submission opportunity
  - □ Search engine
  - □ Survey/poll
  - □ Voting opportunity

☑ **Example**

YouthNOISE contains numerous opportunities for interactivity:
Explore: The 411 on issues affecting teens
Take Action: Ways you can be the solution
click Play Hard and Win Big! This summer help create a safe place for 1,000 kids to play and you could win a free Playstation 2.
Sound Off Poll Do you think juvenile offenders should be put to death?

- Use strong active verbs.

☑ **Example**

KidsCom uses verbs such as Play, Go, Chat, Make Friends, Create, Enter, Check it out.

▷ International Audience

Consider international readers. Because the Web is a global medium, your audience may reach readers around the world. Web pages reach a wide audience; even though you may have targeted a specific audience for your Web page, you may have many secondary readers.

Tips

Introduction

- If your company or organization is global, provide initial links to the country.

☑ **Example**

Select a Country
Home page by country/region and language:
Select one ▼
❑ Remember this choice
Check this box to go to your preferred country homepage every time you visit ibm.com

- If your Web site is available in other languages, provide links.

☑ **Examples**

Translations of Web Content Accessibility Guidelines
Information on translations of Web Content Accessibility Guidelines, including completed translations and translations in progress.

Select Language
Please click a flag or use the drop-down menu below to select your preferred language English ▼

Web sites
English

LIBRARY
English
French
German
Italian
Portuguese
Spanish

- If your topic is specific to a country or region, put that prominently on the home page.

> ☑ **Example**
>
> The purpose of this site is to reveal the hidden world of Britain's inland waterways network. It shows how you can find and enjoy them, how they are managed and how you could get involved in their future conservation.

WRITING STYLE: WORDS

- Use the following:
  - ☐ A controlled vocabulary (word choice).
  - ☐ Consistent, unambiguous words (clarity).
  - ☐ Words with few meanings.
  - ☐ The most common meanings of words.
  - ☐ Generic terms rather than trademarks.
  - ☐ Simple forms of verbs.
  - ☐ Active voice.
- Avoid the following:
  - ☐ Ambiguous verbs, such as *should*, *may*, *could*.
  - ☐ Contractions.
  - ☐ Jargon, abbreviations, and acronyms.
  - ☐ Multiple adjectives.
  - ☐ Slang, puns, idioms, figures of speech, adages, figurative language, and colloquial language.

WRITING STYLE: OTHER TECHNIQUES

- Use the following:
  - ☐ Standard grammar.
  - ☐ Simple sentences.
  - ☐ Specific geographical references.
- Avoid the following:
  - ☐ Humor, wit, and sarcasm.
  - ☐ Non-standard spelling.
  - ☐ Omitting articles such as *the, a, an.*
  - ☐ Prepositional phrases.
  - ☐ Phrases that do not translate well.
  - ☐ Political and religious references.
  - ☐ References to things unique to one country.
  - ☐ Symbols (/ and &) that are difficult to translate.

Layout Techniques

- Use distinct topics (chunking).
- Use graphics.
- Use simple icons. Avoid icons that may be misunderstood or offensive.
- Avoid colors with negative associations in other countries.
- Avoid human figures.

Measurements and Abbreviations

- Use measurements, numbers, and abbreviations independent of country, or make it clear what country you are from. The following types of items may be affected:
    - Metric equivalents.
    - Country names in addresses.
    - Format for large numbers. Commas and periods are used differently.

Currency

- Do the following:
    - State the currency.
    - Use the currency of the country involved.
    - Provide more than one currency, or link to a conversion page.
    - Mention if you accept other currencies.

Addresses

- Give specific addresses, including country.
- Give shipping information.

Telephone Numbers

- Begin a phone number with a plus sign followed by the country code. The plus sign indicates that you first dial the number for international access, which is 011 in the United States.
- Do not include the long-distance access code used within the country to make an international call.

> ☑ **Example**
>
> +011: + means access code needed; 011: AT&T access code within the U.S.

- Next, include the area, province, or city code in parentheses.
- For U.S. numbers, include a (1), the U.S. country code.
- End with the local number.
- Use parentheses to separate the area code from the seven-digit phone number.

> ☑ **Example**
>
> (44)(71) 555-5555 (44 is country code for United Kingdom; 71 is city code for London).

- If an 800 number is provided, indicate which countries it serves.
- For local phone numbers, use spaces instead of hyphens because countries punctuate phone numbers differently.

TIMES

- Remember time zone differences.

> ☑ **Example**
> All times are in the Central Time Zone.

- Use the 24-hour clock.
- State what time measurement you are using.

> ☑ **Example**
> All times quoted are in UT (Universal Co-ordinated Time - for all practical purposes, the same as GMT), unless otherwise stated.

DATES

- Use a date format that positively identifies the month, such as 2-JAN-07 or January 2, 2007 because some countries reverse the order of date and month: 2-1-07 or 1-2-07.
- Spell out names of the month.

▷ Introduction: Site

The introduction to your Web site should get readers immediately interested in reading your page. The introduction is used to

- Get attention by providing a hook.
- Give readers the idea of the purpose, content, scope, and target audience of your site.
- Briefly describe contents/topics (advance organizer).
- Help readers decide whether to continue reading.
- Attract Web crawlers/spiders that create keyword databases for search engines.

Tips

COMPONENTS OF A SITE INTRODUCTION

- Include the following in the site introduction:
  - ☐ State the site's content, goals, and purpose.

> ☑ **Examples**
> -The information in these pages is geared to providing you with a starter's guide to authoring a World Wide Web page. The purpose is to introduce you to various types of pages you might want to consider creating, explain the design elements you will want to consider as you plan your page, and provide a hotlinked list of methods for actually producing your web page. The index below provides a guide to the material available.

-Welcome to Wotsit's Format, the complete programmer's resource on the net. This site contains file format information on hundreds of different file types and all sorts of other useful programming information; algorithms, source code, specifications, etc.

-This site is devoted to links to recipes, food history, and food lore of all regions of the globe, with the goal of discovering, sharing, and appreciating the diverse tastes of all the world's people.

-Our goal is simply to provide an easy to use site to aid you in finding a job by providing direct links to 1683 Higher Education Institutes throughout the USA and their Human Resources and Job Listing webpages if they have any.

- ☐ Tell who you are, what you do, and why the site is important.

☑ **Example**

Welcome to GPO Gate, the University of California's gateway to federal information. GPO Gate is a World Wide Web interface to the Government Printing Office's suite of databases known as GPO Access. GPO Access databases contain the **full text** of selected information published by the United States Government. GPO Gate is designed to help citizens easily access the laws, regulations, reports, data and other information provided through the GPO Access system. Among the growing list of titles available are the *Federal Register*, the *Congressional Record*, *Congressional Bills*, *United States Code*, *Economic Indicators* and *GAO Reports*.

- ☐ Make it clear who sponsors the site.

☑ **Examples**

The Virtual Reference Desk Project is sponsored by the
United States Department of Education.

Page sponsors: Get gigs or buy freelance work on Elance.com -- the new way that freelancers of all kinds connect with markets. Sign up to bid on a variety of technical services or post your job to request bids.
Site sponsors: Elance • Match • SMARTpages • Amazon • The Market

- ☐ Define the scope of your site.

☑ **Examples**

This site contains a tutorial of over 35 printed pages, 80K of readable text, animated illustrations, well over 1 megabyte of images and information. All of this is distributed over about a dozen sections.

Introduction
The National Institute on Alcohol Abuse and Alcoholism (NIAAA) has created this portal to support researchers and practitioners searching for information related to alcohol research. This page includes links to a number of databases, journals, and Web sites focused on alcohol research and related topics. Also included is a link to the archived ETOH database, the premier Alcohol and Alcohol Problems Science Database, produced by NIAAA from 1972 through December 2003.

Find Funding - Search more than 23,000 records, representing over 400,000 funding opportunities, worth over $33 billion -- updated daily.

Site Stats
There are currently 1,017 entries and 506 comments, contained within 32 blog categories.

ScienceDirect offers more than a quarter of the world's scientific, medical and technical information online.
•Over 2,000 peer-reviewed journals
•Hundreds of book series, handbooks and reference works
•Back to volume one, issue one

- Identify the intended audience.

☑ Example

Welcome to the Copyright Website, the ultimate copyright portal providing real world, practical (and some whacky) copyright information. For over twelve years, we've delivered the goods:
❖General copyright information for educators, students, websurfers and confused citizens.
❖Specialized information for webmasters, musicians, moviemakers, screenwriters, programmers and photographers.
❖The web's first online Copyright Registration service.

- Define any terms or acronyms if necessary.

☑ Example

Welcome to the GNU Project web server, www.gnu.org. The GNU Project was launched in 1984 to develop a complete Unix-like operating system which is free software: the GNU system. (GNU is a recursive acronym for "GNU's Not Unix;" it is pronounced "guh-NEW.")

- Include a date and tell how often the site is updated.

☑ Example

Corporate Giving Online is the fastest path to find grants and in-kind donations from U.S. corporate funders. Search companies, grantmakers, and grants to find support that meets your needs. Updated weekly.

TECHNIQUES TO USE

- If your introduction is long, make it a link from the main page. This technique is common for well-known businesses, organizations, and news sources.

☑ Examples

About This Site
Introduction

- Capture readers' attention near the beginning of your page with a graphic or banner and an introduction.
- Welcome readers to the site. This greeting can be on the home page or a link to a separate page.

☑ Examples

These examples tie the welcome to a summary of the site's contents.

-Welcome to Multimedia & Internet Training Newsletter's Online Resource Center, your best source of business-critical information for technology-based training.
-Welcome to BrowserWatch, the leading site for information about browsers, plug-ins and ActiveX controls.

This example ties the welcome to the scope and an invitation to participate.

Welcome to the **Canon World Wide Network**: Gateway to all Canon Web sites and the vast global map of Canon research, manufacturing and sales companies. **Begin your journey** into the world of Canon now!

Site Welcome & Overview

☒ **Example**

Welcome and thanks for stopping by my little spot on the World Wide Web.

- Invite readers to explore your site.

☑ **Example**

Take a moment to explore our web site and the possibilities of interactive multimedia as a cost-effective solution to your communication needs.

- Make it clear who is responsible for and maintains the site.

☑ **Examples**

Welcome! You are now connected to a wide range of accurate, credible cancer information brought to you by the National Cancer Institute (NCI). CancerNet™ information is reviewed regularly by oncology experts and is based on the latest research.

Welcome to the GERD Information Resource Center, sponsored by AstraZeneca LP

- Motivate readers to read, interact, and participate.

☑ **Example**

Feel free to read from any of our collections, search our entire site for topics of interest to you, send comments (or new texts) to our editors, talk on our conference line, and join our public mailing lists.

- State the site's benefits.

☑ **Example**

We'll help you save money, and provide insights and advice that will improve **your** child's chances for success! *(SuperKids site).*

- Use lively text at beginning of site or people will not stay.
- Use links immediately.

☑ **Example**

The World Wide Web Consortium (W3C) develops interoperable technologies (specifications, guidelines, software, and tools) to lead the Web to its full potential. W3C is a forum for information, commerce, communication, and collective understanding. On this page, you'll find W3C news, links to W3C technologies and ways to get involved. New visitors can find help in *Finding Your Way at W3C*. We encourage you to learn more about W3C and about W3C membership.

- Keep the introduction short.
  - Avoid long paragraphs that make the page look too dense with text.
  - Be concise.

☑ **Example**

This site links to details to keep the introduction short.
Welcome to World Lecture Hall, your entry point to free online course materials from around the world. Please browse, search, learn and enjoy.
Tell me more about WLH >>

☒ **Example**

With a wide area of experience in internet technologies, we feel that we are helping to bring the truth about website production to you, the customers.

- Consider using a brief slogan that summarizes the site's purpose or theme.

☑ **Example**

A simple quotation is a method of stating your purpose.

PC Help Online: The PC Information you **need** to get the answers you **want**
Learn the Net: Knowledge when you need it.
Native American Rights Fund: Standing Firm for Justice

Avoid a vague slogan: Internet Underground Music Archive: It's Fun! It's New!

TuneUp.com ties their slogan with their site name and theme: **TuneUp Your PC For Maximum Performance!** Regular tune-ups protect your PC from viruses, crashes and more. Get tuned and stay tuned with TuneUp.com to keep your PC working at peak performance.

- Get right to the point.

☒ **Examples**

Having had the pleasure and experience of setting up a few Web sites of my own, I realized that it would be even more gratifying to share my knowledge and experiences with those who wish to learn. This site is full of information and links that are related to **Designing a Web Site**. Whether you're interested in creating a Homepage, a multiple page Web site, or just want to spice up an existing page or site, then this is the place for you.

I set out with the intention of writing an entire guide to good Web design. Then I looked around on the Web and realized there were already plenty of these, so I decided instead to just give my list of the ten commandments of Web design.

☑ **Example**

This site contains a Weekly Magazine on New Technology, Business Opportunities on Technology Transfer, Licensing, Financing of Research etc., Sections on Intellectual Property and Patents, Trade Shows and Conferences, Science and Technology Policy, Venture Capital & Finance, with VC funding opportunities. The link section provides several useful links to International Technology and Professional Contacts in many areas.

- Consider using the introduction for contact information.

☑ **Example**

Some Web sites put the following types of information at the top of the page.

Last update:	Official home (URL):	Sponsored by:
Edited by:	Comments/suggestions to:	Copyright:(date)

- Use the introduction to direct readers to the appropriate section:

☑ **Examples**

For those of you that are completely new to the web, go to our Intro to the Web. Others will want to start with the site design link.

For first time visitors, please take the time to read the ***t-faq Background Topic***.

LexisNexis™ provides authoritative legal, news, public records and business information; including tax and regulatory publications in online, print or CD-ROM formats. View our products and services by occupation, industry or task to see how LexisNexis unites innovative technologies and authoritative content to create solutions to support critical decisions.

The U.S. National Institutes of Health, through its National Library of Medicine, has developed ClinicalTrials.gov to provide patients, family members and members of the public current information about clinical research studies. Before searching, you may want to learn more about clinical trials and more about this Web site. Check often for regular updates to ClinicalTrials.gov.

- Use a confident style.

☒ **Examples**

Many Web sites begin with language such as the following:
I hope you find something of use in here. Please suggest any improvements at all that occur to you.

If you're interested in Web publishing, this is a great place to begin your adventure. I hope you find this site helpful.

- Use headings.

TECHNIQUES TO AVOID IN THE INTRODUCTION

- Avoid using the following techniques in the introduction:
  - ☐ Advertisements.
  - ☐ Apologizing for the site and offering excuses why your site has not been updated.
  - ☐ Clichés.
  - ☐ The following over-used words and phrases:

    I hope that . . .
    This site focuses on . . .
    This site is designed to . . .
    This site is intended to . . .
    This site is meant to . . .
    Take a look around.
    You've come to the right place.
  - ☐ Dull and over-used statements or facts.

> ☒ **Examples**
>
> -We don't need to tell you what an ***exciting*** time we live in.
> -In our fast moving world, change is inevitable.
> -There are thousands of companies that are already on the Internet and hundreds joining each day.

- ☐ Boring text.

> ☒ **Example**
>
> Energizing our people resources promotes an organization's growth and productivity to new levels of prosperity. We're talking the bottom line – people power for your future. In today's competitive business climate, companies can't afford to settle for anyone but the most qualified professional to manage project-related employment opportunities.

- ☐ Generalities.
- ☐ Instructions on how to use the site.

> ☒ **Example**
>
> To browse the site either jump to a section using the pull-down menu or scroll down the home page.

- ☐ Introducing yourself with *Hi, I'm . . .*
- ☐ Jumping right into the content.
- ☐ Lists (with logos) of all the awards the site has received.
- ☐ Long lists of quotations or long poems. It takes too long to get to the point of the Web page.
- ☐ Requiring the reader to immediately register, sign in, or download a plug-in.
- ☐ Stating the obvious.

☒ **Examples**

-This page and related links will take you to several topics related to **Web pages design, authoring and style**.
-It looks like you are looking for graphic or web design. Well you've come to the right place.

- Too many links.

☒ **Example**

The school is a community-based medical school that emphasizes primary care and enriches the profession through a diverse student body to address national needs. Its 16 academic departments include basic science departments on the Wright State University campus and clinical departments in seven teaching hospitals in the Dayton, Ohio, area.

▷ Introductions

Write concise introductions for each Web page, not just the introduction to the site. Each introduction should explain what the page is about and summarize the contents. The introductory sentence should concisely state the purpose of the page and contain keywords.

The introduction to each Web page is used to

- Help readers who scan pages quickly decide whether to continue reading.
- Serve as a contextual clue for readers who may not have entered at the home page. Many Web pages omit good introductions and jump right into the material without orienting readers.

Tips

- Introduce the subject and preview the contents (advance organizer).

☑ **Examples**

This document links you to reference information and demonstrations of hypertext markup language (HTML). You'll find information about all levels HTML, as well as examples, tag summaries, and supporting and reference information.

Welcome to Sierra's Customer Service area. Here you'll find answers to our customers' most frequently asked questions (FAQs), product refund and exchange information, sign-up forms to become a Sierra beta tester, and a Customer Service email form so we can address your specific suggestions and concerns.

- Relate this topic to other topics by presenting the context.

☑ **Example**

Once you've organized your site and designed some pages, it's time to produce the individual graphics that will go on each page.

- Give an overview of the topic's scope.

☑ **Example**

> The next three pages cover the entire process of creating an animation.
> First, the preparation of individual frames and what to consider
> Second, putting together an animated GIF file
> Third, looping animated GIFs and other information

- If appropriate, provide links to necessary background, definitions, or conceptual information.

> ☑ **Example**
>
> Free Advice's starting a business law legal information helps individuals and small businesses to understand their legal rights. To use Free Advice effectively, we recommend you read our Q&A's, post on our legal advice Bulletin Boards, and/or visit our state resource center! Click if you need assistance in locating an attorney, or finding starting a business law legal forms.

- Use the introduction to help readers choose where to start.

> ☑ **Example**
>
> **CGI Documentation**
>
> If you have no idea what CGI is, you should read this introduction. Once you have a basic idea of what CGI is and what you can use it for, you should read this primer which will help you get started writing your own gateways. If you are interested in handling the output of HTML forms with your CGI program, you will want to read this guide to handling forms with CGI programs. When you get more advanced, you should read the interface specification.

- Check that all your introductions are worded and formatted consistently.

☑ **Example**

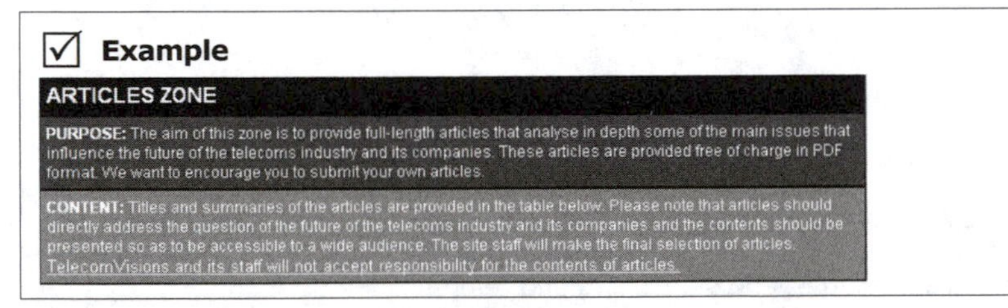

ARTICLES ZONE
PURPOSE: The aim of this zone is to provide full-length articles that analyse in depth some of the main issues that influence the future of the telecoms industry and its companies. These articles are provided free of charge in PDF format. We want to encourage you to submit your own articles.
CONTENT: Titles and summaries of the articles are provided in the table below. Please note that articles should directly address the question of the future of the telecoms industry and its companies and the contents should be presented so as to be accessible to a wide audience. The site staff will make the final selection of articles.
TelecomVisions and its staff will not accept responsibility for the contents of articles.

▷ Inverted Pyramid

In the inverted pyramid structure, you put the most important information (*who, what, where, when, why, how*) first, then details, and finally background information. While this technique is used often in journalism, it is also an effective technique for Web pages. The inverted pyramid writing style lets busy readers get key information up front.

Tips

MOST IMPORTANT

LEAST IMPORTANT

- Use the following organization:
  - Main point: who, what, where, when, why, and how
  - Summary or conclusion
  - Details

- ☐ Background information (or a link to background and conceptual information) needed to understand the topic

- ▸ Arrange points in descending order of importance.
- ▸ Link to the following:
  - ☐ Background
  - ☐ Conceptual information
  - ☐ Details and examples

> ☑ **Example**
>
> **Nature's Own: Ocean yields gases that had seemed humanmade**
> Chemical analyses of seawater provide the first direct evidence that the ocean may be a significant source of certain atmospheric gases that scientists had previously assumed to be produced primarily by industrial activity.
> References & Sources

- ▸ Do not require that readers scroll to view the key points. Put key points in the visible area of the screen.

▷ Italics

Italic letters slant to the right. Italics are a subtle way to create emphasis.

HTML Code: The italic tag <I> </I> is the *physical* style. Emphasis <EM> </EM> is a *logical* style that usually results in italics. Logical styles are used when browsers don't display italics.

Tips

- ▸ Avoid italics because they are difficult to read on a computer screen.

> ☒ **Example**
>
> *Long passages of italicized body text can be difficult to read online. Pixels are square, but italics are angled. So it is best to avoid italics except for small amounts of text.*

- ▸ If you do use italics for subtle emphasis,
  - ☐ Avoid small italicized fonts.
  - ☐ Use italics sparingly; avoid large blocks of italicized text.
- ▸ Use asterisks (*) to represent italics in online documents such as e-mail.

▷ Jargon

Jargon is technical language used by particular professions or subject-matter experts. You cannot assume everyone who reads your Web page will be familiar with technical words or Web terminology.

Tips

- Define terms and acronyms.
- Create a glossary available as a link from every page.
- Know your audience before you use technical jargon. Make sure your readers will understand the meaning of technical terms.

☒ **Examples**

Java is a simple, robust, object-oriented, platform-independent multi-threaded, dynamic general-purpose programming environment. It's best for creating applets and applications for the Internet, intranets and any other complex, distributed network.

The following is an explanation of hosting services provided by a Web design company.

The host system facility is comprised of a Primary DS3 via Lucent Technology, utilizing a Bay Networks BLN2 Router, and a Secondary T1 backup via UUNET NAP, utilizing a Cisco 2501 Router; redundant telecommunications with primary DS3 and secondary T1 backup.

- Link to background information.

☑ **Example**

Cascading Style Sheets (CSS) is a simple mechanism for adding style (e.g. fonts, colors, spacing) to Web documents. Tutorials, books, mailing lists for users, etc. can be found on the "learning CSS" page. For background information on style sheets, see the Web style sheets page. Discussions about CSS are carried out on the (archived) www-style@w3.org mailing list and on comp.infosystems.www.authoring.stylesheets.

- If you must use technical jargon for a non-technical audience, explain it.

☑ **Examples**

The definition of cookies (paragraph 1) is explained in simpler terms (paragraph 2).

Cookies are a general mechanism which server side connections (such as CGI scripts) can use to both store and retrieve information on the client side of the connection. The addition of a simple, persistent, client-side state significantly extends the capabilities of Web-based client/server applications.

In human terms this means that Web servers now have (and have had for a long time) the ability to customize a Web site on a person-by-person basis. Imagine how hard it would be to keep preferences for every browser that has ever visited Yahoo. If the preferences had to be kept on the Web server, it would amount to billions of bytes of data. A much better way to do this is for each browser to keep its own preferences. That's what cookies do.

▷ Journal (Electronic)

A Web journal (electronic journal or e-journal) is similar to a print journal. It is a scholarly journal distributed over the Web. A journal's Web site may exist solely to promote the print version only. The journal may also be available online or available in both print and online versions. It may be free or require a subscription or registration fee. An online journal saves money from printing and distribution costs and can take advantage of hyperlinks and multimedia. An online promotional site for a print journal can provide publication information, tables of contents, sample issues, and can solicit subscriptions and articles.

Tips

Home Page

- Keep the home page simple and quick loading. People access the site to obtain information.
- Put the name of the journal and a short description at the top of the page.

> ☑ **Example**
>
> **The Writing Instructor**
> TWI: A networked journal and digital community for writers and teachers of writing.

- Summarize the most important features of the journal.

> ☑ **Examples**
>
> The Writing Instructor is a blind, peer-reviewed journal, publishing in print since 1981 and on the Internet since June, 2001. Its distinguished editorial board consists of over 150 scholars- teachers- writers representing over 75 universities, community colleges, and K-12 schools.
>
> *This journal immediately identifies the sponsors, publication information, purpose, and contains links to the current issue first, then other important sections. Contact information and a date stamp complete the page, which is about one screen long.*
>
> The Online Chronicle of Distance Education and Communication is the electronic source for information about distance education produced by Nova Southeastern University. The Chronicle appears quarterly and provides an information exchange related to distance education and online communication.
> Current Issue
> Past Issues
> Conferencing Area
> Editors, Staff & Advisory Board
> Editorial & Copyright Policy
> Distance Education Resources, Providers, Publications

- Put all pertinent information within the focal point.
- Make it easy to jump to the current issue.

Items to Include

- Include the following:
  - ☐ Abstracts

☑ **Examples**

*The **Journal of Artificial Intelligence Research** provides abstracts with choice of downloadable Postscript or .pdf (Adobe Acrobat) files.*

Ginsberg, M.L. (1993) Dynamic Backtracking, Volume 1, pages 25-46.
PostScript article ginsberg93a.ps (211K)
PDF article ginsberg93a.pdf (244K)
Online appendix ginsberg93a-appendix.lisp (7K) containing crossword data
Abstract: Because of their occasional need to return to shallow points in a search tree, existing backtracking methods can sometimes erase meaningful progress toward solving a search problem . . . .

- ☐ Annotations or summaries
- ☐ Announcements
- ☐ Archives (links to old issues)
- ☐ Articles
- ☐ Citation information for quoting from the journal
- ☐ Columns
- ☐ Community features such as discussion forums, ability to post comments and contact authors with e-mail links
- ☐ Contact information
- ☐ Description of the journal's purpose, audience, scope, and content
- ☐ Download information (if articles are available in PDF or other formats)
- ☐ Editor's report
- ☐ Editorial board information and links
- ☐ Editorials
- ☐ Employment information
- ☐ FAQs
- ☐ Features
- ☐ Index
- ☐ Keywords for each article
- ☐ Letters to the editor
- ☐ Links to information about the organization (mission, history, etc.)
- ☐ Meeting and conference information
- ☐ News
- ☐ Publication information
- ☐ Resource lists (related to the subject)
- ☐ Reviewers of articles
- ☐ Sample (free) issue if subscription is required
- ☐ Sponsors
- ☐ Submission/editorial guidelines
- ☐ Subscription information for both the online and print versions
- ☐ Upcoming articles
- ☐ What's New section

LINKS

- Avoid simply dumping articles online or, worse, presenting entire pages as graphic files. Take advantage of linking.

> ☑ **Example**
>
> *The **World Wide Web Journal of Biology** uses frames containing links to sections of each article and back to the table of contents. The article contains buttons navigating up and down and separate links to figures. There is also an e-mail link to contact the author and a keyword list.*

- Provide internal links within articles to aid navigation.
- If articles are chunked across several pages, provide navigational aids.
- Provide cross-references to related articles.
- Establish editorial/submission guidelines for hyperlink references/footnotes, and links to figures and tables.

> ☑ **Example**
>
> *The **Journal of Technology Education** uses links within articles to end-of page bibliographies.*
>
> Glaser and Strauss (1967) and Strauss and Corbin (1990) refer to what they call the theoretical sensitivity of the researcher.

- Provide a search engine and a variety of ways to browse articles (by author, date, volume, topic, etc.).
- Use layering to give readers options on viewing articles.

> ☑ **Example**
>
> *The **Journal of Biological Chemistry** provides links to both abstracts and full text versions.*
>
> Zheng Wang and Douglas M. Templeton **Induction of c-*fos* Proto-oncogene in Mesangial Cells by Cadmium**
> 1998 273: 73-79. [Abstract] [Full Text]

- Make articles available in a variety of formats.

> ☑ **Example**
>
> *The **Journal of Infectious Diseases** provides several options for viewing articles:*
> **Abstract Full Text PDF PostScript**

▷ Keywords

Keywords are descriptive adjectives and nouns that describe your site. Search engines find documents by using keywords or phrases readers enter. Search results are then usually ranked according to relevancy. Search engines use indexes generated by Web spiders or crawlers. These spiders determine keywords for Web sites based on the titles, headings, introductions, link text, alternative (ALT) text, META information, and the home page text. There are many tools available to help you generate keywords and determine what words users use to make search queries.

HTML Code: <META name=keywords content=>

Tips

Writing Keywords

- Make a list of all the keywords (at least ten) that describe your site. Include core keywords for the site in general and keywords specific to each page.

> ☑ **Example**
>
> *Keywords for the Home Internet Entrepreneur's Newsletter:*
> newsletter, entrepreneur, entrepreneurs, home-entrepreneurs, home success, business, Internet, Marketing
>
> *Keywords for a company offering full service firm offering complete web design services for both large and small businesses:*
> web site development, web design, web site design, web designers, business web site design, site evaluations, web graphics, scanning services, search engine information

- Use your keywords and search on the Internet to find similar documents. Look at keywords others, particularly any competitors, use.
- Use your theme or focus, important information, and audience to help generate keywords.
- Include concept words, product names, brand names, and company names.
- Describe your topic in as few words possible. Omit all unnecessary words.
- Consider these categories of keywords: single words, phrases, synonyms for words, words that both generally and specifically describe the topic, both plural/singular, misspellings. Do not worry about capitalization because search engines are not case sensitive.

> ☑ **Example**
>
> Introduction to ISO: international, worldwide, standards, standard, organization, guidelines

- Avoid common words or words that may have more than one meaning.

> ☑ **Example**
>
> *In the Introduction to ISO example above, the word "convention" may bring up words having to do with meetings rather than standards.*

- Test your keywords with search engines.

Using Keywords

- Use keywords in the following locations:
  - □ <META> tags

- ☐ The URL (Web address)
- ☐ Title for browser
- ☐ Page title
- ☐ Introductions
- ☐ At the beginning and end of a page
- ☐ Headlines, headings, and subheadings

> ☑ **Example**
>
> Using Interlaced Graphics
> What Are Interlaced Graphics?
> Why Use Interlaced Graphics?
> How Are Interlaced Graphics Created?

- ☐ Repeated throughout the document and on every main page
- ☐ Bold and large text
- ☐ Filenames (e.g., interlace.htm)
- ☐ Alternative (ALT) text for graphics

- ▶ Do not repeat invisible keywords on your Web pages, a method of spamming. Some search engines will reject pages with excessive keywords.

▷ Large Initial Cap

An *initial cap* is an enlarged first letter that extends above the baseline: Initial cap

A *drop cap* extends below the baseline: Drop cap

Large initial caps, which are usually used in the first word on a page or article, have the following purposes:

- ▶ Draw the eye to key points and paragraph beginnings.
- ▶ Capture attention.
- ▶ Add visual interest to the Web page.
- ▶ Contribute to the overall design scheme and tone conveyed.
- ▶ Create unity with other graphical elements used on the page.

HTML Code: Use Cascading Style Sheets or the <FONT> tag with the SIZE attribute. You can also use a graphic.

Tips

- ▶ Use enlarged initial and drop caps sparingly.
- ▶ Use the same look and type of graphics that you use in graphical bullets, icons, and lines.
- ▶ Use color for additional emphasis and unity.

> ☑ **Example**
>
> *This Web site uses large red initial caps throughout the site.*
>
> **Welcome to the Web Site Design Guide.**

- Provide ALT (alternative) text if you use a graphic.

▷ Layering Information

Layering is the technique of beginning with general information, then providing links to more detail and supplementary information. Layering is also called *progressive disclosure*. You essentially "zoom in" to details. Taking advantage of layering is one of the benefits of hypertext.

Layering has the following advantages:

- Keeps the page simple.
- Provides minimal information up front.
- Accommodates different audience levels from experts to new users.
- Allows readers to choose the type of information to read.
- Reduces information overload.

Tips

- Use a hierarchical structure, which supports main topics and sub-topics.

> ☑ **Examples**
>
> *Many sites that use browsing let users "drill down" through topics:*
>
> **Computers & Internet**
> Internet, WWW, Software, Games...
>
> Home > Computers and Internet > Internet > World Wide Web > **History**
>
> **Getting Started — Planning the Web Site**
> Planning the Site
> Planning Usability Activities
> Links to Related Articles
>
> **Planning the Site**
> Planning is critical because it helps you focus your objectives. It also helps you plan for usability activities that are part of the process of developing a successful site.
> Before you design, you must think about:
> 1. Why are you developing a Web site?
> 2. Who should come to your site?
> 3. When and why will they come?
>
> • **Why Are You Developing a Web Site?**

- Use the inverted pyramid method of providing key information up front with links to detail.
- Provide high-level summaries or general overviews (advance organizers); then link to details.

> ☑ **Examples**
>
> By identifying your business problem, developing a strategic solution, executing a plan and measuring the results, we can optimize your marketing spend and ensure your success.

Getting Started. There are four basic approaches to starting your own web pages.
Use a *template*.
Use a *word processor* to create your page and convert it to web format.
Use *WYSIWYG software* to create your page.
Learn *HTML coding* to completely control your page and to enhance the above methods.

Sample forms
Permission to link: View sample | Explanation

- Use layering to accommodate different reader levels and backgrounds.
- Provide sections for experts and new users.

☑ **Example**

The ZDNet Web page on how to create links begins with a 30-second link lesson:

We're going to dig deeper into links in just a few moments, but first we thought we'd throw you the bare-bones how-to. Here's the fast food version:
How to make a basic link:
<A HREF=insert_URL_here>Click me</A>

*Their section on **Link basics** lets users skip the information.*

Link basics: If you already know how to create a basic link, you can skip ahead to the next lesson. Otherwise, read on to refresh your understanding of how links work.

- Use layering in the table of contents to help readers navigate quickly to the information they need.
- Use links such as More, Learn More, Full Story.

☑ **Example**

Surf School provides levels for readers to learn about the Internet. Each level provides more advanced information. Readers click the DEEPER icon to go down another level.

Animated GIF - What Is It?
Most of the pictures you see on the Web use the GIF format. Now there's a way to string several GIFs together to make simple animations.
Slang Watch: The version of the GIF file format that allows this is called GIF89a, but most people just call them animated GIFs. DEEPER

- If you want to keep all information on the same page, separate detailed information.

☑ **Example**

This site lets readers move to the next topic or view more details (In-depth).
You've completed the basic drawing step. If you're making your first few animations I'd recommend Fast Forwarding to the next step -- Assembling the animation. But if you've successfully made a few and would like to read about transparent backgrounds, anti-aliasing, tips to minimize file size and more, read on. Fast Forward >>

V In-Depth V
Optimization starts here
What you do in this step has a big effect on our two optimization goals -- 1) selecting colors that look good on all video displays and 2) minimizing the file size of the animation.

- Avoid too many layers because they require more navigation and require readers to wait for pages to load.

▷ Line Length

Line length is the width of text. Browsers will size lines of text to fit the screen. The line length affects the following:

- Readability. Long lines are tiring to read online. It is difficult to follow long lines across, and it is easy to lose your place.
  Short lines require the eye to constantly read back and forth.
- The amount of white space on your page. Long lines make the page look too dense with text.

☑ **Example**

Many Web pages use multi-column tables to control line length (left). Others use tables with a graphic or design in one column (center). Long lines make the page look dense with text and difficult to read (right).

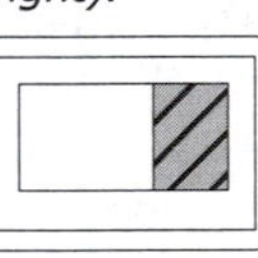

HTML Code: HTML defaults to placing text across the entire width of the screen. However, you can control length of lines with line breaks using the following techniques:

- Cascading Style Sheets
-
 tag
- Frames
- <Blockquote> tag, which gives a margin on both sides of the page. You can also nest Blockquotes to vary text width.
- Tables. A table lets you control text width in pixels or a percentage of the screen.

Tips

- Keep line length short (about 65 characters, or about 10-15 words, or two alphabets).
- Design your layout for the average screen resolution of 800 x 600. Make sure that the width of page elements is under 595 pixels so readers will not have to scroll horizontally.
- For text that can be skimmed quickly, use a wide left margin containing headings.

▷ Lines (Rules)

Lines are graphics used to separate and emphasize information. You can create lines by

- Using the horizontal rule tag.
- Inserting simple, textured, and 3-D graphics.
- Using table borders.

You can create vertical rules by inserting solid color graphics that are only one pixel wide and the height of the screen.

HTML Code: <HR> creates a horizontal rule.
The width (WIDTH=), size (SIZE=), and align (ALIGN=) attributes allow you to vary the length, thickness, and position.

Tips

- Use horizontal rules to
  - Separate and divide information.

> ☑ **Examples**
> *Use rules to separate your footer, pull quotes, and definitions.*

  - Provide emphasis.
  - Capture attention.
  - Anchor text.

> ☑ **Example**
> *Use rules to highlight and anchor headings.*

  - Help readers find information.
  - Direct eye movement.
  - Tie in with your page design.
- Use horizontal rules sparingly because they interrupt reading.
- Remember that graphical rules may not appear in some browsers and will decrease loading time.
- Use ALT tags.

▷ Link Title

A link title is text that pops up when the cursor is placed over a link. The text describes where the link leads.

HTML Code: A link title is an HTML attribute:
<A HREF="/link text goes here/" TITLE="description goes here">.

Tips

- Include any of the following types of information:
  - Site name
  - Type of page (e.g., Registration)

> ☑ **Example**
>
> **This link will open a new browser window which contains a list of products covered by Hewlett-Packard Instant Support – online diagnostic and repair.**

- Prerequisites (e.g., Registration Required)
- Type of information and how it relates to the context

- Keep text short (under 25 words and 60 characters)
- Do not repeat the same text as the link itself.
- Do not use link titles as substitutes for clear links and supporting context.
- Recognize that link titles may display differently on various browsers.
- Use only when necessary, such as when your link text has limited space.

▷ Links

A link is a connection between two nodes (locations in a document) that allows a user to jump from one location to another. Two parts of a link are the *source* or trigger (what is clicked) and the *destination* (where it goes). The source can be text (short text or embedded text in paragraphs), lists of links, or graphics. Links provide navigational aids for readers to move from one location to another. *Links are the primary advantage of hypertext.*

Types of links include the following:

- *Internal links* help users navigate within the same document. They are usually used on long Web pages.
- *Intra-site links* go to other pages in the same Web site.
- *External links* go to other Web pages.
- *Media links* go to graphics or multimedia files (sound, animation, video).
- *E-mail links* can open an e-mail program.

HTML Code: Links use anchor tags <A> </A>
Within these tags is the link's destination (a URL or filename) specified by the HREF= attribute. (The NAME attribute is used to link to a target within a document.) Text typed between the tags becomes the visible hyperlink text that users click.

Tips

When to Use a Link

- Within the body of the topic, link only to topics that directly contribute to what you are saying. In general, link to relevant, valuable, accurate sites. Don't create a link to every possible related topic.
- Link to the following:
  - Archives
  - Background and conceptual information
  - Closely-related topics
  - Companies mentioned
  - Cross-references
  - Definitions
  - Details
  - Footnotes
  - Information for different audience levels
  - Large graphics

- Options
- Sites on similar topics
- Sites with a tone similar to that of your own site
- Software required to view files at your site
- Support information
- Tools

How Often to Link

- Use links early on your Web page.
- Use links sparingly.
    - Use the seven plus or minus two rule. Too many link choices may confuse and overwhelm users.
    - Provide a short path to all information; information should be only one to three mouse clicks away.
    - Use no more than about five links per topic and no more than three levels of jumps.
- Use each link once per page. Don't make every occurrence of the word a link–only the first one on a page or major text block.
- For internal links, use links for about every 10 to 20 lines.

Necessary Links

- Home page, table of contents, or main menu
- A way to go from every page back to the main page
- Forward and Back
- Feedback
- Frequently used or critical tasks (preferably locate at the top of the page)

Links to Consider Including

- Consider linking to the following:
    - Related sites
    - Search engine
    - What's New

Links to Avoid

- Avoid the following:
    - Dead-end links: pages with no navigational options.
    - Links to pages with little content.

☒ **Examples**

A What's New link on a Web page links to a page with this sentence:
Everything is new right now, but as I add new tips, links, and information, you can check here to see the latest additions.

A Services link on every page links to a page containing this statement:
Due to this being a very competitive market this area has been temporarily removed to protect our clients.

Do not make available a link that goes to the current page. Omit it or dim the text to avoid sentences such as the following:
You really don't want to click that link since you'll just get cycled back to this page!

- ☐ Links to someone else's pages unless you first get permission.
- ☐ Links to redundant information.
- ☐ Links that are not obvious. These range from text with underlining removed to buttons that do not look like buttons. However, on sites with lots of links, remove underlining to avoid clutter.
- ☐ Links that are inconsistently formatted on the same page.

☒ **Examples**

News headline (no emphasis) **News headline** (bold) **News headline (bold, blue)**

- ☐ Links that are too small, such as symbols or characters.

☒ **Examples**

≥ >> *

PLACING LINKS

- ▶ Place links consistently along the left, right, top, and/or bottom.
- ▶ Put links in a logical order.
- ▶ Use *Related Topics* at the end of the topic.

CHECKING LINKS

- ▶ Regularly test and check your links.
- ▶ Keep links current.

LINKING TO YOUR SITE

- ▶ If your want readers to link to a certain page on your site, give the URL.

☑ **Example**

If you wish to link to this site, please link to *http://www.newbie.net/*

▷ Links: Embedded Text

Embedded links are contained within sentences and paragraphs. Alternatives include menus or bulleted lists of links. There are pros and cons of embedding links.

PROS

Embedded links have the following advantages:

- ▶ Links within text are more conversational.
- ▶ Surrounding text provides a context and explanation.
- ▶ Links are in the same location as the text rather than separated.
- ▶ Usability studies show that users prefer embedded links.

Cons

Embedded links:

- Are more difficult to search and skim quickly.
- Interrupt reading.
- Look cluttered and distracting.
- Are often unclear about where they lead.

Tips

- Consider using both embedded and listed links to accommodate both readers who skim and search and readers who need the initial explanation. This technique is often useful on the home page.

☑ **Example**

Take advantage of the available resources to learn more about the Earth Sciences!
...Review the latest Earth Science news in the News Room...
...Get current information on Earth and its resources at Earth Data...
...Search the Book Center for Earth Science-related books by Topic, and learn which books are most popular...
...Participate in Earth Science Week and share your enthusiasm for the Earth Sciences with others...
...Investigate Careers in Geoscience and learn about job opportunities, degrees programs, and enrollment statistics in the Earth Sciences...

- Use embedded links to preview organization.

☑ **Examples**

Inktomi® information retrieval products, including search, categorization and advanced XML technology, give you the power to find the right information and profit.

On this site you can find information **ABOUT the collection**, a selection of online **EXHIBITS** illustrating visual phenomena, a set of pages exploring novel and thought-provoking **IDEAS** about seeing, and many **LINKS** to other sites where you can find more information about the science and mystery of how we see the world.

- Avoid too many embedded links within a paragraph.

☒ **Example**

The fruit information link page includes links to such plants as apples, strawberries, pears, plums, blueberries, apricots, cherries, citrus, grapes, peaches, nectarines and other fruiting plants. The Vegetable and herb information links about such crops as corn, carrots, peppers, tomatoes, onions, herbs, beans, cabbage,, cucumbers garlic, lettuce, and other vegetable crops.

- Do not embed general navigational aids that are not also available on a navigation bar or menu.
- Write the hyperlink text so readers do not have to read the surrounding text.
- Make linking words or phrases part of a meaningful sentence so readers have a clear understanding of where they are going.

- Write as though there were no links.

☒ **Examples**

For more information on how we work, please go to info. For more information on the services we offer, including our web site packages please go to services. *("please go to" calls attention to the link)*

Just another reason to make The Ultimatorium your start-up page so you can start searching immediately when you open your browser program (click here to find out how) *("click here" calls attention to the link)*

- Make no mention of the links themselves in the text.

☒ **Examples**

See the Download section in the FAQ for some sound players.

If you don't already understand HTML basics, here are some links that will help you in your education.

- Create context for a link. Use surrounding text/context to clarify the link text.

☑ **Example**

Whether you want to know how to search the Net, build your first Web page, or become a better browser, Web Basics' highlights will lead you to the best beginner's resources on the Net.

- Highlight the important information.

☑ **Examples**

The Purdue Online Writing Lab contains useful handouts for various types of writing.
Answer questions based on Gallup poll survey findings.

☒ **Example**

Ohio Senator Wants Answers Senator George Voinovich is calling for hearings on why the military is considering cuts. *Make the article title the link.*

- Highlight the word that best indicates the type of resource it links to.

☑ **Example**

A noun link would lead to a definition. A verb link would lead to a procedure. A name would link to a biography.

-Ted Nelson first coined the phrase hypertext in his non-linear books of the sixties. *(links to a biography)*

-Ted Nelson first coined the phrase hypertext in his non-linear books of the sixties. *(links to a definition of hypertext)*

-Ted Nelson first coined the phrase hypertext in his non-linear books of the sixties. *(links to a bibliography)*

- Allow readers to jump to the section they are interested in.

☑ **Example**

Sun's cookbook for Web creation addresses several writing-and content-related topics, including identifying your audience and identifying its needs, page length and how to segment sections of text, and Web cliches.

- Don't put hotspots too close together, or readers may accidentally select the wrong hotspot.

☒ **Example**

Home pages are used to provide information on individuals, enterprises, organizations.

- Design the links to form a pattern. Look at just the links and imagine the reader reading only the links. Are they informative and unified? Do they tell a story?

☑ **Example**

Wired charts the impact of technology on business, culture, life. Browse selections from archived issues. Then feel free to talk online with our editors and readers. (Or get a free issue to check out the real thing.)

▷ Links: External

External links jump to other Web pages. They may be located within your own site (intra-site) or another Web site. External links allow you to cross reference other Web sites that are related to your topic. They also add credibility to your content.

HTML Code: <A HREF=[URL]>[text to be linked] </A>

Tips

Types of External Links to Include

- Link to sites that
  - Are geared to the same audience.
  - Have similar content.
  - Contain background or conceptual information.
  - Contain related support information, such as vendors.
- Make your lists of external links more than just a list of other addresses. Provide your own unique contribution to the subject.

Writing External Links

- Distinguish between internal, intra-site, external links. Make it obvious when links are external:
  - Warn readers that the link will take them outside your Web site.

☑ **Examples**

Thank you for visiting OhioTourism.Com! Come back soon! To transfer to the Ohio Hotel & Lodging Association's online reservation system, **click here**.
You are exiting The White House Web Server. You will now access **http://www.opm.gov**

Thank you for visiting. We hope your visit was informative and enjoyable.
When you visit our client's Web sites, you will be leaving the Peak Web Design site. To return to Peak Web's site, click your browser's Back button until you see this page again.

- ☐ Use clear headings and labels such as [off site].

☑ **Examples**

Links to External Sites
Resources: External
Featured Net Links
On The Web
Related Web Sites

HOWTO documents (off-site)

- ☐ Use icons (also called "glyphs").

☑ **Example**

Note: ⇨ leads to an external site

- ☐ Warn readers if a link goes to another protocol, such as an ftp or gopher site.
- ☐ Use phrases that give clues.

☑ **Examples**

More information about the drill ship may be found at the Ocean Drilling Program's web site.

This page uses "our" to indicate pages within the site and "site" to indicate an external link.

To download a free clipart, visit the Netscape site. To create it yourself, look for the tools at our Web Graphic Tools page. To learn more about the graphic file format, visit our GIF or JPEG page.

- ☐ Show a thumbnail of the page the link leads to.

☑ **Example**

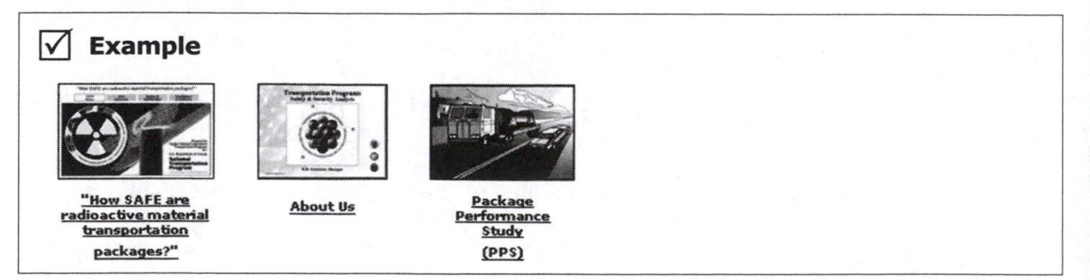

- ☐ Explain how links are formatted.

> ☑ **Examples**
>
> This page contains links to the World of Advertising. (items in **bold italic** are located on this Web site)
> All external links in the tutorial are identified by a yellow dot.

- Provide useful annotation. Describe what information is available at the other site. If you believe it is worth including a link, state why you like the site.

> ☑ **Example**
>
> A Beginner's Guide to URLs: If you are interested in developing a greater understanding of the elements in a web address, this is a good place to start.

Creating External Links

- Link to the original Web page rather than mirror or referring sites.
- Link to the actual location in the document rather than the home page. Be careful of *deep linking*—links to a site's internal pages but not the home page. This type of link takes readers to the information they need. However, it can lead to copyright issues.
- To avoid making errors with URL addresses, copy and paste the address. Some sites even help you link to their site.

> ☑ **Example**
>
> Click here to generate a link to this page

- Check links periodically to verify that they are still working.

Formatting External Links

- Avoid too many external links, or readers may not return. They may forget about your site or lose interest in your topic. Each external link encourages readers to leave your site.

> ☒ **Example**
>
> *Each link in the following excerpt goes to a new page. Will the reader remember to return?*
>
> Setting up a Web Site can be fun and easy. However, you should consider several aspects of a Web page design before you get started. For instance, you should consider
>
> Presentation Content Location

- Avoid external links on the first page of your Web site, or readers may leave without reading on.
- De-emphasize external links by placing them at the end of important text.

> ☑ **Example**

For more information about URLs, read NCSA's Beginner's Guide to URLs or the World Wide Web Consortium's Fact Sheet on URLs.

☒ **Example**

This sentence would be unnecessary if the links were at the end of the article:

A good example of this is the Geocities homepage. You can go look at it, but you'd better hurry back.

- Avoid numerous external links within paragraphs. They detract from your message. Instead, consider using one of the following options:
  - Link to an *Other Links* page of external links.
  - Box a list of jumps and give it a title, such as *Find Out More*.
  - Put links to external sites in a separate frame, column, or window.

☑ **Examples**

According to the Skin Cancer Foundation (opens in new window)
▪ For young women 15-29, rates of melanoma have grown more than 60 percent since the mid-70s.

  - Place the links in a list at the bottom with a heading.

☑ **Examples**

Related articles:
For further reading:
More information on URLs can be found at:
References:

- Group numerous external links under headings.

▷ Links: Internal

Internal links (intra-page) connect locations inside one document. They are most often used to link items in a menu or table of contents to a heading and then return to the menu. The link points to a page location called an *anchor* or *target*. You can also link to an anchor within another document. (For example, you could link directly to the definition of a term within a longer glossary page.) Internal links help readers navigate long documents without scrolling. They also help chunk information.

HTML Code: An anchor uses the <A> tag.
HREF= indicates where to jump.
NAME= indicates the internal label. You create the anchor: <A HREF=#keyword> anchor text for current document </A>. Next, you create the text you want to jump to: <A NAME=keyword>Text to jump to</A>

Tips

- Use internal links to chunk information on long scrolling pages.
- Use internal links as advance organizers about what will be covered.

☑ **Example**

What You'll Learn On This Page...
Electrons
Conductors & Insulators
Voltage & Current
See also...
Follow the Power Trail – How electricity gets to your home

- Always provide links back to the top.
- Make sure that the jump to an internal link will be visible (sometimes internal links are not obvious when the page is short). Otherwise, the reader will think the link did not work.
- Make it clear that the link is internal rather than a link to another page.
  - Use words such as "this page" or instructions.

☑ **Examples**

Topics on this page include:
Interpretation of Food Labels by Parents of Food-Allergic Children
Study on the Genetics of Peanut Allergy
Archive of Research Summaries

On this page: • New • Services for Faculty and Staff • Academics • Administration • Jobs • News & People

  - Use small icons such as arrows.

☑ **Example**

▼ *A "down" arrow can indicate an internal link.*
▶ *A "forward" arrow can indicate a link to another page in the site.*

☒ **Example**

It is not obvious what this icon represents; it is also too small.
How to make your Web site fast and usable *

▷ Links: Lists

A Web page can contain a list of jumps (these pages are sometimes called jump pages, list pages, or navigation pages). An advantage of this technique is that links incorporated into a regular Web page may distract the reader, who may follow an external link and never return to your page.

Tips

Using Lists of Links

- Consider using both embedded and listed links to accommodate readers who skim and search and readers who need the initial explanation. This technique is often useful on the home page.
- Begin with an overview (advance organizer) of the type of information on the page, especially if you have several list pages.

☑ **Example**

Internet is about the Internet in general, including history, structure, applications, email, protocols, commerce, Web-related topics, organizations, statistics, intranets, e-zines, and pointers to further information.

- Don't require that users jump down more than three levels.
- When using nested (levels of) links, provide a link back to the main list.
- Word links consistently; for example, use the page title as the link.

☒ **Examples**

The links in this list are inconsistent.

WWW Homepage Access Counter has . . . *(uses the name of the counter rather than the page)*

Here's a page about counters that makes use of NCSA's . . . *(vague link)*
In-Line Counters contains . . . (*uses the name of the page*)

- Use annotation or list the topics covered so readers know where the link leads and why it is useful.

☑ **Example**

About Kregel Publications
Mission Statement
Editorial Focus
Other Information

- Indicate how many items are available for each category.

☑ **Example**

The Dostoevsky Resource Station contains a menu of categories, with jumps to specific topics. Parenthetical numbers show readers how many items are available for each topic. (Partial list shown here).

200 Relevant Links
Artwork Inspired by Dostoevsky (16)
Critical Essays (31)
Discussion Groups & Mailing Lists (5)

- Avoid beginning lists with the same word. Skimming is difficult.

☒ **Example**

- links to "Global Studies" learning objects
- links to other learning objects
- links to collections of learning objects
- links to key organizations associated with learning objects and metadata

- Use parallel construction.
- Make sure headings following a lead-in sentence complete it correctly.

☒ **Example**

In the following, only the second heading (Use) correctly completes the lead-in sentence, which requires a verb.

What would you like to do?
What is the Personal Page Builder? What does this tool do? How do I use it?
Use the Template-based Page Builder Don't know HTML? Don't want to know? No problem! Our Templates-based Page Builder will help you build a page in just minutes!
File Upload Upload HTML pages and images to your Netcom Personal Web Page.

- Keep lists updated.

Formatting Lists of Links

- Arrange link lists in a logical order.

☒ **Example**

Get SAT dates and fees
View and send your SAT scores
SAT Registration FAQs
Changing test center, date, or test
SAT Registration terms & conditions
SAT Prep Center

- Group long lists under meaningful headings to categorize related links.

☑ **Example**

Services
- Email services
- Headline service
- Text alerts and PDA

- Use a grid or tabular arrangement. Readers can find the appropriate category, then skim quickly down the list.

☑ **Examples**

LINKS	LEGISLATION	*CONGRESSIONAL RECORD*	COMMITTEE INFORMATION
About THOMAS	Bill Summary & Status 93rd - 108th	This Congress by Date	Committee Reports 104th - 108th
THOMAS FAQ	Bill Text 101st - 108th	Text Search 101st - 108th	House Committees Homepages
Congress & Legislative Agencies	Public Laws 93rd - 108th	Index 104th - 108th	Senate Committees Homepages
How Congress Makes Laws: House | Senate		Roll Call Votes 101st - 108th	
Résumés of Congressional Activity			

Links to statistics　　Links to statistical agencies

- Tell the reader what order links are in.

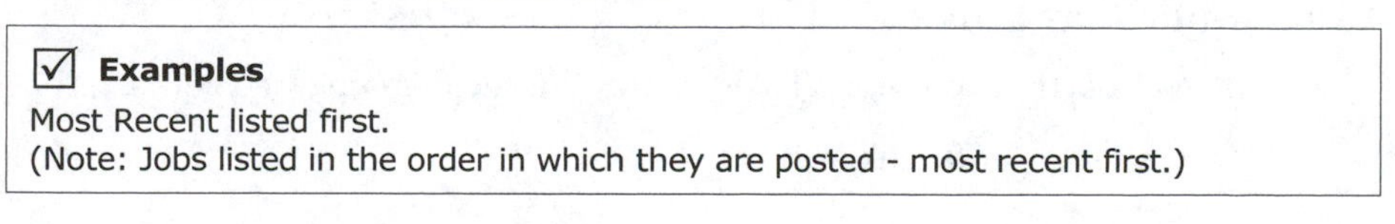

☑ **Examples**

Most Recent listed first.
(Note: Jobs listed in the order in which they are posted - most recent first.)

▷ Links: Text

A text link is a word or phrase that jumps to an internal or external location. A text link provides a simple navigational aid or accompanies graphical links so your site is accessible. It also helps readers predict where the link leads and decide where they want to go.

HTML Code: <A NAME=keyword>Text to jump to</A>

Tips

When to Use Text Links

- *Always* provide text links, even when you provide graphical buttons. This technique supports users who do not view graphics. In addition, users can begin using the links and do not have to wait for the graphics to load.
- Provide links to large graphics.

> ☑ **Example**
>
> *The link in the following excerpt goes to this picture.*  *Readers can choose whether to download it.*
>
> Picture of the letter A as an array of dots (left). In reality the dots overlap (right), as they are round or oval this is necessary to avoid small gaps between dots.

Types of Information to Include

- Include the following information:
  - ☐ Site name
  - ☐ Type of information on the destination page
  - ☐ Warnings (file size, file format, registration required, etc.). If you are linking to files to download (download menu), provide all necessary information.
- Don't explain how to use links unless your audience is new to the Web or the information appears on a Help page.

> ☒ **Examples**
>
> Follow each of these links to learn more.
>
> This page is designed to act as a jump point to specific information of interest to site developers, marketers and promoters. All topics are targeted links to allow you to find what you want quickly and easily.
>
> To the left, you will notice two selection lists; under each one is a variety of topics related to web site design.
>
> Click an underlined word to enter.

Explaining Text Links

- Always make it clear where a link goes by using the following techniques.
  - ☐ Clear, specific link text.

☒ **Example**

Spiders crawl around the Net to collect information about what's available online.
(Is this a link to a definition? This link actually leads to an article about Webcrawlers.)

- ☐ Context.
- ☐ Explanation/description of the link and why readers should go there.

☑ **Example**

Find out how we are working to provide upgrades to customers, when they will be available, and how to get them.

- ☐ Link titles (pop-ups that give more information about links).
- ☐ Warnings about what will happen.

☑ **Examples**

(The examples used in this article are in popup windows).

Warning: If you select this link, you will leave this site and go to a new browser window. You will automatically return to this page when you close the new browser window.

I analyzed 24 Websites (open the list in a new window).

Messages are loading to the left.
Click on any link, and the message will appear in this frame.

- ▶ Always provide information about the contents of the page the link goes to.

☑ **Example**

The Web	Learning about the Web
Starting	Browsing the Web
Browsers	Using Browsers
HTML	Using HTML
Servers	Running a Web Server
Searching	Searching the Web

- ▶ Consider listing the topics found on the linked page so readers know the contents.

☑ **Example**

Creating an image map
  Create a .gif image
  Create a .map file for the image
  Edit the .map file

- ▶ Before a list of bulleted links, provide a lead-in sentence.

Writing the Link Text

- ▶ Choose meaningful, informative words or phrases for links.

☒ **Examples**

Additional Information
Miscellaneous - lots of links and other info
Ooops

☑ **Examples**

-Who We Are
-Find everything you need to know about HTML
-Download great tools for building Web sites
-Jazz up your site with Java and JavaScript
-New to Web publishing? Start with the basics

- Avoid ambiguous terms.

☒ **Examples**

Tour *(Is this an interactive lesson or a site map?)*

Web Graphics *(Does this link contain links to graphics sources? Does it give tips on using graphics?)*

Free Images *(Does this link go to a page with free graphics? This link actually goes to a list of Web sites that contain graphics.)*

Help *(Does this link go to a page providing help on using the site? It actually goes to a page with links to sites to help teachers use technology in classes.)*

Catalog *(Is this an online catalog? It actually goes to a form to request that a catalog be mailed.)*

- Avoid clever link text. It does not make clear where the link leads and makes it more difficult for search engines to index your site.

☒ **Example**

The Science Fiction Channel site uses the following names for links:

Trader (for the catalog), Orbit (for Web resources), and Pulp (for fiction).

  - Use clever link text only when you want to convey a light tone or use humor.
  - Explain what the humorous link means.

☑ **Example**

Avoiding Conversion Aversion: Conversion of Familiar Units
I've Got Logarithms...Who Could Ask for Anything More? Reading a logarithmic scale.
Pull up a Chair (or a Web Browser) to the Periodic Table! Choose your favorite on-line periodic table and click away!

- Use terms most users will understand or that are not dependent on a certain context.

☒ **Examples**

The terms "push," "appliances," and "shopping" carts have unique meanings here:

Push Technologies
Internet Appliances
Secure commerce and shopping carts are available.

- Avoid abbreviations.

☒ **Examples**

ISP locator   Important #s
Comm Effectively   Brd. Of Elections – Voter Registration

- Avoid mismatched text. Match the link text with the title or heading it links to.

☒ **Example**

Site Building Blocks *links to a page called*

THE BEGINNING
Building a Good Site Foundation

- Edit links so they are concise and to the point.
- Put keywords at the beginning of link text.
- Use titles of documents and people's names.
- When possible, use active verbs to encourage reader interaction.

☑ **Example**

The task-oriented verbs used in these sites emphasize reader actions and benefits:

discover   visit   teach   join   shop   search

Download Free Software   Get The Latest News   Buy Internet Products
Read It | Discuss It | Fix It | Cite It

I need to. . .
I need to. . .
Roll my 401(K) to an IRA
Find a mutual fund
Get help from an advisor
Get advice on my portfolio
Save & invest for retirement

BREAST CANCER ACTION
Get Informed!
Do Something!
Support Us!
Learn About Us!

- Consider using questions when you know the type of information visitors to your site will be looking for.

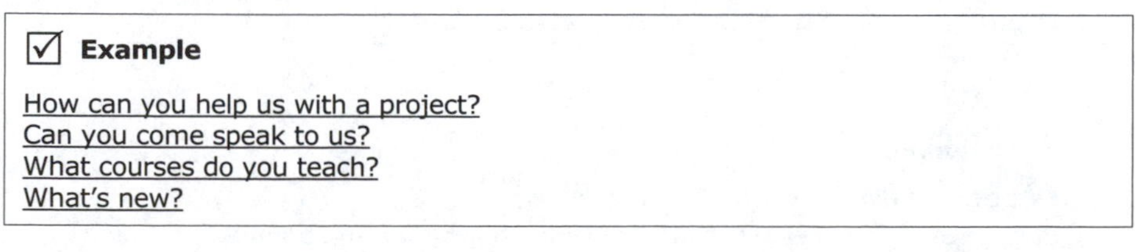

☑ **Example**

How can you help us with a project?
Can you come speak to us?
What courses do you teach?
What's new?

- Word links to show what you are emphasizing, preferably reader benefits. The emphasis, in turn, can affect the tone and image you project.

> ☑ **Example**
>
> *The links you choose can show what your site emphasizes: products, company, support, etc.*
>
> ***Poor:*** What We Offer ***Better:*** Customer Services *or* Customer Care

- Don't repeat the same link within a few sentences.

> ☒ **Example**
>
> Our Web Page Enhancers provide companies with new and innovative ways to make their Web sites more effective. From an on-line multi-media presentation manager, to order form validation and automation, we have a wide selection of affordable Web Page Enhancers to choose from.

- Don't repeat the same link with different names.

> ☒ **Examples**
>
> *Both links go to the story:*
>
> Sector outlook: Consumer Discretionary and Financials
> "The m100 analyzes their first and third largest sector weightings: Consumer Discretionary and Financials." ... click here to read more
>
> *The graphic and both links all go to the same page:*
>
> 
>
>
> This month marks the one-year anniversary of our Magazine Subscriptions Store, and we're inviting you to celebrate with us by discovering our everyday low prices on your favorite titles. Visit Today's Deals to see the huge savings.

- Avoid simply listing the URL; instead, provide the name of the site with the URL as the link destination. However, provide addresses for readers who print out the page.

> ☑ **Example**
>
> ***Poor:*** The Allyn & Bacon site at http://www.abacon.com contains . . .
> ***Better:*** The Allyn & Bacon site contains . . .
>
> *The following description should make* The Children's Partnership *the link text. The Web address should be available if readers will print the page.*
>
> **The Children's Partnership**: The Parents' Guide to the Information Superhighway. Rules & tools for families online. Comprehensive look at the information superhighway and what parents should know to help their children use it safely and wisely.
> http://www.childrenspartnership.org/parentguide/parentguide.html

- Use explicit rather than implicit links.

> ☑ **Example**
>
> *Explicit links directly state where link leads:*
>
> Back to Using Frames Forward to Using Tables Up to Contents

Implicit links imply where link leads:
Back Forward Up

Formatting Text Links

- Choose between embedded text links (links within the paragraph), link lists, or a combination of both.

☑ **Example**

Links are embedded within paragraphs and available in a boxed "Story Links" sidebar.

Story Links

- UPS Warehouse
- list of USP terms
- Year 2000

The UPS Warehouse is a British site that contains detailed information on power supplies, including definitions of the common terms . . . For a quick overview of UPSs, refer to Windows Magazine's list of UPS terms, which explains everything from runtime to data line protection.

- Distinguish types of links (e.g., internal vs. external).
- Make links recognizable.
  - Avoid removing underlining. Some sites remove underlining for links, making it difficult to distinguish colored text from actual links.

☒ **Example**

Every bullet item in this list is a link.

MSNBC QUICK LINKS

- Free video
- Nightly: Iraq Watch
- Hardball tour
- Crossword
- Comics
- Horoscope
- Sports scores
- Yellow Pages
- Week in sports
- Gossip
- Stock quotes
- Meet the Press
- MSNBC interactives
- Week in Pictures
- MSNBC Alerts
- Breaking news email

  - If you do remove underlining from links, format all links consistently.
  - Avoid changing the default link colors because some people change the ways visited/unvisited links will appear in their browsers.

- Choose an appropriate length for the link text. Avoid making the hotspot too long or too small.

☒ **Examples**

A single word may be too small for readers too click on and may not be informative.

A long link like this is not necessary and makes the link text distracting.

These links to states are difficult to click because they are so small:
AK | AL | AR | AZ | CA | CO | CT

All links in this Web site are too long:

Need to create a new logo for your site but don't know how to use the image tools such as PhotoShop? No problem, with the help of this cool site's tools, you could create a cool logo in a matter of seconds. The powerful technology behind this tool is a program known as GIMP, yet another great Linux tool!

- Use position and/or emphasis to distinguish types of links.

☑ **Example**

The following example shows a page grid using zones of links. Links are both vertical and horizontal and a grouped and are positioned differently. Their position (left or right) and emphasis (top level, secondary level) give readers clues about the types of links.

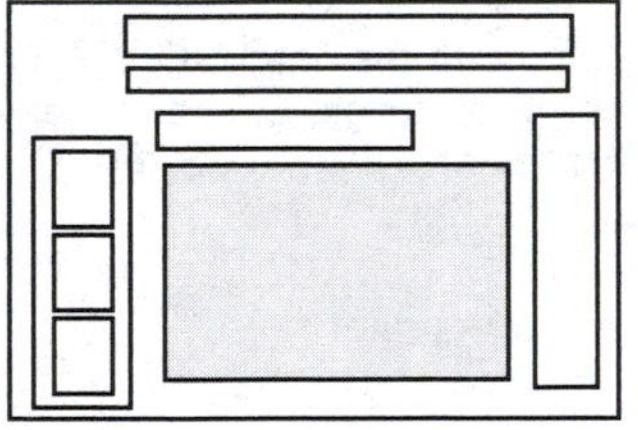

- Show the active page by deactivating the hotspot in the navigational aid: italicize, gray, or remove the link.

☑ **Example**

[Home] [Topic 1] [Topic 2] [Current Page] [Topic 3]

- Avoid lines that wrap to two lines and look like separate links.

☒ **Example**

Listen to Sounds *is one link but is on two lines.*

Press Room
Meetings
Listen to
Sounds
Jobs

Links to Avoid

- Avoid wording links as follows:
  - □ *Back* or [←] because these may mean something different than *Previous*; you never know where a reader has been.
  - □ *Click here* (not everyone is using a mouse or looking at the screen).

☒ **Example**

Click here to check out the many options that are available to you. Still want to know more? Click here to find out how information moves through the Internet.

☒ **Before**

For more about the selection criteria click here.

☑ **After**

Read our selection criteria *or simply* Selection criteria

☒ **Before**

Click here for previous page Click here for home page Click here for next page

☑ **After**

Previous Home Next

- Go to
- Here is
- Link to
- Next (unless you are using a linear structure). Use a phrase describing where your link leads.
- Press this button
- Select this button
- Select here
- This link will take you to

Should you include a *Coming Soon* heading? You may show planned topics in the form of links that appear underlined but are not yet functional. They do help users see topics you have planned and may encourage them to check back. But they may also frustrate some users.

▷ List Page

A list or resource page contains a list of useful Web sites. The list may be links to sites in a specific field or discipline or may be restricted to a narrower category such as companies, businesses, services, software, freebies, etc. A resource list may be a separate page in your Web site. However, the sole purpose of some Web sites is to provide resource lists ("megalinks" or "megalists" sites). A list site provides a list of useful links related to a particular topic.

☑ **Example**

The Nuthin' but Links site is a list of sites offering information on learning HTML, web page design tips, and creating web pages.

Tips

Creating a List Page

- Clearly state the audience, focus and purpose of the resource site or list.

☑ **Examples**

Webmasters: Welcome to Reallybig.com!

We offer more than **5000 Webmaster** resources for web builders including free clipart, CGI scripts, hit counters, fonts, HTML tutorials, javascripts, clipart animation, free backgrounds, icons, HTML editors, buttons, photographs, SEO and site promotion tips and tricks, and much more. Reallybig.com is The Complete Webmaster Resource for ANYONE who wants to create Web pages on the internet, from beginner to experienced Webmaster.

- Indicate the scope.

☑ **Example**

264,800 + links for family history!
254,380 in 180+ categories
10,400 + new and uncategorized

- Put useful resources on a separate Web page.
- Include a clearly labeled link from the main menu or table of contents.
- Make it clear that the resources are external links.
- Provide a variety of ways to access information, including browse, search, and menus/drop-down menus.
- Annotate each link.
- Provide information about who has compiled the site.
- State how often the site is updated and links verified.
- Avoid creating a Web page that is simply your list of favorite Web sites.

FORMATTING A LIST PAGE

- Divide resources into categories separated with headings. These can be internal links or links to separate pages, depending on the length.

☑ **Example**

WRITING AND RESEARCH SKILLS
WEB WRITING - MULTIMEDIA
WEB WRITING - COLLABORATION

- Use a logical organization for the links. Tell readers if links are in a special order, such as most recent first.

☒ **Examples**

This accounting site arranges links into categories. However, the order chosen for the information is not obvious.

Destinations: [Accountants - Misc resources related to] [Accountants - Lists] [Accounting Profession] [Accounting Software - Dealers] [Accounting Software for CPAs] [Banking sites] [Accounting Software for Businesses] [Accounting Organizations] [Tax and Law links] [Bar Code] [Book and Magazines] [Business] [Computers] [Consulting] [Databases] [Electronic Commerce] [Government Online] [Internet: General] [Internet: Oddities] [Internet: Security] [Internet: Web specific] [Investments] [Law] [ISO 9000 & Manufacturing] [Construction] [Newspapers and news sources] [Online service providers] [Finding people] [Shareware] [Commercial software] [Spreadsheets] [Contributor list] [INTERNET CPE] Accountant's Bulletin Board

- Use the Web site name as your link text rather than the address itself.
- For annotated items, use any of the following formats:
    - Place the name of the site in bold and hyperlink.
    - Indent the annotation describing the site.
- If appropriate, use symbols to flag certain sites.

☑ **Example**

A dollar sign **($)** is placed next to those sites that charge for access.

▷ Lists

Lists itemize or prioritize items. You can create the following types of lists: bulleted, numbered, and definition.

HTML Code: Ordered (numbered) list: <OL> or unordered (bulleted) lists <UL>tags, along with list items <LI> </LI>

Definition list: <DL> </DL>, along with definition term <DT>, and definition description <DD>.

Nested list: You can create a nested list by including the list tags inside list items (<LI>).

Tips

- Use lists to do the following:
    - Emphasize.
    - Condense and summarize information by omitting transitions and connectives.
    - Make information easy to scan.
    - Create lists of links.
- Avoid over-using lists.

Introducing the List

- Avoid drawing attention to the list.

☒ **Examples**

Here is a list . . .
The tasks are listed below.

▶ Introduce the list with a sentence or phrase.

☑ **Examples**

Organizing your Portfolio
Although your portfolio should stress concrete training and experience in your vocation, three other important items will help organize it effectively:
Academic Skills
Personal Management Skills
Teamwork Skills

▶ Make sure each list item grammatically follows the lead-in.

☑ **Example**

Let us help you ...
REGISTER your domain name
DESIGN your web site
ARRANGE for hosting service
SET UP your e-mail forwarding

☒ **Example**

Your hosting account gives you all of the following benefits:
- 24/7 access (barring emergencies) and a fast T-1 connection
- access to secure page serving and form submission
- name your main page whatever you want *(Does not follow the lead-in correctly.)*

▶ Pick a logical organization for the list items. State how the list is organized or what it contains (advance organizer).

☑ **Example**

The software used to create multimedia Web pages can be placed into three distinct categories:

▶ Cut extra words by using them in the lead-in.

☒ **Before**

The following items are repetitious:

Some browsers can do things that other browsers can't do.
Some browsers are better at doing certain things.
Some browsers display the same things differently.

☑ **After**

Revise the previous list by creating a lead-in sentence:

Some browsers
- can do things that other browsers can't do.
- are better at doing certain things.
- display the same things differently.

- Use the appropriate punctuation:
    - With a sentence, use a colon after *as follows* or *the following*.
    - With an incomplete sentence, use no punctuation.

FORMATTING THE LIST

- Determine if bulleting, numbering (for sequential items), or using neither is appropriate.
- Use the seven plus or minus two rule (i.e., use no more than about seven items in a list). Break up long lists by using headings.
- Keep lists items short. Keep list items about the same length and of equal importance.
- Show the hierarchy of topics by using indentation.
- Put keywords at the front of each list item and bold them for emphasis.

WRITING THE LIST

- Check lists for parallel construction (e.g., begin with same part of speech, such as a verb).

> ☒ **Example**
>
> - The Web Site is similar to your business card, resume, or brochure
> - Can be your Customer's first impression of you
> - A place for your Customers to find you

- Omit articles (*a, an, the*) from the beginning of each list item.
- Avoid beginning each list item with the same word. The list is not easy to skim.

> ☒ **Example**
>
> Introductions to Mineralogy
> Introductions to the History of Mineralogical and Geological Sciences
> Introductions to Petrology and Rock Identification
> Introductions to Fluid Inclusion Research
> Introductions to the Solid Earth and Plate Tectonics

- Begin the first word of a list with a capital letter.
- End each list item with a period unless it contains only a few words.
- Check lists for consistency in the following:
    - Punctuation
    - Capitalization
    - Use of either complete sentences or fragments

> ☒ **Examples**
>
> *The following list of topics uses capitalization inconsistently.*
>
> about the author   Form processing

About this document	Image-maps
Incomplete information	input
Character map	Internal images
Comprehensive listing	Obsolete features
Documented features	Undocumented features

▷ Location Identification

Always let users know their location: where they currently are, where they can go, and where they have been. Location identification helps orient readers and provides contextual clues.

Tips

- Use the following methods to show users their location:
  - Repeat your header and footer on all pages.
  - Use a site map.
  - List the path as a heading at the top of the page.

> ☑ **Examples**
>
> **Top**: **Computers and Internet**: **Software**: **Internet**: **World Wide Web**:**Browsers**: Plug-Ins
>
> You are here: Software / Home

  - Highlight, dim, or disable the navigational aid for the current page.

> ☑ **Examples**
>
> Indexes: Topics | **Destinations** | Authors | Site Index
>
> Web | Images | **Groups** | Directory | News

▷ Logos/Trademarks

A logo is a graphic used to identify an organization, institution, or company. It is often used in the banner and header; a smaller version often appears in the footer. A trademark is a name, word(s), or symbol identifying a company or product. Logos and trademarks

- Identify the organization/group/business behind the Web page.
- Add identity and thus reader trust.
- Add visual interest to the page.
- Help readers recognize each page as part of the same site.

Tips

- Use the logo throughout your site to establish an identity and contextual clues. Position it consistently.
- Consider using a small version of the logo in the footer.
- Keep the file size small (under 12 KB) to reduce loading time.

- Keep the logo one-fourth the size of the screen or smaller. Do not allow the logo to be so large that the readers must scroll to read the important information.
- If you use the logo as a link to the home page, provide alternative links as well. It is often not obvious to click a logo.
- Protect the logo/trademark by listing acknowledgments and copyright information. Clarify who is the owner.

▷ Long Description

A long description is a detailed explanation of any graphics that convey information. This text is important for making your site accessible. The description is provided on a separate Web page. You link to the description using the LONGDESC attribute, which is a pointer to the file's address. This technique is used when you do not want to put all the information about the graphic in the surrounding text.

Because not all browsers support this attribute, a [D] link is used as well. Because it is small, the link provides minimal distraction on the page.

HTML Code:
LONGDESC = "URL" (of the document containing the long description of the image). It is an attribute of IMG (image) and is used with the ALT attribute.

Tips

- Link to the long description page.
  - Link to the description using the *longdesc* attribute.
  - Provide a link *back* to the page containing the graphic.

> ☑ **Example**
>
> Link to a version of the Interactive Mortality Charts and Graphs section, which has been optimized for the visually-impaired

- Describe or summarize the content of each important graphic (graphs, charts, photos).
- Include all the information a person who cannot see the image would need to understand it.

> ☑ **Examples**
>
> Figure 2 shows an iPAQ handheld computer with an extension card for wireless communication via 802.11b. The iPAQ screen shows the buttons "Power", "Mute", "CC" (for closed captioning), "Volume Up", "Volume Down", "Channel Up", and "Channel Down". Text labels provide the current setting for the volume and channel: "Volume: 5", and "Channel: 1".
>
> Statistics are shown as time period, race / gender, Mortality Rate per 100,000 person-years, number of deaths.
> 5-year Cancer Mortality Rates per 100,000 person-years,
> Age-adjusted 1970 US Population
> All Cancers, 1950 to 1994, All Ages
> X Y Line graph with 4 lines.
> The x-scale starts at 1950.0 and ends at 1990.0.

Line 1, US White Male.
Line starts at 1950, 174.1499.

- Consider providing the long description in the context of the document itself or using a normal text link to link to it.
- Until most browsers support long description attributes, also use a D-link (description link) or invisible D-link.

▷ Magazine or E-Zine

An online magazine is an electronic magazine that may or may not have a printed counterpart. An *e-zine* or *webzine* is only distributed electronically and is usually considered to be informal. Unlike a print magazine, an online magazine can provide more interactivity, take advantage of links to support information, allow searching, and make archives available. If a printed counterpart is available, the online magazine also can be used for promotion and to solicit subscriptions.

Tips

- Make it immediately clear what your magazine is about.

☑ **Example**

Welcome! Netsurfer Science is an e-zine bringing neat science and technology sites directly to your mailbox. Subscribe and we will bring you a hot-linked HTML gateway to a selection of great online science related sites.

- Include a main page containing attention-getting headlines, summaries, and links to stories.

☑ **Example**

■ **Business**

Air Travel Gets a New Model
Going broke fast, the major airlines will have to change the way they operate. Here's what it means for you
• **Technology**: Windows vs. Linux?
• more stories

- Organize departments and stories into categories.

☑ **Example**

HOME, NATION, WORLD, BUSINESS, ARTS, SCI-HEALTH, PHOTOS, COLUMNISTS

Items to Include

- Include the following types of information:
  - ☐ Advertising information
  - ☐ Archives/back issues
  - ☐ Articles
  - ☐ Columns
  - ☐ Community features
  - ☐ Contests

- ☐ Current issue
- ☐ Description of the purpose, audience, content of the publication
- ☐ Downloadable articles
- ☐ Editorials
- ☐ Editorial guidelines
- ☐ Features
- ☐ Feedback
- ☐ Forum
- ☐ Help
- ☐ Links to information about the sponsoring organization
- ☐ Links to primary sources and related articles
- ☐ Links to technical papers
- ☐ Masthead
- ☐ Multimedia
- ☐ News
- ☐ Polls
- ☐ Press releases
- ☐ Publication information
- ☐ Reprints and permissions
- ☐ Resource lists (related to the subject)
- ☐ Reviews
- ☐ Search
- ☐ Site guide
- ☐ Sponsors
- ☐ Subscribers' services
- ☐ Subscription information (if applicable)
- ☐ Table of contents of current stories
- ☐ Tips/how-to information

Formatting: Home Page

- ▶ Include a logo or banner.
- ▶ Annotate headlines.
- ▶ Use an attractive design.
- ▶ Include attention-getting graphics.
- ▶ Chunk information into categories.
- ▶ Use columns and tables to create a grid layout. Place changeable, updated information in one location.

Formatting: Articles

- ▶ Chunk the story into short pages and provide links.

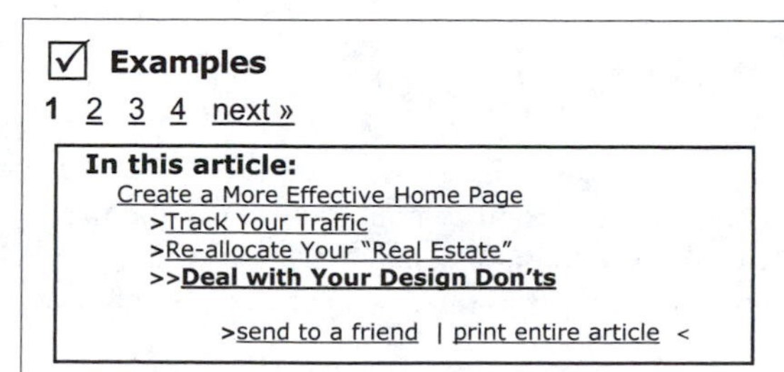

Examples
1 2 3 4 next »
In this article:
Create a More Effective Home Page
>Track Your Traffic
>Re-allocate Your "Real Estate"
>>Deal with Your Design Don'ts
>send to a friend | print entire article <

- ▶ Use small paragraphs with headings.
- ▶ Link to author information, details, sidebars, support information, and other Web sites that expand on the story.

☑ **Examples**

AUTHOR INFORMATION

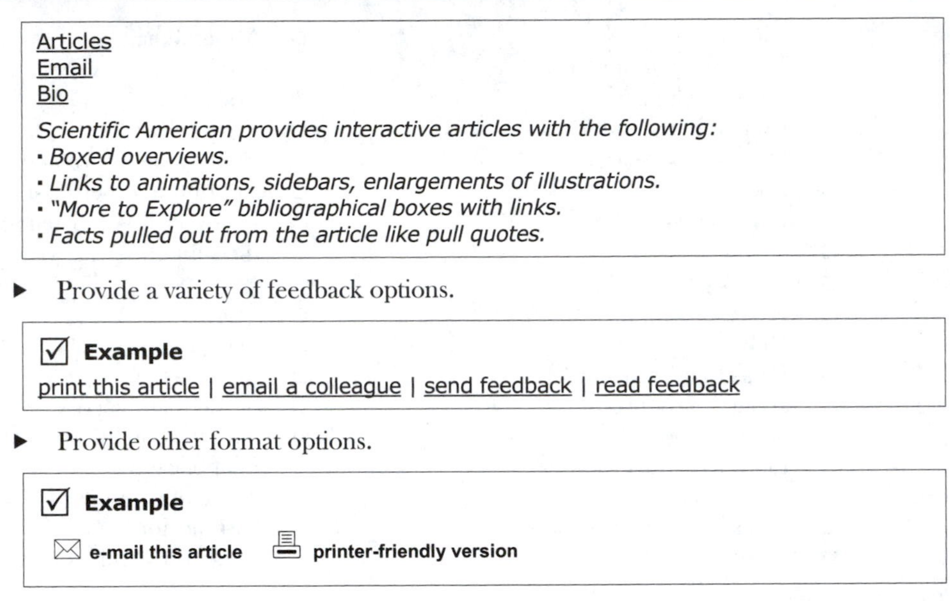

Articles
Email
Bio

Scientific American provides interactive articles with the following:
- *Boxed overviews.*
- *Links to animations, sidebars, enlargements of illustrations.*
- *"More to Explore" bibliographical boxes with links.*
- *Facts pulled out from the article like pull quotes.*

- Provide a variety of feedback options.

☑ **Example**

print this article | email a colleague | send feedback | read feedback

- Provide other format options.

☑ **Example**

e-mail this article **printer-friendly version**

▷ Margins

Margins add white space to your Web page. Margins will vary depending on the browser. White space around text

- Makes line length shorter and thus easier to read.
- Emphasizes headings.
- Adds contrast.
- Reduces eye strain and fatigue.

There are several tricks used to create margins:

- Tables
- Transparent GIF graphics used as spacers
- A graphic of the entire page, including text
- Cascading Style Sheets margin-property

Tips

- Use wide left margins with headings in the margin so readers can scan the page.
- Avoid using too many unique margins on the same page.
- Remember that wide margins are usually unnecessary on a monitor because there is a frame around the screen.

▷ Marquee

A marquee is scrolling text that moves across a Web page without using animation. You can make the text scroll or slide, enter from the left or right, and loop. Scrolling marquees are used for announcements, to attract attention, and to fit a large message in a small area.

HTML Code: <MARQUEE> </MARQUEE>

You can also use JavaScript to add scrolling text.

Tips

- Use marquees sparingly because they are distracting.
- Provide a text version of the message on an alternate page.
- Never use marquees as links.

▷ Menu

A menu is a list of links to topics on a Web page or site. It is usually less detailed than a table of contents. A menu can be arranged vertically or horizontally. It may also consist of text in one or more columns, graphics, or both. You use a menu to

- Provide "entry" into your site.
- Let readers go directly to specific pages.
- Show the document's contents and scope.
- Provide a navigational aid.
- Layer information.

HTML Code: <MENU> </MENU>
(You can create special effects menus with JavaScript and Dynamic HTML.)

Tips

General

- Determine the main information readers look for on your site; then make this information obvious on the menu.
- Weigh the pros and cons of using a broad, shallow menu (many categories with less levels) or narrow, deep menu system (less categories with more levels). With many categories, readers must spend more time searching. With fewer categories, readers must spend more time clicking and making decisions.

> ☑ **Example**
>
> *Nissan USA's site has only four main categories:* VEHICLES, BUYING, OWNERS, and ABOUT NISSAN.

- Group related items into categories; then write clear labels to describe them.

> ☑ **Example**
>
> **mobile products**
> → calculators
> → handhelds and palmtops
> → notebook PCs

- Separate information *about* your site from information provided *in* your site.

> ☒ **Before**
>
> EPA Newsroom
> Browse EPA Topics
> Educational Resources
> About EPA

Laws, Regulations & Dockets
Where You Live
Information Sources

Programs
Business Opportunities
Jobs

☑ **After**

Information about the EPA should be grouped:
EPA
About EPA
Programs
Business Opportunities
Jobs

The following menu items are unclear:
Where You Live
Information Sources
Laws, Regulations & Dockets

- Put items in a logical order (e.g., alphabetical, most important to least important).

☑ **Example**

The order of the following menu items is not logical.

Resumes
Business/Professional Writing
Research Paper Writing/Citing Sources
Spelling
English as a Second Language

General Writing Concerns
Punctuation
Sentence Concerns
Parts of Speech

- Put the menu on every page.
- Use care with redundant menus (two versions of the same menu on the same page), because they can be confusing. You can provide an abbreviated menu on the home page. This menu will appear on every page. On the home page provide an annotated version of this menu as well.

☑ **Example**

IBM's Ease of Use site repeats the same links on the left and on a main annotated menu.

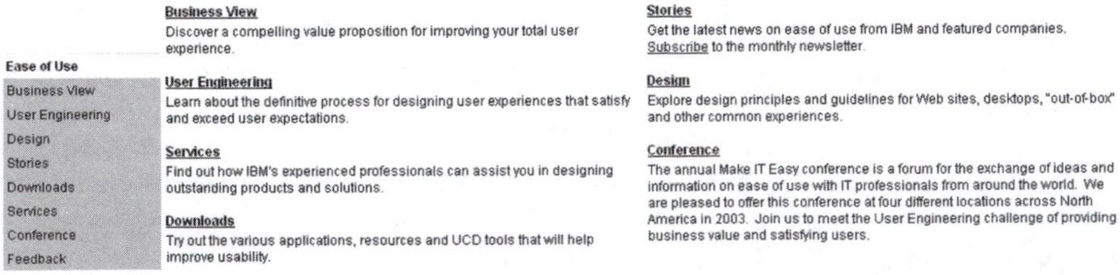

Ease of Use
Business View
User Engineering
Design
Stories
Downloads
Services
Conference
Feedback

Business View
Discover a compelling value proposition for improving your total user experience.

User Engineering
Learn about the definitive process for designing user experiences that satisfy and exceed user expectations.

Services
Find out how IBM's experienced professionals can assist you in designing outstanding products and solutions.

Downloads
Try out the various applications, resources and UCD tools that will help improve usability.

Stories
Get the latest news on ease of use from IBM and featured companies. Subscribe to the monthly newsletter.

Design
Explore design principles and guidelines for Web sites, desktops, "out-of-box" and other common experiences.

Conference
The annual Make IT Easy conference is a forum for the exchange of ideas and information on ease of use with IT professionals from around the world. We are pleased to offer this conference at four different locations across North America in 2003. Join us to meet the User Engineering challenge of providing business value and satisfying users.

- Provide a site menu (*global* navigation) and a menu for sub-topics (*local* navigation).

☑ **Examples**

MM & Web Page Design Principles: [Simplicity] [Consistency] [Clarity] [Balance] [Harmony & Unity]
Page Design:[Multimedia & Web Page Design Principles] [Screen Design Research] [Screen Resolution & Size] [Writing Style]
Main Level: [Home Page] [Design Theory][Site Design] [Page Design] [MultiMedia] [Teacher Resources] [Table of Contents]

This Site
home page | table of contents

This Page
Color Help | Counters | Graphic Information | Graphics / Icons
Imagemaps | Interesting Ideas | Intranets | Site Design / Information | Sound and Video

- Consider putting the phrase "main menu" in the page title that appears in the browser bar.

☑ **Example**

<title>Virtual University Site Map (Main Menu)</title>

Writing Menu Items

- Write menu text that is specific and informative and that helps readers find information quickly. Avoid categories that are too general.

☑ **Example**

Alzheimer's Disease
Caring for Someone with Alzheimer's
Exercise for Older Adults

- Use terms that your audience will recognize and understand.
- Keep items short (less than five words) and descriptive. They will be easier to scan and repeat on every page.

☒ **Example**

These chapters need titles to be useful.

Section 1: The HTML Language
Chapter 1   Chapter 2   Chapter 3   Chapter 4   Chapter 5

- Annotate the main menu.

Formatting

- Format the menu so readers can skim it quickly.
- Format the menu differently from content.
    - Use a different font size/color, or use white space, lines, or boxes to separate it.
    - Avoid removing underlining from links so readers recognize menu links.
- Make text highly legible.
- Make menu items easy to click.

☒ **Example**

This site requires that you click small buttons.

Click on the [button] below or use the new Website Search Engine:

- Avoid making readers scroll to view menu items.
- Avoid a cluttered look by providing too many menu items.
- Use vertical or horizontal menus.
    - Vertical menus are easier to scan quickly.
    - Horizontal menus are easier to add to later and take up less room.
- Place menus at both the top and bottom of the page.
- Use headings to organize menu items into categories.
- Make all menus in your site consistent.
- Avoid solely graphical menus, such as image maps. If you do use graphics:
    - Make sure they load quickly.
    - Use textual alternatives as well.
- Use JavaScript or DHTML (Dynamic HTML) to create special types of menus (pop-up, cascading, expandable/collapsible, slider, and drop-down).

> ☑ **Example**
>
> *To keep the interface simple, you can use a drop-down menu (also called a pop-up or pull-down menu) that allows readers to select an item from a list:*
> *This type of menu is created by using a form or JavaScript.*

▷ Meta Information

Meta tags (metadata) are part of the HTML code that may appear at the beginning of the document inside the <HEAD> tags and after the <TITLE>. They are not required tags.

There are several meta tags; two of the most important are description and keyword tags. Both tags were designed to help search engines find and summarize your site, but many search engines do not use them any more. The *description* tag is used for the site summary. The *keyword* tag provides keywords associated with the site. It is especially useful for pages without text or with frames.

HTML Code: <META NAME=>

Tips

- Identify what your site is, what it does (functions), what readers can learn or what tasks they can perform.

> ☑ **Example**
>
> <META Name="description" Content="Meta tag tutorial overview of the web site promotion skills needed to build high traffic counts to your site. Webmaster tools for promoting your site. A free service for all web site developers.">
>
> <META Name="keywords" Content="search engine secrets, meta tag, meta tags, web site traffic, web site promotion, meta tag, meta tags, search engine spamming">

- Provide keywords for your site.
- Provide a short, concise description of the site.

> ☑ **Example**
>
> <META Name=description Content=This site contains the names and descriptions of key writing-related Internet sites.>
>
> <META Name=keywords Content=writing, research, college students, college writing, writing, editing, researching, library skills, Web resources>

▷ Metaphor

A metaphor is a comparison to an object. It is usually used to help readers understand something new by comparing it to something they are familiar with. Metaphors are common on the Web, as seen in terms such as "home page," "shopping cart," and "table of contents." A metaphor might also describe the way the site is

- Organized (e.g., building, store, library, file cabinet)
- Navigated (remote control, map, buttons, tabs, book, travel)
- Designed (newspaper, radio)
- Written (terms used for links, extended comparisons)

Using a metaphor for the interface of your Web site can

- Make it easier to use and navigate.
- Inform users about the type of information to expect.
- Create a graphical model of the site's organization or the functions of various parts of the page.
- Create a mood or atmosphere.
- Make the site consistent.

Tips

- Use a metaphor that is related to the topic.

> ☑ **Examples**
>
> *The Disney vacation planning site called Magic Alley uses the metaphor of a neighborhood with shops, exhibits, passageways, and tour bus.*
> *The Dory Kanter virtual art gallery lets you pick rooms from a floorplan. You can visit rooms where various types of art (e.g. watercolors, oil pastels) are displayed, a gift shop, and an information booth.*

- Use a metaphor only when it supports the reader's goal, not just for looks.
- Develop the metaphor through images (background, image maps, icons, navigation bars), the color scheme, and names for links.

> ☑ **Example**
>
> *Fact Monster, a free reference site for students, teachers, and parents, uses two metaphors for the site design: monster and reference. The logo is a red monster; the*

title font, background colors, and navigational icons all fit the "monster" look. There is also a Reference Desk with a more formal design.

- Explain the metaphor.

☑ **Example**

Here is the introduction to the PC Lube and Tune site, which contains suggestions for making system repairs:

PC Lube and Tune is a Service Station and convenience store at Exit 130.132 on the National Information Highway.

- When using the metaphor with your links, provide descriptive explanations.

☑ **Example**

The Backyard: Gardening and Outdoor information
The Back Porch: Current Events poll
The Kitchen: Recipes and Household Hints
The Playroom: Everything about the Kids, Toys, Homework, Software & the Web

- Use the metaphor with consistency throughout the Web site.
- Keep the metaphor simple.
- Avoid multiple metaphors.
- Consider using a metaphor to write about your topic.

☑ **Examples**

The Web Kitchen uses a culinary metaphor:

The browser is like the cook, the information you want to provide is like the ingredients, the HTML document like the recipe, the HTML language like the instructions, the server where your web page lives like the dinner table, and the word you get out about your site like ringing the dinner bell. And if you get the hang of all that, the more advanced things you can do with a web site are like gourmet.

▷ Misplaced Modifier

A misplaced modifier is a word, phrase, or clause that is not next to the word it should modify.

Tips

- Check that words such as *only, just, nearly, barely* are in the correct position.

☒ **Example**
These costs are for the web address (domain name) SOLELY.

☒ **Before**
Some tags don't need an ending tag because they only perform one function.

☑ **After**
Some tags don't need an ending tag because they perform only one function.

- Check that clauses and phrases are in the correct position.

☒ **Before**

We are dedicated to providing creative web sites for businesses that intrigue their clients.

☑ **After**

We are dedicated to providing business Web sites that intrigue clients.

☒ **Before**

A promotions manager is able to identify prospective customers for campaigns with an easy-to-use interface.

☑ **After**

Using an easy-to-use interface, a promotions manager is able to identify prospective customers for campaigns.

▷ Mission Statement

A mission statement summarizes the philosophy of your business, organization, institution, publication, or Web site. A mission statement lets readers know what type of business, organization, or site it is, and how it differs from any competition.

Tips

- If your mission statement is brief and central to the purpose of your site, put in on the home page.
- If your mission statement is long, put a highly-visible link to the mission statement. It may be called *Mission Statement, Our Mission, About (Organization or Company Name).*

☑ **Example**

The first two mission statements appear on the home page because they are short and tied in with the introductions.

Our mission is to provide clients worldwide with state of the art World Wide Web related services including but not limited to Web development, Web hosting, multi-lingual communications and more.

Our mission is to preserve the nation's network of estuaries by protecting and restoring the lands and waters essential to the richness and diversity of coastal life.

The EPA mission statement appears prominently on the home page:

Our Mission: . . . to protect human health and to safeguard the natural environment.

This site contains a link to the mission statement from the home page:

Mission Statement: Read the IPAC mission statement and learn more about the objectives of the organization.

Home Page > Mission

National Weather Service Mission Statement

- Clearly state the organization's purpose, functions, what you do, how, and why.
- Emphasize reader benefits.

☒ **Example**

We want to make money.
There's another reason we exist:
Passion.
We love design.
We love creativity.
We love identifying problems.
We love solving problems.

- Include specific details rather than vague generalities, obvious statements, or clichés.
- Consider grouping your mission statement with other information, such as history and background.

▷ Monospaced Type

Monospaced type uses characters that take up the same horizontal space. Examples are fonts such as `Courier` or `Typewriter`. Monospaced type is usually used for computer documentation, to illustrate HTML code, or fit a design theme.

HTML Code: Monospaced text is created from the following HTML tags:
<CODE> </CODE> (code) is for programming code.
<KBD> </KBD> (keyboard) is for text typed on a computer.
<SAMP> </SAMP> (sample) is for computer-displayed text.

Tips

- Use monospaced type to format calculations, computer code, commands, and messages.
- Use monospaced type to fit a design theme.
- Use highlighting for additional emphasis. For example, use shading and separate the text from the paragraph.

☑ **Example**

Shading and monospaced type provide emphasis.

The following font style elements are available:

```
<TT>teletype or monospaced text</TT>
<I>italic text style</I>
<B>bold text style</B>
```

▷ Multimedia

Multimedia includes video, animation, and sound. On the Web, multimedia can be used for entertainment, training, samples, demonstrations, and news. It is also a means of adding interactivity to your Web site.

Locations for Multimedia

Multimedia can be inline and external.

Inline media include words and pictures browsers display in place within the browser window. Browsers use plug-ins to access and execute the embedded files. A *plug-in* is a small program that attaches itself to a Web browser and allows it to play one type of file.

External media include files that a Web browser does not directly load. When the user clicks on a link to the media file, the browser requests the file from the host server. The server then determines the kind of data in the file. It sends the media file and the file type to the browser. The browser stores the media file and determines the type of *helper application* needed to play the file. The helper application then displays it in a separate window. The browser launches a helper application when it downloads a file it cannot process.

Helper applications include viewers and players.
Viewers display static graphics files.
Players display sounds and dynamic pictures, such as video, animation, sound clips, and virtual reality.

Types of Multimedia

Sound files include voice, music, and sound effects. Popular sound formats include AU (Sun Microsystems), AIFF (Apple's Audio Interchange Format), MIDI (Musical Instrument Digital Interface), MP3 (MPEG audio), RA (Real Audio), SND (Sound), VOX (Voxware ToolVox), and WAV (Windows). A podcast (iPOD broadCAST) is an audio broadcast that has been converted to an MP3 file or other audio file format for playback in a digital music player or computer.

Streaming audio lets the user listen to a file as it downloads.

A method of putting sound in the background of your page is to use the EMBED tag:
<EMBED SRC=sound.mid HIDDEN=TRUE AUTOSTART=TRUE>

Video formats include ASF (Advanced Streaming Format), AVI (Microsoft Audio/Visual Interleaved), MPEG/ MPG (Moving Picture Experts Group), MP4 (MPEG-4), MOV (Apple QuickTime), RM (Real Media), FLA/FLV (Flash), and WMV (Windows Media Format).

Streaming video lets users with fast connections view video as it is downloading. QuickTime files can use a technique called Fast Start that let them begin playing soon after the page loads.

Adobe **Shockwave (SWF)** applications combine interactive video, streaming audio, 3-D animation, and graphics. These files are created with a variety of software products and are compressed for use on the Web. Plug-ins are available for Netscape and Microsoft Internet Explorer.

Animations are created in programs such as Macromedia Flash and Shockwave and include animated GIFs or Java applets.

Java is a programming language developed by Sun Microsystems. It is used to create applets distributed on the Web with HTML documents. The browser downloads applets to your computer. These applets may incorporate text, audio, animation, image, and video objects.

JavaScript is code included as part of the HTML document rather than a separate file. It allows embedding of small programs in an HTML document to control actions.

Tips

- In general, aim for simplicity in your Web page design. Avoid jazzy special effects and enhancements.
- Use multimedia only when it supports your content.
- Don't rely only on multimedia alone to convey information. Not all readers will choose to view it.
- Recognize that files are large and require long download time.
- Do not force reader to use multimedia; provide links to files.

> ☑ **Example**
>
> Example 11
> Audio:
> [Modem] [Broadband]

- Preview what the multimedia file contains through a summary or still photos.

> ☑ **Example**

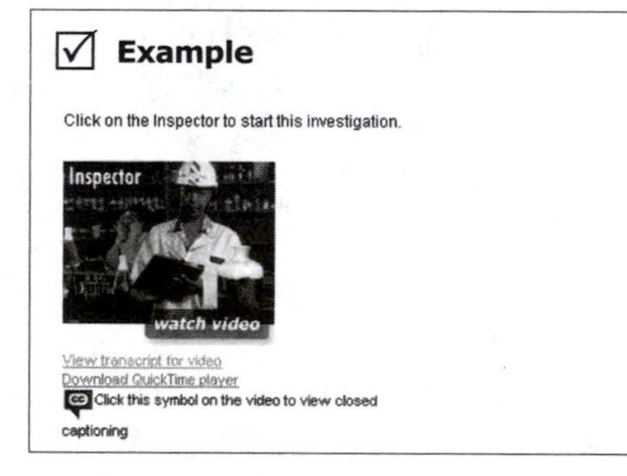

Click on the Inspector to start this investigation.
Inspector
watch video
View transcript for video
Download QuickTime player
Click this symbol on the video to view closed captioning

- Warn readers about file format, file size, required bandwidth, and required plug-ins.
- Provide necessary plug-ins.
- Use icons to show the type of media.
- Know copyright issues before you use multimedia on your Web page.
- For accessibility, provide the following for sound, video, and animation:
    - Transcript (text version)
    - Description

▷ Navigation

Navigation is method by which users move around your site. Navigation is one of the most important components of a Web site because it helps readers know

- What is available
- Where they are and where they can go

- How to get to information
- How the site is organized

Types of Navigational Aids

The *navigation system* is a combination of all of the following types of navigational aids:

- Browse feature
- Clickable "breadcrumb" path at the top of a page showing the reader's location
- Drop-down menus/shortcuts
- Frame for navigation
- Index
- Links lists
- Links such as Previous, Next
- Menus
- Navigation bars
- Page numbers
- Search engine
- Site map
- Table of contents

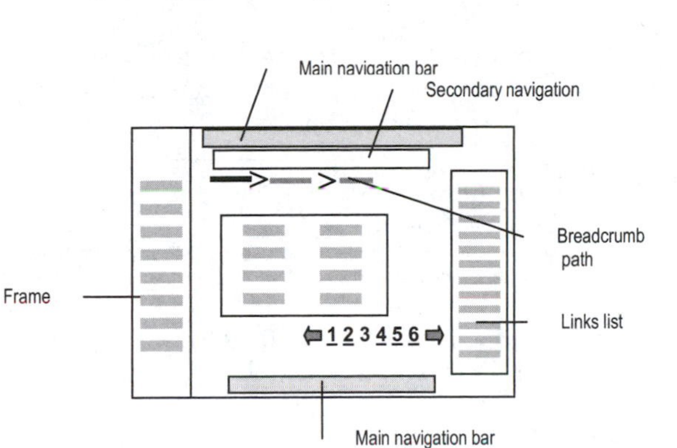

STYLE	DESCRIPTION	PROS	CONS
Text	Text, possibly separated by [ ] or | |	Simple, has quicker download time, supports all browsers	Not as visually interesting as graphics
Icon/button	Representational graphic	Adds visual interest, can fit with theme	Not visible in all browsers, adds download time
Navigational aids in a frame	Icons or text displayed in horizontal or vertical frame	Stays in place	Not supported in all browsers, slower download time
Image map	Graphic with hotspot links	Has graphic appeal, can fit with metaphor	Requires programming, hotspots often not obvious

Elements in your navigation system are related to your site's structure, content, and types of readers.

There are two "types" of navigation:

- *Site-wide (main)* navigation used for the entire site.
- *Local (secondary)* navigation used for specific topics, modules, or sections within the site.

Tips

TYPES

- Select the appropriate type of navigation system for your content and audience.
- Provide several methods of navigation and multiple paths to the same information to accommodate different experience levels.
- Always let readers know where they are (contextual clues), where they should go, and how to get there.
- Distinguish categories for navigation: site-wide, local, content (topics), background (mission, company information), tools (search, site map, contact information, ordering information).

> ☑ **Example**
>
> **site** Sign In **help** Site Map · Site Help
> **quicken** Quicken Web Entry · Buy Software Online · Download Free Updates · Quicken Software Support
> **fyi** Privacy Statement · Information for Advertisers · Free Newsletters · Intuit.com

- Provide shortcuts to frequently accessed pages.
- Make navigation accessible by providing textual alternatives.

FORMAT

- Make navigation prominent.
- Make navigation consistent by using a standard set of buttons and words.
- Dim/gray the navigational aid for the current page, or use a folder tab graphic.
- Balance flexibility with overwhelming readers with too many options (information overload). Make paths to information relatively short; don't require numerous clicks to information.
- Use unusual patterns for the links if text is legible and if it fits your design scheme.

> ☑ **Examples**
>
> Library
> Traveller's Handbook
> Internet Guide
> Traveller Magazine
> Healthbook
> More...
>
> **faq**
> **code**
> **awards**
> **journals**
> **subscribe**
> **older stuff**
> **rob's page**
> **preferences**
> **submit story**
> **advertising**
> **supporters**
> **past polls**
> **topics**
> **about**
> **bugs**
> **jobs**
> **hof**

POSITIONING NAVIGATIONAL AIDS

- Place the navigational links in a logical order.
- Place navigational aids in any of the following locations:

- ☐ At the top and bottom of each page. Navigational aids at the *top* allow readers to immediately see the choices. Navigational aids at the *bottom* allow readers to navigate once they have finished the page.

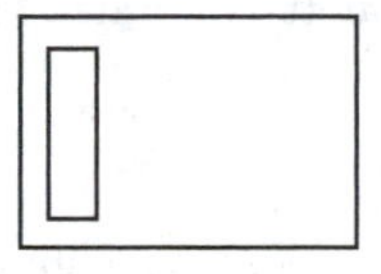
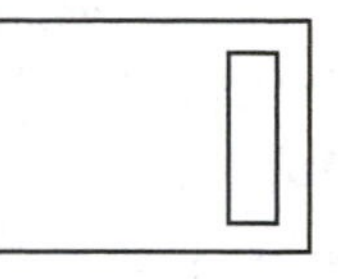

- ☐ On the left (most prominent)
- ☐ On the right
- ☐ In a frame

☑ **Example**

Main chapter links appear along the right and chapter sections appear across the top of Jan's Guide to HTML.

Style Sheets

Contents: Introduction - Using An External Style Sheet - Using A Style Tag - Using A Style Attribute - The Formatting Model - Background Colour - Background Image - Foreground Colour - Borders - Padding - Margins - Font Family - Font Weight - Font Style - Font Variant - Font Size - Letter Spacing - Line Height - Text Align - Setting Link Styles - Forcing a Page Break When Printing

Introduction

Under HTML 4, everything to do with lay-out has to go into separate styles. There are three ways to combine HTML and styles. You can load an external style sheet, you can include a STYLE element inside the HEAD element of your document or you can use a STYLE attribute with tags in the body of your document. You can also combine the three methods in one document. This chapter first shows the 3 ways to include style information and then gives examples of what you can do with it.

Chapters
Table of Contents
Introduction
Basic HTML
Text
Images
Lists
Anchors
Tables
Frames

- ☐ After each topic or screen of text (for long pages)

☑ **Example**

This site repeats the navigation after each topic on the page.

Deliverable: Beta Version of Project

The major deliverable at this stage of your project development process is the beta version of your project - i.e. a full version of your site, ready for user testing. Have a look at the criteria that your testers will apply when evaluating your site.

Concept Flowchart Text Interface Test Publish

- ▸ Group navigational aids into distinct areas of the screen.
- ▸ Repeat navigation on every page.
- ▸ Do not require that readers scroll to view navigation.

▷ Navigation: Types of

There are several types of navigation:

- ▸ *Site-wide navigation* helps users navigate the entire site.
- ▸ *Local navigation* helps users navigate within a sub-site or subject area.

Navigation can also be *hierarchical* or *sequential*. The type of navigation depends on the site's structure.

NAVIGATION: HIERARCHICAL

A hierarchical navigation scheme is similar to an organizational chart. It allows readers to move from general topics (top level), to specific categories (down) and back up, and to related topics (across). Hierarchical navigation is appropriate for topics arranged in a branching structure.

Tips

- Consider labeling document chunks, such as chapter, section, and sub-section.
- Include both *vertical* and *horizontal* navigational aids:
    - Home link (to main menu, table of contents, home page)
    - Down ("child")
    - Up ("parent")
    - Across to topics at the same level
- If you use terms such as *Up a level/Down a level* or *Parent/Child* (or icons representing this type of navigation), be sure your audience will understand their meaning.

☒ **Example**

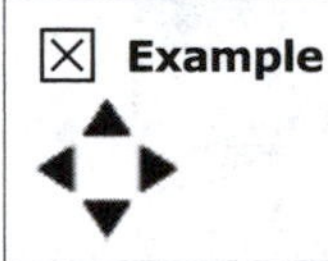

Navigation: Sequential

Sequential navigation forces readers to navigate forward and backward through topics in a linear structure. The most common sequential navigational aids are *Forward/Next* and *Previous/Back*. In general, *Back* or is confusing because many browsers include a BACK button. Because you do not know where a reader has been, *Back* may mean something different than reading the previous topic in your Web site. Use sequential navigation when sections should be read in a linear order, such as tutorials. However, you cannot predict the exact order readers will read Web pages.

Tips

- Use a browse sequence with linear navigation for topics in a sequential order or training/tutorials.
- Consider allowing readers to read out of sequence.

☑ **Example**

Or Jump Ahead
If you're in a hurry, you can skip directly to the section of the tour that interests you. Choose a section from the list to the below, or use the navigation to the left.

- Place sequential navigational aids at the beginning or end of a section or near the last line of text.

- Use any of the following methods of providing sequential navigation:
  - Forward to (topic title), Back to (topic title). This technique is preferable because it lets readers know the names of specific topics.
  - Next Page, Previous Page
  - Next (Page 1 of —)
  - More
  - Buttons
  - A list of all available pages

> ☑ **Examples**
>
> [PREVIOUS] 1 2 3 4 5 6 7 [NEXT]
>
> (Up to overview , back to Introduction , on to: writing each document)
> Next: Starting and Maintaining a Website.
>
> General Design Options
> Consider these choices, not rules. It's up to you, the page creator... (Contains three pages and a summary)
>
> 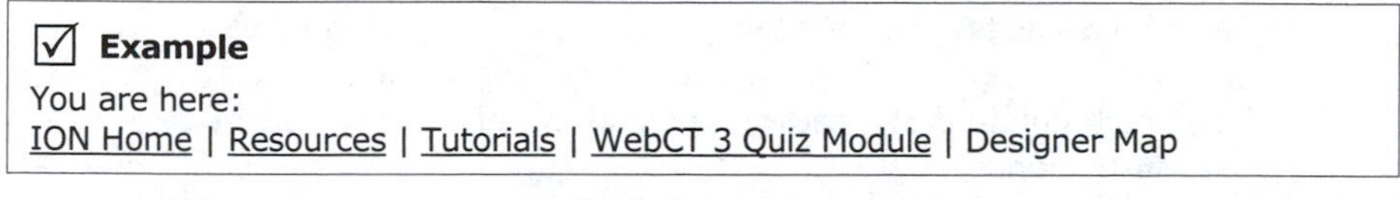
>
>
> Page 4 of 4
>
> **Next page** ... Previous page ... Core Rules ... Netiquette Contents
>
> Several points were raised on preceding pages, so here's a summary of the general design options that I covered.
> Make the site navigable by every browser. (Page 1)
> Make the look, feel and function consistent. (Page 2)
> Make links to every section from every single page. (Page 3)

- Let readers know the following:
  - Where they are located (contextual clues).

> ☑ **Example**
>
> You are here:
> ION Home | Resources | Tutorials | WebCT 3 Quiz Module | Designer Map

  - How many pages are available in a topic.

☒ **Example**

This Web site does not indicate how many pages are in a topic. It contains only forward and backward buttons and a menu on the left. The layout section has eight pages.

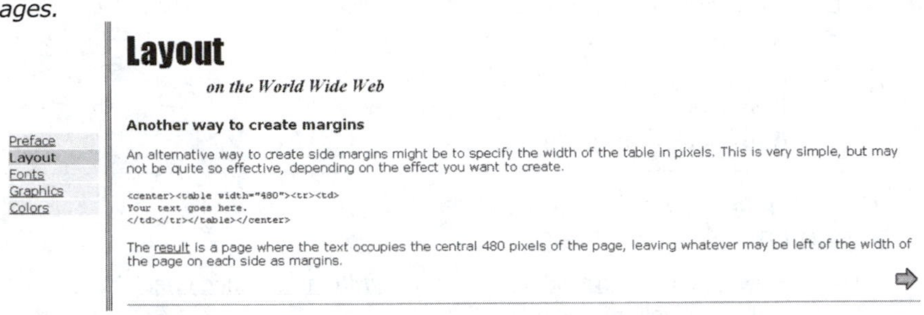

Preface
Layout
Fonts
Graphics
Colors

Layout

on the World Wide Web

Another way to create margins

An alternative way to create side margins might be to specify the width of the table in pixels. This is very simple, but may not be quite so effective, depending on the effect you want to create.

```
<center><table width="480"><tr><td>
Your text goes here.
</td></tr></table></center>
```

The result is a page where the text occupies the central 480 pixels of the page, leaving whatever may be left of the width of the page on each side as margins.

▷ Netiquette

Netiquette, or "net etiquette," refers to good Web manners: the do's and don'ts of being a good netizen (Internet citizen). Using good netiquette affects the way others perceive you and receive your opinions. It also affects a Web site's credibility.

Tips

Some examples of good netiquette for *writing* include the following.

Site Netiquette

- Consider the following:
  - The time it will take to download your page. Keep file sizes small and allow users to link to large images.
  - Consider users who use different browsers and platforms (platform-independent terminology).
  - Copyright issues for graphics or information you use. Ask permission to link to someone's site.
  - Accessibility issues by providing alternative text for graphics.
  - Privacy issues by not publishing private information.
  - Web publishing policies and procedures of your own organization or institution.
- Identify your organization and sponsors.
- Provide the following:
  - Contact information
  - Date stamp
  - Copyright, privacy, and disclaimer information
  - Updated information
  - Information about file types and sizes so readers can decide if they want to download them.

Discussion Group Netiquette

- Read archives, FAQs, posting rules, and discussions–"lurk"–before posting.
- Don't insult (flame) people.
- Post messages that are well edited, clear, concise, relevant to the list, and knowledgeable.
- Include short edited summaries of messages you are replying to so readers remember the context.
- Have responses directed to you by e-mail; then post summaries to the group.

▷ News

News information appears on many kinds of Web sites. It may include news stories and announcements. It is, of course, the sole focus of online newspapers. Distributing news over the Internet is an ideal way to distribute it quickly to millions of readers. It is also one way to encourage readers to return to your site.

Tips

Items to Include on News Sites

- Include the following types of items:
  - Advertiser information
  - Archives
  - Articles
  - Background information
  - Breaking news/top news
  - Columns
  - Community features
  - Download options for articles
  - Editorials
  - Entertainment
  - Features
  - Links to primary sources and related articles
  - Multimedia links, such as photos, video, and animation
  - News stories
  - Resource lists (related to the subject)
  - Reviews
  - RSS feeds
  - Search engine
  - Subscription information (if applicable)
  - Table of contents of current stories
  - Tips/how-to information
  - Translation options

Formatting News

- Make content the priority over layout.
- Emphasize the news by using a variety of techniques:
  - Use design elements, such as headings, boxes, white space, color, and large fonts.
  - Put important information in the focal point of the screen, usually the top and slightly to left of center.

☑ **Examples**

Bloomberg.com scrolls headlines in a shaded box across the top of the home page:

U.S. Treasuries Rise After Drop in New Home Sales; Yield Curve Steepens

Boeing puts the stock price on the top of the home page:
BA stock price 90.83 [- 0.15] at 4:09 PM ET on Mar 26

- ☐ Arrange stories under familiar headings. Label with captions, such as News, This Week, Today's News, What's New, Top Stories, Lead Stories, Headlines, What's Happening, Latest Headlines, Just In, Breaking News.
- ☐ Put news in a frame so that it remains visible.

- ▶ Use a grid for different types of information so readers can skim quickly. Use a layout similar to print newspapers. However, keep the layout uncluttered.
- ▶ Put changeable (updated) information in one location for easier updates.
- ▶ Provide other formats or methods of using the information.

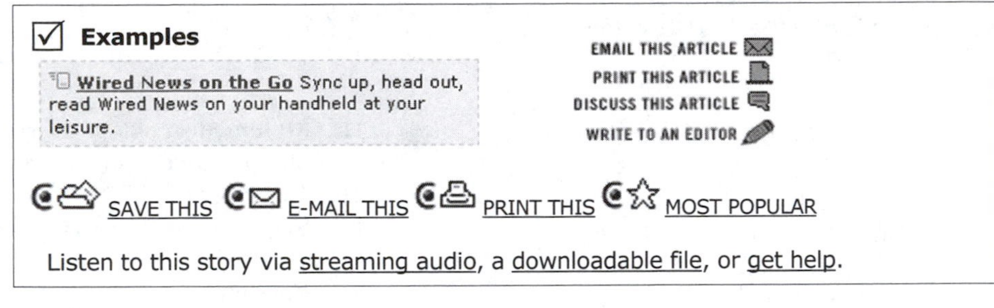

Using Links

- ▶ Provide browsing and search options.
- ▶ Use any of the following techniques:

  - ☐ Tabs that link to separate categories.

☑ **Example**

HOME PAGE	MY TIMES	TODAY'S PAPER	VIDEO	MOST POPULAR	TIMES TOPICS

- ☐ Provide a link to news from the home page.
- ☐ Place the headlines on the home page itself.
- ☐ Place both headlines and summaries on the home page. Link to more headlines and details.

☑ **Example**

U.S. U.S. NEWS
- MLK plot back in court
- Iraq, Venezuela factor in U.S. gas hike
- Nasty weather wreaks havoc

WORLD WORLD NEWS
- Serb hardliner arrives in Hague
- Venezuela strike leader's house arrest
- **TIME.com:** Saddam: Threat or pest

- ☐ Link to a separate *Top Stories* page.

☑ **Example**

Top Stories
Want to see what the top-read stories on ScienceDaily are? Click on one of the links below to see the most popular stories, with the number of hits for each story:
Most popular stories in the past WEEK.
Most popular stories in the past MONTH.
Most popular stories in the past YEAR

- Provide alternative ways to view stories.

☑ **Examples**

News Pages
Summaries
Headlines
News by topic
Find stories

click to see the headlines Show All Top Stories
To add a Clipping click on the Clipping icon:

- Link to details, sidebar, more stories in the category, related stories, and supplementary information that expands on the main stories (layering). Place links within the text or separate from the story.

☑ **Examples**

COMPLETE STORY

Company MultiLink™

Find out more about these companies.

RADIO EXPRESS
Coca-Cola Company
PepsiCo, Inc.

RELATED INFORMATION

Related Stories
- Lieberman Criticizes Bush Domestic Defense Spending
- President Outlines New Terror Analysis Agency
- States Struggle to Pay for Homeland Security
- Ridge Faces Early Test of Power as Head of Homeland Security

Background Info
- **Fast Facts:** Dept. of Homeland Security
- **Fast Facts:** Explaining the Alert Codes
- **Bio:** Tom Ridge

- Link to graphics and multimedia files (video, sound, interactive version, slides). Provide icons that represent the type of file:
- Chunk long articles and provide either internal links or links to separate pages.

☑ **Example**

[Click here to move to the next part of the article]

Writing Web News Information

- Organize stories into categories.
- Write clear, concise, descriptive headlines.

- Provide the date, author, and source.
- Provide a concise summary and lead-in.

▷ Newsletter

A Web newsletter (e-newsletter) is similar to a print newsletter. A Web newsletter may be

- Available in print version only, online, sent by e-mail, or downloaded.
- Free or require a subscription.
- Available as a link from the main page of a company or organization site or as the entire focus of the Web site.

A newsletter can be used to market, advertise, or promote an organization or business and provide information about a group, issues, and events. Distributing it online or electronically reduces printing and distribution costs and can take advantage of hyperlinks and multimedia.

An *e-mail newsletter* is sent to readers to present news, promote new services or products, and encourage them to visit your Web site.

Tips

Home Page

- Clearly state the purpose, scope, audience, and cost.

> ☑ **Example**
>
> The VUG is a FREE, monthly newsletter covering the Internet University movement. We serve over 30,000 distance learning professionals and students at the adult, post-secondary levels.

- Include a table of contents.
- Provide attention-getting headlines.

Items to Include

- Include the following types of information:
  - ☐ Announcements
  - ☐ Archives (links to back issues)
  - ☐ Articles
  - ☐ Columns
  - ☐ Community features/discussion area
  - ☐ Contact information
  - ☐ Current issue
  - ☐ Description of the purpose, audience, and content
  - ☐ Download information (if applicable)
  - ☐ Editorials
  - ☐ Employment information
  - ☐ Features
  - ☐ Links to information about the organization (mission, history)
  - ☐ Meeting and conference information
  - ☐ News
  - ☐ Officer reports
  - ☐ Profiles about members
  - ☐ Publication information
  - ☐ Reports and research

- Resource lists (related to the subject)
- Reviews
- RSS feed
- Sample issue if subscription is required
- Sponsors
- Submission/editorial guidelines
- Subscription information for online and print versions
- Table of contents of current stories
- Tips/how-to information
- Update information

Formatting

- If your newsletter is a Web page format, put articles in one column. Using multiple columns (like traditional newsletters) requires that readers scroll up and down to read a story.
- Make the newsletter available in various formats, such as Adobe Acrobat.

E-Mail Newsletter

- Target a specific audience, such as by locale or what they've bought.
- Use the following techniques to provide an overview of contents:
    - List headlines at the beginning of the newsletter. Repeat the headlines with short, concise summaries of stories.
    - List headlines with brief summaries and link to articles.

> ☑ **Example**
>
> **News**
>
> **Stanford To Host Extreme Makeover: Internet Edition**
>
> A team of Stanford researchers is mulling what the Internet should look like if it could be rebuilt from scratch....

- Choose news and articles that emphasize reader benefits.
- Use a skimmable format.
- Link to your Web site.
- Give users the option to subscribe/unsubscribe.
- Provide masthead information (title, date, volume, publisher, copyright, contact information).

▷ *Next* Links

Next links are used to direct readers to the next page in a sequence. By providing links to suggested pages readers should visit, a Web page can provide guidance and direction. The top of a Web page should provide the context, and the end of a Web page should let readers know which page to read next. This technique is especially important for documents with sequential navigation, training/tutorials, and educational or informational Web pages.

Tips

- Determine whether you want to force readers to read sequentially or allow them to choose topics. Sequential information is important when readers must read information in order. Consider your audience and subject matter.

> ☑ **Examples**
>
> *This site on bicycle safety links only to the next principle.*
> Next: Principle #3, Be Visible & Ride Alertly
>
> *This site on pollution links to the next principle and also lets readers view the entire list.*
> Next Principle
> See List of Principles

- In general, place navigation at the bottom of the page.

> ☑ **Example**
>
> **Where can you go from here?**
> - School for Champions
>   - Curriculum Outline
>   - Succeed in Training
>   - Training Your Customers
>
> **Other good stuff:**
> Succeed in Business
> e-Learning, CBT and WBT
> Succeed in Education
> Total Quality Management (TQM)

- Use either text (e.g., *Next*) or arrows (▸).
- For topics that require scrolling, consider placing a *Next* link at both the top and bottom of the information.
- Make it clear where the *Next* link leads.

> ☑ **Examples**
>
> **Now move on to Step 2:** Tune ClearType Settings
>
> ◂ PREVIOUS: Page length | NEXT: Page headers and footers ▸
>
> **Next**: Using "See also" links in Web pages
> **See also**: Chunking for easy Web navigation
> **Previous**: Designing the end links of a Web page
>
> ☒ **Example**
>
> Continued

- Use a sentence recommending where readers should go next.

☑ **Example**

Now that you have a good idea of what should be on a school Web site, it is time to get ready to build your own! Chapter Three takes you,step-by-step through the process of planning your site.

☒ **Example**

There is no link taking readers to the next page.
So how do you even get on the Internet in the first place? Let's take a look at the different ways to connect to the Internet.

- Place the Next link on the right.

☒ **Example**

Next * Previous

▷ Noise

Noise is any element of a Web page that distracts from the main message. Noise confuses readers, clutters the page, and detracts from the content. It also causes information overload.

Tips

- Avoid the following causes of *visual* noise:
  - Techniques such as animation, blinking, useless graphics
  - Overusing devices of emphasis (color, fonts, bold, boxes, lines)
  - Too many links
  - Irrelevant information
- Avoid the following causes of *verbal* noise:
  - Errors in accuracy, grammar and punctuation
  - Wordiness
  - Jargon
- Use the following to keep your design simple:
  - Layering
  - White space

▷ Non-Sexist Language

Non-sexist language is gender-neutral and avoids offending readers.

Tips

- Use the following:
  - You, one, we
  - Plural rather than singular

☑ **Example**

If you use singular, you must use he/she. Using plural eliminates this problem.
Corporate sites have the corporate logo prominently displayed on all pages lest the viewer forgets where he/she is in cyberspace. *(Viewers forget where they are in cyberspace.)*

☒ **Example**

In most text formatting, the author describes how he wants each part of the document to look.

- Do not try to avoid *his/her* by using *their* with a singular noun.

☒ **Example**

The reader can, if they wish, modify the text to try out their own ideas of style and formatting.

- Avoid pronouns by
  - Eliminating them
  - Using titles (manager, employee)

☑ **Example**

Our membership includes professionals, students, and hobbyists; programmers, artists, writers, educators, and entrepreneurs; beginners and experts. The HTML Writers Guild is a non-profit educational corporation with the goal of helping our members improve their skills in the craft of web design through cooperation and sharing of experience.

  - Rewording the sentence

☒ **Before**

A table of contents at the beginning and a site map will help the user find his way around the Web Site.

☑ **After**

A table of contents at the beginning and a site map will help the user navigate the Web Site.

- Avoid generic references to "man" or "mankind" by substituting words such as "human" or "humanity."

▷ Numbered (Ordered) Lists

Numbered lists are ordered lists used when the order of points is important. Numbered lists show priority and sequence.

HTML Code: <OL> </OL>tags mark the beginning/end of the list. Each item is preceded by <LI>.

Tips

- Use numbered lists to format a list in which the *order* is important (e.g., instructions).
- Use bulleted lists to format a list in which the order is not important.
- Use nested lists (lists within lists) to show hierarchy or sub-points.
- Check lists for parallel construction.
- Break up long lists
    - With headings
    - By using sequential pages

> ☑ **Example**
>
> NOW GO TO STEPS 13 to 24
> Instructions steps: | 1 to 12 | 13 to 24 | 25 to 35 |

- Use vertical or horizontal format.

> ☑ **Example**
>
> | Step **1** >> | Step **2** >> | Step **3** >> | Step **4** >> | Step **5** >> |
> |---|---|---|---|---|
> | Find out more about AIESEC and this site... | Apply for an AIESEC exchange... | The selection process... | Find a Traineeship... | Preparing for your experience... |

▷ Numbers

There are rules for writing numbers as words or numerals, although they vary and depend on the type of writing and situation.

Tips

- Use numerals for the following:
    - Decimals
    - Dimensions
    - Hyphenated adjectives: (e.g.,12-volt battery)
    - Large numbers (millions)
    - Line numbers
    - Measurements (when measurements form a modifier)
    - Page numbers
    - Percent
    - Room numbers
    - Table numbers
    - Temperature
    - Weight
    - Years and dates (non-military style)
- In sentences following weights and measures, spell out the noun, such as pounds, inches, meters, etc.
- Do not begin a sentence with a numeral. Either write it as a word or rearrange the sentence.

☒ **Examples**

10MB is optimal.
(A file size of 10 MB is optimal)

270 different communities to come explore, populated by well over 3 million members to meet -just a click away
(the first number is incorrect; the second is correct).

- Write out numbers 1-10 as words; use numerals above ten (the exception is the first rule listed above).

☒ **Examples**

In my opinion, the best 2 HTML editors are HotDog Pro™, and Microsoft FrontPage. *(two)*

If you are not sure which is the best website design solution for your needs, you can follow 4 simple but important steps. *(four)*

- Use numerals with *a.m./p.m.* but not with *o'clock.*
- Write out periods of time.
- Write related numbers in the same sentence as numbers, even if some are ten or less.

☑ **Example**

Scanned and edited 15 photos, 1 map, and created 9 thumbnails.

- For consecutive numbers (e.g., 10 twelve-foot beams), there are two versions of the rule:
  - Write the first as a numeral and the second as a word.
  - Write out whichever word is shorter.
- When possible, write fractions as decimals. Write fractions as numerals when written with a whole number ($4\frac{3}{4}$).
  - Write out when alone (three-fourths: modifier; three fourths: noun).
  - In the text, write the fraction as a decimal preceded by a zero (0.53).
- Because different countries write large numbers differently (using commas or space), identify the country you are from.

▷ Objectivity

Objectivity is presenting information without bias or emotion. It is especially important for news and educational/informational sites.

Tips

- Make it clear when information on your site is expressing a certain point of view.

> ☑ **Examples**
>
> The Ohio Travel Association's mission is to lead the Ohio travel and tourism industry to promote hospitality and economic development.
>
> Guerrilla web site hosting provocative debates on safety issues, with a bias against the air companies and regulators.

- Present sources, evidence, statistics, and examples.

> ☑ **Example**
>
> Ohio Travel and Tourism Statistics

- Present or link to alternative opinions.
- Separate advertising from important content.

> ☑ **Examples**
>
> ▾ advertisement Advertiser Spotlight

- Avoid exaggeration and promotional and marketing language.
- Make it clear if your site has any sponsors or advertisers so readers can determine if there is a conflict of interest.

> ☑ **Examples**
>
> -Presentersuniversity.com brought to you by INFOCUS
> -**SPONSORED LINKS**
> - WirelessAdvisor.com® is a free unbiased service started to help consumers make wise choices among the many cellular, digital PCS and wireless phone companies that serve each area of the U.S. . . . we do not sell wireless services or cell phones.

▷ Organization

Organization is the structure or outline of information. A key step in developing a Web page is identifying information categories. You organize information at all levels: from the content of the entire Web site to the order of points in a bulleted list.

Tips

- Determine the most appropriate organization for the information and your audience:
  - ☐ Alphabetical
  - ☐ Cause and effect
  - ☐ Chronological or reverse chronological
  - ☐ Classification (categories)
  - ☐ Comparison
  - ☐ General to specific
  - ☐ Hierarchical
  - ☐ Least important to most important
  - ☐ Most complex to least complex

- Most important to least important
- Most popular
- Partition (parts)
- Product type
- Sequential
- Spatial
- Specific to general
- Task-oriented
- Topical

> ☑ **Examples**
>
> *The W3C (World Wide Web Consortium) arranges topics alphabetically (left). Microsoft arranges topics in order of most to least important (right).*
>
> **W3C A to Z**
>
> - Accessibility
> - Amaya
> - Annotea
> - CC/PP
> - CSS
> - CSS Validator
> - Device Independence
> - DOM
> - HTML
>
> Products & Related Technologies
>
> Downloads & Trials
>
> Customer & Partner Solutions
>
> Security & Updates
>
> Training & Events

- Put the most important information early in your Web page. Because readers are busy and often browsing, their first few minutes at your site will determine if they stay or return.
- Let readers know how information is organized through:
  - Explanations

> ☑ **Examples**
>
> Public Notices that are active are listed below in reverse chronological order (the most recent are listed first).
>
> Welcome to my guide to HTML and beyond. Regular visitors may want to check out the recent updates page. An alphabetical index is now also available.
>
> The Topics are arranged in order beginning with "General" topics, covering motorcycles in general, and then proceeding to topics about individual models. These are arranged in order by date with the newest appearing at the top of the list. Topics about the same model of the same date are placed in alphabetical order.
>
> **1 How this Document is Organized**
> Section 2 of this document reproduces the guidelines and checkpoints of the "Web Content Accessibility Guidelines 1.0" [WCAG10]. Each guideline includes:
> The guideline number.
> The statement of the guideline.
> A list of checkpoint definitions. Checkpoints are ordered according to their priority, e.g., Priority 1 before Priority 2.
>
> *This site contains a link,* Site Organization.
> **Site Organization**
> There are three main technical categories of information on this site . . .

  - Visual clues

☑ **Example**

This site's "How This Site is Organized" page shows the logic behind the navigational structure. Information is arranged with introductory information first. Site help is last. The grouping is conveyed by differences in shading.

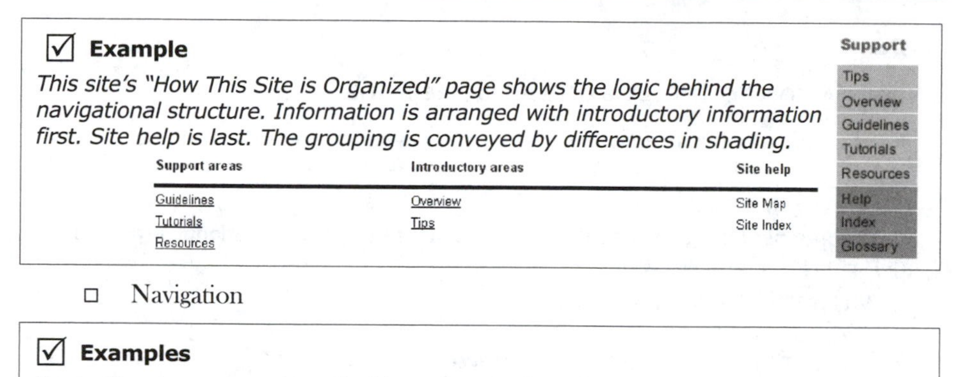

- ☐ Navigation

☑ **Examples**

Navigation menus can show the hierarchy of information. The following expandable and collapsible menu lets readers see how information is organized.

SPBs Website
- Home Page
- Jobs Area
  - Getting Started
    - How to Get A State Job
    - Frequently Asked Questions
    - General Information
    - IT Careers
    - Students/Interns
    - PT/Seasonal
    - Transfer/Reinstate

Department of Development
- Building & Zoning
- Community Dev.
- Economic Dev.
- Planning

- ☐ Site map and site guide

- ▶ Select an organization that is easy to update.
- ▶ Provide readers with a variety of organizational methods:

☑ **Examples**

Sort By: Position | Company | Posting Date | Location | Job Type |

Usable Web provides multiple organizational schemes (by date, topic, destination, author, popularity).

Search: [ ] Search

Indexes: Topics | Destinations | Authors | Site Index

Home
Popular
Books
About

▷ Organization Page

An informational page provides factual information and details about an organization. The URL usually ends with *.org*. Creating a Web page for an organization allows you to

- ▶ Establish a presence.
- ▶ Focus on the purpose, mission, and history.
- ▶ Promote your group.
- ▶ Answer reader questions.

Tips

Information to Include on the Home Page

- Provide a tagline that clearly describes who you are and what you do.

> ☑ **Examples**
>
> The Internet Society is a non-profit, non-governmental, international, professional membership organization.
>
> WILLIAMS-SONOMA *a place for cooks*

- Answer the questions *who, what, where, when,* and *how.*
- Make the important information easy to find on the home page.

Links to Provide

- Provide a high-level overview of the organization and link to more in-depth information.

> ☑ **Example**
>
> The World Wide Web Consortium (W3C) develops interoperable technologies (specifications, guidelines, software, and tools) to lead the Web to its full potential. W3C is a forum for information, commerce, communication, and collective understanding. On this page, you'll find W3C news, links to W3C technologies and ways to get involved. New visitors can find help in Finding Your Way at W3C. We encourage organizations to learn more about W3C and about W3C Membership.

- Provide links to other information that will be of interest to readers.

Items to Include

- Include the following types of information:
  - ☐ About us/background
  - ☐ Address
  - ☐ Awards
  - ☐ Calendar
  - ☐ Chapters/societies/divisions
  - ☐ Conferences
  - ☐ Constitution/bylaws
  - ☐ Contact information
  - ☐ Dates/currency of information
  - ☐ Description of organization
  - ☐ Directions/map
  - ☐ Donation information (if applicable)
  - ☐ E-mail list information
  - ☐ Employment information
  - ☐ Fact sheet
  - ☐ FAQs
  - ☐ Financial information
  - ☐ Governance
  - ☐ Grants
  - ☐ History
  - ☐ Hours of operation
  - ☐ Member services
  - ☐ Membership information
  - ☐ Mission statement and goal
  - ☐ News and announcements
  - ☐ Newsletter
  - ☐ Philosophy
  - ☐ Phone numbers
  - ☐ Press information
  - ☐ Policies

- ☐ Product information
- ☐ Professional articles
- ☐ Professional development
- ☐ Publications
- ☐ Request for information section
- ☐ Sales section for products
- ☐ Services
- ☐ Special interest groups
- ☐ Staff information/organization
- ☐ Student groups and activities
- ☐ Support information
- ☐ Volunteer information
- ☐ What's new/announcements

Formatting

- Use formatting techniques such as tables, headings, bold, and lists to make the information easy to read.
- Provide a variety of ways to access the information.

☑ **Example**

The National Science Foundation site contains a wealth of information that is easily accessible through shortcut menus, a Site Map, search engine, A-Z Index, navigation bar, and NSF at a Glance menu.

▷ Page Length

The size of the page displayed in the browser is determined by space required to display the text and images on your page. If your document fits within one screen, no scroll bars will appear at the side or bottom of your browser window. The *vertical length* of your Web page is determined by

- The amount of text and graphics in the file.
- Screen resolution.
- The browser and the display options selected.

Tips

Deciding Between Long and Short Documents

Should you create several short documents that require more navigation, or create longer documents that require scrolling? Each option has pros and cons.

Short Pages

PROS	CONS
▸ Fit on one screen. ▸ Require less scrolling. ▸ Are modular; easier to provide updated information. ▸ Make it easier to find information quickly. ▸ Provide simpler pages that are less dense with text.	▸ May require more loading time overall (but load more quickly than long pages). ▸ Are more difficult for author to keep track of topics. ▸ Require more navigation, so readers may become confused.

Long Pages

PROS	CONS
▸ Are easier to print. ▸ Are easier to scan the topic all at once. ▸ Help reader concentrate on one concept. ▸ Support visual and conceptual continuity.	▸ Take more time to read. ▸ Make it more difficult to find information. ▸ Require scrolling. ▸ Have a more rigid structure.

Tips

- Use the following guidelines:

TO DO THIS	USE THIS PAGE LENGTH
Attract attention	One window
Present short discrete topics	One window or less
Maintain pages easier	Short pages
Provide quicker loading pages	Short pages
Present simpler pages with less information overload	Short pages
Present text user must read	One document Short pages with sequential navigation
Let user print or save	One document
Show contiguous topics and unified material	One document
Require less clicking and retrieval time overall	One document
Allow faster skimming	One document

- Make some pages short and some long, depending on their function:
    - Pages or sections at the top level should be short.
    - Navigation pages and pages lower in the hierarchy can be longer sections.
- In determining the length of a Web document, consider the bandwidth and types of browsers your readers will use.

- Be sure that each Web page has the following characteristics:
  - Contains something worthwhile.
  - Is on one topic (chunking). Don't split topics across pages.
  - Is, on average, one to three screens long.
  - Places important information above the scroll line (also called "above the fold"—a newspaper term.
- Consider making the document available two ways: short separate documents and one long document for printing. This option allows readers to obtain chunked text or text on one topic in one document. However, using Cascading Style sheets can help you avoid the need to provide "printable page" links.
- For text that should remain together (e.g., an essay or article) or a document that users will want to download and print, create one large file.

Creating Long Documents

- For long documents, use internal links.
  - Provide a table of contents or a list of topics at the beginning.
  - Provide links such as *Back to Top* or *Return*.
  - If appropriate, use chapter and section headings in long documents with several parts.
- Put important information on the visible portion of the screen.
  - To keep your text within a screen area without requiring the user to scroll, keep it within a *600 pixel wide by 250 pixel high* area. This guideline accommodates various monitors, resolutions, platforms, and browsers.
  - Avoid placing key navigational aids and small amounts of text hidden below the visible portion of the screen. Readers may never see them.
  - If necessary, use visual cues or break text to encourage readers to continue below the edge of the screen.
  - Repeat navigational aids at both the bottom and top of the document.

▷ Page Numbers

Using page numbers on a Web page is similar to using them in print documents. Page numbers are often used to help readers navigate documents organized in a sequence, such as articles, handbooks, manuals, or online books. They are used because page numbers are familiar to readers (book metaphor). However, page numbers have less meaning on Web pages because Web pages can be any page length, may be read in any sequence, and are separate HTML documents.

Tips

- Break long articles into shorter segments. However, also provide the ability to view or print the article as one long document.

☑ **Example**

Printer Friendly Version

- Provide navigation forward, backward, and to any page.
- Disable the link for the current page number.

☑ **Example**

Previous 1 2 3 4 5 6 7 8 9 10 11 12 13 14 Next

- Always let readers know what page they are on and how many total are in the series (contextual clues).

☑ **Examples**

Currently viewing page 1 of 9

page 1 of 4   1 | 2 | 3 | 4   Next > >

You are here: OPM Home > Insurance > FEGLI > Handbook > Eligibility Page 1 of 2

☒ **Example**

next >>

- For numerous pages, group links.

☑ **Example**

Page: 1 **[2]** 3 4 5   6-10 >>

- If pages are long, place page numbers at both the top and bottom of the page.
- In general, avoid page numbers. Instead, use a text link that names a Web page title or that links to a specific location on that Web page.

☑ **Example**

Next: 5 Creating a Source

▷ Page Size

The width of a Web page is important because it affects whether you require readers to scroll horizontally. The phrase "graphic safe area" is used to describe page dimensions that you can be sure display and print correctly. The dimensions for this "safe area" are determined by two factors:

- The average monitor size and screen resolution. Monitors are now usually 17 inches or more and at least 800 by 600 pixels.
- The width of standard paper used to print Web pages.

Pages in a site will vary in length. The page length is determined by the amount of text and graphics on the page. Page size is important because it affects what information readers see when they load your page. The important information should be in the visible portion, or focal point, of the screen.

Tips

- Develop guidelines for page size to accommodate small monitors, as well as navigational bars at the top and sides. The average screen resolution worldwide is now 800 by 600. The following are guidelines for 640 x 480 resolution.

Page Designed For	Max. width	Max. length
Maximum screen use	595 pixels	295 pixels
Printing	535 pixels	295 pixels

- Never require that readers scroll horizontally.
- Use a page grid to standardize page dimensions.
- Keep the line length short.
- Use Cascading Style Sheets.
- Use tables to control width. Two options for sizing tables are
  - Fixed width (fixed pixel width)
  - Variable width (percent of screen)

▷ Paragraphs

In Web documents, a paragraph (or group of several short paragraphs) covers a specific type of information. Web paragraphs should be short and focused. Web readers tend to scan documents, looking for a few sentences that contain the information they want. Text in short paragraphs is easier to read and skim quickly and adds more white space.

HTML Code: Paragraphs use <P> </P> tags. You cannot control the leading of text (space between lines). However, the paragraph tag is a method of providing white space between sections. Use extra paragraph space (line break <BR>) between paragraphs.

Tips

Writing Paragraphs

- Put one type of information per paragraph.

TYPE OF INFORMATION	WHAT IT PROVIDES
Introduction	Overview
Definition	Explanation of terminology
Evidence	Specifics
Examples	Concrete illustrations
Fact	Statements, data, statistics
Principle	Guidelines on how to handle situations
Procedure	How to do something
Process	How something works
Reference	Supplementary information
Summary	Review

- Discuss only one topic per paragraph.

- Use strong topic sentences. The most significant part of the paragraph is the first sentence.

> ☑ **Example**
>
> **Internet connection**: Internet connections come in two forms: direct and dialup. Direct connections are always on and require a physical wire connected to the Internet topology. Dialup connections are established only when the user connects to the Internet with a modem or terminal adapter over phone lines.

- Put the main point first (use the inverted pyramid method). Some readers read only the first sentence.
- Provide coherence within paragraphs by using the following:
  - Transitions.
  - Repeated keywords and phrases.
  - Synonyms.
  - Pronouns and words such as *this, these, those.*

> ☒ **Example**
>
> *The following is too chunked, with no relationship among ideas:*
> Putting together a web site is a unique blend of publishing, user interface design, and technology.
> The three main activities of visiting a web site are reading text, viewing images, and interacting with its interface.
> Web publishing is not an opportunity to show off your technical prowess. Use the technical aspect to support and enhance, but don't let it overpower the other aspects of your work.
> Web publishing is not an opportunity to show off your graphic arts skills. Use the graphical aspect to support and enhance, but don't let it overpower the other aspects of your work.

- Use occasional one-sentence paragraphs for emphasis.

> ☑ **Example**
>
> *This excerpt uses a number of techniques that make the information easy to scan quickly. An overview sentence is separated with bold and centering. It summarizes the point of the next two paragraphs. The next sentence, which is also separated, expands on the overview. The two paragraphs are short and indented. The first sentence of the first paragraph provides a transition and repeats the keyword "interactivity." The first sentence of the second paragraph provides a transition and contains the keyword "content."*
>
> **The second most important trait a Web site should have is interactivity.**
>
> **Be interactive**; good interactivity engages the user and makes your site memorable.
> After original content, the second most important trait a Web site should have is interactivity. The Web is an interactive hypermedia communications medium that your Web site should reflect. Sites that involve the user and have a sense of fun or adventure will get more hits and can charge more for ad space.

Another advantage of interactivity is self-generating content. If your visitors interact with your site, they actually create content for you. Script-driven user surveys and forums allow visitors to share information with others and can help shape your site to better serve their needs. Forum or chat software is a great way to do this.

Formatting Paragraphs

- Keep paragraphs short: 4-8 lines long.
- Use extra paragraph tags to create more vertical space between paragraphs and sections of text. Some sites separate each sentence with space.
- Use headings to label the type of information the paragraph contains (labeling).

☑ **Example**

Another method of chunking is categorizing links as most useful/important to read, less useful or important.
Principle All information should be presented in small digestible units.
Digestible unit defined A digestible unit of information contains no more than nine separate items of information.
Rationale Research suggests that human beings can understand and remember no more than seven plus or minus two items of information at a time. This phenomenon is called the "chunking limit". Further, as the complexity of the information increases the chunking limit decreases.

- Ask yourself if a paragraph is the best format for the information. Use formatting techniques such as lists, tables, and bold to help readers scan the information more quickly.

☒ **Before**

A paragraph is not the best form for this information:
The Style Guide is organized by three views to accommodate the different preferences for accessing information. This page is the Index View. To use it, find the topic you are looking for in the index list at left. (You may click the alphabet letters at the top to jump to the categories alphabetically, or simply scroll down.) The topic you select will then appear in this window. Continue using the index on the left to select new topics. The other two views are Book View, for those who prefer to go through the style guide page-by-page; and Frequently Asked Question (FAQ) View, for those who seek answers to a specific question.

☑ **After**

You can view the information in this Style Guide in three ways:
Index View, **Book View**, and **Frequently Asked Questions (FAQ) View**.

Index View (this page) lets you find topics alphabetically.
To view the topic in the main window.....Click an index entry on the left.
To jump to categories...........................Use the alphabet at the top.

Book View lets you browse page-by-page.
FAQ View provides questions and answers.

▷ Parallel Construction

Parallelism is using similar grammatical structure for words, phrases, and clauses at the same level of importance. Using parallelism helps you

- Be concise.
- Clarify the meaning and relationships of key elements.
- List multiple ideas with symmetry.
- Help readers to identify, compare, and remember the listed elements.

Tips

- Check the following for parallelism:
    - ☐ Words and phrases in lists or a series

☒ **Before**

Discusses all aspects of Web page design, including content selection, organization, and advertising.

☑ **After**

Discusses all aspects of Web page design, including selecting content, organizing, and advertising.

- ☐ Bulleted lists

☑ **Example**

- Create a consistent look and feel for the web.
- Separate information into manageable page-sized chunks.
- Provide cues for the reader about the web's information structure and contents, context, and navigation.
- Use links to connect pages along the routes of use and user thinking.

☒ **Example**

- A full service company
- Places your interest first
- Known for fairness and integrity
- Reasonable fees without hidden charges
- Experienced in web development and design
- Knowledgeable about e-commerce and Internet law

- ☐ Numbered lists
- ☐ Headings
- ☐ Lists of links

☑ **Example**

<u>What we provide</u>
<u>Why we exist</u>
<u>Where we are going</u>
<u>How you can help us get there</u>
<u>Who we are</u>

☒ **Example**

Find
Air Tickets
Auto Price Quotes
Find a Job
Get an Apartment
Maps & Directions
Old Friends

▷ Parentheses

Parentheses enclose words, phrases, and sentences that are not essential for the text's meaning.

Tips

- Avoid parentheses because they are difficult to see online and interrupt reading.
- Use parentheses to enclose digressions and elaboration.

☑ **Examples**

-Use this box to use a picture (JPEG or GIF) as your background.
-That's all we need to build HTML files (i.e., Web pages).
-The title (as we mentioned earlier) is surrounded by the <title> and </title> commands.

- Use parentheses sparingly for definitions of acronyms or for brief definitions.

☑ **Example**

HTML (HyperText Markup Language) is fundamentally different from most text formatting.

▷ Patterned Information

An information pattern is a method of using the same wording, the same information, and a similar amount of detail on each page. Pages within a Web site should contain similar information patterns for consistency. This technique is especially important for reference information, product specifications, course or book details, and resource guides.

Tips

- Present similar information in a similar manner.

☑ **Examples**

A troubleshooting section would include standard information:
Model Affected: Symptoms: Solution:
A review of search engines might include
When to use Strengths Weaknesses
This site uses the same links to discuss tornadoes and other storms:
Introduction The Effect The Phenonmena The Science

- Use similar format (headings, subheadings, lists, devices of emphasis) and wording.

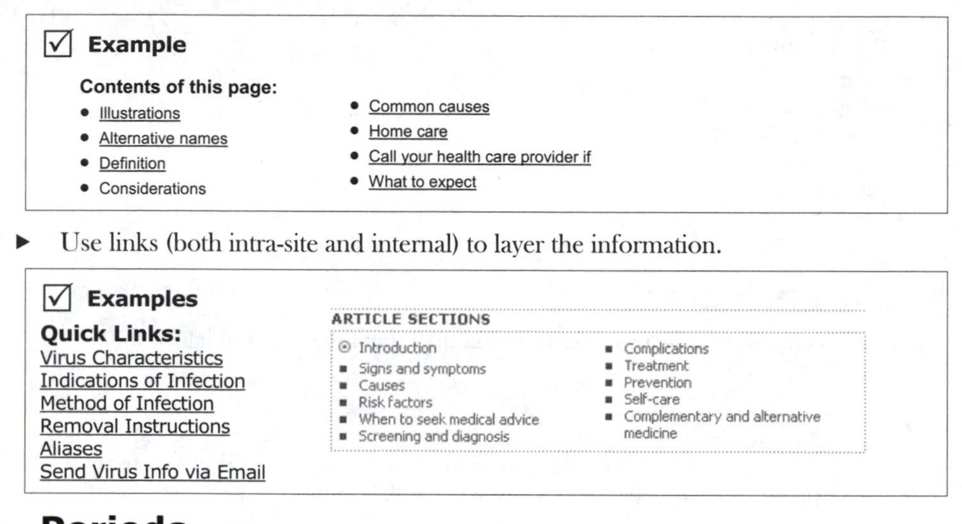

Example
Contents of this page:
Illustrations
Alternative names
Definition
Considerations
Common causes
Home care
Call your health care provider if
What to expect
Examples
Quick Links:
Virus Characteristics
Indications of Infection
Method of Infection
Removal Instructions
Aliases
Send Virus Info via Email
ARTICLE SECTIONS
Introduction
Signs and symptoms
Causes
Risk factors
When to seek medical advice
Screening and diagnosis
Complications
Treatment
Prevention
Self-care
Complementary and alternative medicine

- Use links (both intra-site and internal) to layer the information.

▷ Periods

Periods are crucial in writing Web addresses and filenames.

Tips

- Use a period at the end of a sentence.
- Put periods inside quote marks.
- In general, omit periods after letters in acronyms.
- Use a period at the end of list items to help readers using screen readers.
- If a Web address (URL) is at the end of a sentence, include the period.
- Avoid periods after computer code or user input because readers may misunderstand exactly what to type. Instead, rearrange the sentence.

☒ **Before**

The name of the file should end with `.html`.
Tags are given between the brackets < and >.

☑ **After**

The filename should end with an .html extension.
The brackets < and > enclose the tags.

▷ Personal Page

A personal home page (PHP) is created by an individual who may or not be associated with an institution. Many people publish a personal home page to share a hobby or interest, express an opinion, display art or other talents, present their credentials, or simply for fun and to learn how to create a Web page. Personal pages often have a tilde (~) in the URL (address).

Tips

- Include the following types of information:
  - Annotated lists of links on a specific subject. However, avoid random lists of your favorite Web sites.
  - Articles/essays
  - Biography
  - Blog
  - Clear statement of any biases
  - Currency of information
  - Hobbies and interests
  - Resume and your qualifications
  - Sources of factual information
- Try to include worthwhile content and think about reader benefits. Simply writing about yourself will not be interesting for readers (unless you are a celebrity!).

> ☑ **Example**
>
> Tim Berners-Lee (Internet founder) provides a short and long biography, e-mail information, talks and articles, press interviews, a FAQ, etc.
>
> ☒ **Examples**
>
> -This is my site, about me, my family, what I like, dislike etc. I am building it as an expression of self, I hope you like it but cannot guarantee that you will!
> -I have designed this site as a showcase of my Web Design ability.

- Remember to protect your own privacy.

▷ Personal Tone

A personal tone is friendly and reader-oriented. Many Web pages tend to be more personal, direct, and conversational than most printed documents. In addition, many readers like Web sites with personality rather than cold, impersonal sites. Readers also like to feel that they can

- Provide feedback and communicate with others.
- Respond to the site individually.
- Tailor the look and feel of the site to their tastes.

Tips

- Analyze your audience and solicit feedback about your site so you understand your audience and their interests.
- Use a personal tone if it is appropriate for the audience.

☑ **Example**

Newbie.net is designed for readers new to the Internet, so it conveys a helpful and informal tone.

If you're looking to learn how to get the most out of the *real* Internet—oh yeah, there's ****lots**** more to the 'Net than just web browsing!—then you've found the right place to start. Our mission is to help newbies become knowbies.

- Use informal writing if it is appropriate for the content and purpose.

☑ **Example**

This is Primer #1 in a series of seven that will calmly introduce you to the very basics of HyperText Mark-up Language. I suggest you take the Primers one at a time over seven days. By the end of the week, you'll easily know enough to create your own HTML home page. No really. You will.

- Avoid the following:
  - A technical tone and jargon.
  - Promotional and sales language.

☒ **Examples**

This excerpt comes from a Web design service. The tone may sound too technical and formal for most customers.

The content at this website provides description of services and related information pertaining to Web development. Interactive pages are located within the various areas supporting inquiry to services and information.

- Use personal pronouns such as *I* and *you.*

☑ **Example**

THIS SITE IS FOR YOU the purchaser of speech technology products. It is also for developers of software using speech products as part of the design. You will find reviews of both retail level products and speech products for programmers. You can also record product bugs you find and wish list features.

- Use a conversational style.
- Define terms and acronyms.
- Use examples readers can relate to.

> ☑ **Example**
>
> *This tutorial from the WebReference site uses personal pronouns, a helpful tone, definitions, and examples.*
>
> In order to be able to write good HTML, you must first understand exactly what HTML is. The most obvious place to look for this information is in its name: HTML stands for ***H**yper**T**ext **M**arkup **L**anguage*. If this doesn't make much sense to you, don't worry, because that's what I'm here to explain. Let's take those initials apart one by one, starting from the right: HTML is a **language**, but not the kind you tell your kids to watch. It's a computer language, and as such it has some specific rules that must be followed. In other words, it has a defined **syntax**, a strict way in which it must be written, and, when the time comes, read. But we'll get to that later.

- Use active verbs, which emphasize tasks and benefits.

> ☑ **Example**
>
> We have compiled the best resources to help you quit smoking, lose weight, eat better, get in shape, and reduce stress.

- Use the following methods of personalizing your Web page:
  - E-mail address
  - Personal pictures (if appropriate)
  - A persona

> ☑ **Example**
>
> **Ask Auntie Nolo**: Auntie Nolo is a plainspoken, mostly patient soul who can help you unravel your knottiest legal problems.

- Solicit feedback and provide surveys and other methods of interactivity.

▷ Platform-Independent Terminology

Browser and platform-specific terminology includes references to browsers, menu items, buttons, and actions specific to a type of hardware or software. Because people access the Web in different ways, you should avoid platform-dependent phrases. A good Web site is accessible to all audiences:

- Browsers: graphical and non graphical
- Operating systems (Windows, Mac, UNIX, Linux)
- Devices: PDAs (Personal Digital Assistants), WAP (Wireless Application Protocol)-enabled phones

Many Web sites now include statements such as the following:
"This site's design is only visible in a graphical browser that supports Web standards, but its content is accessible to any browser or Internet device." (See the Web Standards Project at http:// webstandards.org.)

Tips

- Avoid referring to features of specific browsers. Readers are confused when there are references to functions they cannot see. If you do require certain features, be careful not to use an insulting or sarcastic tone.

> ☒ **Examples**
> -Best viewed with . . .
> -This site requires JavaScript to be enabled to be fully functional.
> -You will need .... plug-in to view this presentation.
> -Open your browser window to here →
> -If your display resolution is set to 640x480 pixels, you may find it easier to view this section without the side-menu.
> -If you don't see animated graphics above, or hear music, you need to upgrade to a better web browser, like Netscape.
> -Best viewed full screen at 800x600.
> -If This Page Seems To Be Not Loading Correctly, It Is Probably Because You Are Not Using Netscape.
> -This site utilizes Macromedia Flash and some other cool tools. If you are browser challenged, you will not be able to experience the full site.

- Do not assume that all readers
  - Are using a specific type of computer or operating system.
  - Have graphical browsers.
  - Are viewing graphics.
  - Are using a mouse. Avoid phrases like *click here*.
  - Are reading the page. Users can listen to Web sites.
  - Have a certain screen resolution, color depth, font size, or window size.
  - Have accepted the default colors and other options in their browsers.
  - Have the appropriate plug-ins loaded for multimedia and other special effects. (*Plug-ins* are components that allow browsers to execute files embedded in a Web page.)

> ☒ **Examples**
> Site best viewed in FireFox - download FireFox + Google toolbar for better browsing!
>
> PLEASE NOTE: This is a computationally *demanding* site. Fast connections and up-to-date software and hardware are critical to seeing it at its best.
> This site uses some advanced techniques like Cascading Style Sheets, Java, and JavaScript.

- Use advanced features sparingly, or make a simpler version of your site available.
- Use standards for both code (HTML, XHTML, Cascading Style Sheets) and accessibility.
- On the home page, specify any requirements for viewing your site.

> ☑ **Example**
>
> This site uses fully compliant Cascading Style Sheets (CSS). Older browsers should display text in their default fonts, while more recent browsers will all display fully formatted text. (However, the styles sheets will look best viewed in Internet Explorer 4.0 or above.) The site also complies with major accessibility standards.

- Use validation tools and test your site.

▷ Position/Placement

Position shows the organization and structure of your Web page. Placement shows levels of importance. It can also affect the direction that the eye moves to view page elements. Key positions are

- Center
- Upper left
- Above the scroll line

Tips

- Place information where it will be most useful for your readers. Ask yourself what you want readers to do.
- Direct the eye from top to bottom or left to right. (However, remember that Western readers read from left to right and top to bottom.)
- Position elements to minimize eye movement.
- Position elements to show
  - Hierarchy
  - Related items (proximity)
- Place important information in the focal point of the screen. Do not require that readers scroll to view it.
- Use consistency in placing items.
- Use a grid to control position of elements.
- Use standard Web conventions to conform to readers' expectations about where items are located:
  - Navigation: left, top/bottom
  - Logo: top left or center
  - Header and footer

▷ Positive Language

Positive language avoids negative words and phrases. Positive statements are easier to read and understand quickly and are less wordy. Negative words, especially two in a row, require more effort from the reader to understand, causing information overload.

Tips

- Avoid words such as the following:
  - can't
  - do not/don't
  - never
  - no
  - none
  - not
  - not different
  - not many
  - not the same
  - not unlike
  - not unless
  - not until
  - nothing
  - unable to
  - won't
- If possible, use positive rather than negative statements. Tell what to do or what can be done.

Examples

-This is not to say there aren't useful Web sources.
-When it comes to designing your website there is very little that cannot be achieved on the Internet nowadays.
-With the Internet's tremendous growth and reasonably inexpensiveness, there is no reason why your small business or organization should not be online.
-We are not the typical web design house that stops working for you after the initial design of your website.
-Don't have long Web pages.
-Consider your background color. The first step is not to use the standard grey one.
-Be careful not to confuse tags with elements.
-GIF images should be no bigger than 20K.
-Never forget to include target=_top on all outgoing links when using frames.
-Don't have extremely long pages, especially the home page.
-A web site is of no value until customers and prospects see it!

- Begin your introduction with positive statements.

Example

A Web site begins with this statement:

If you've been here before, I'm sure that you are at least slightly taken aback by the new site, but don't fear, we're still the same.

- Use positive wording in tips, instructions, and procedures because clarity is crucial.

Example

If you don't want anything to get into that menu in the first place...
... select Open Location from the File menu and type there the address you want to visit, instead of typing it directly into the Location field.

Preformatted Text

Because HTML gives you little control over text spacing, preformatted text allows you to bring in text with exact formatting. The text will be monospaced (fixed-pitch) font.

Preformatted text formats the font and spacing exactly as you typed it. It also saves time required to reformat the text. You can use it for text such as computer code, poetry, etc.

HTML Code: <PRE> </PRE>

Tips

- Use preformatted text when you want to maintain the format of the original, especially tabs, spaces, and character returns.

▷ Press Release

A press release conveys company announcements to the media and the public. It announces a product, rollout, service, business venture, personnel change, etc. The audience for most press releases is journalists who usually have tight deadlines and need to find information quickly.

Tips

Using Links

- Provide a clearly visible and labeled link (such as *Press Releases, For the Media, Media Information,* or *Journalists*) on your home page.
- List the titles and dates with links to the articles.

> ☑ **Example**
>
> Friday, August 23, 2002
> Microsoft and Access Softek Align to Simplify Financial Data Transfer

- Arrange releases in reverse chronological order. Allow readers to change the sorting order.
- Make navigation flexible and easy to use.
- Provide links to supplementary material, such as company facts, financial information, other companies mentioned, statistics, media FAQ, previous press releases, reviews, external press coverage, glossary, images that can be used for stories, PR contacts, and related topics.

> ☑ **Examples**
>
> **Note to editors**: If you are interested in viewing additional information on Microsoft, please visit the Microsoft Web page at http://www.microsoft.com/presspass/ on Microsoft's corporate information pages. Web links, telephone numbers and titles were correct at time of publication, but may since have changed. For additional assistance, journalists and analysts may contact Microsoft's Rapid Response Team or other appropriate contacts listed at http://www.microsoft.com/presspass/contactpr.asp.
>
> *The American Cancer Society News Room page provides the following links:*
> Media Contacts, Media FAQ, Radio News Service, News Archives, Rumors, Myths, and Truths, News Releases, ACS News Today

- Make archives available.
- Provide a search engine with advanced search options.

WRITING THE PRESS RELEASE

- Give a concise headline that summarizes the information.
- Match the headline with the link text.
- Begin with a summary of the author, organization, source, date, time, location, and headline.
- Explain *who, what, where, when, why* (inverted pyramid).
- Put key information in the first paragraph.
- Provide PR and company contact information: name, address, phone number, e-mail (*mailto:*), and Web site address.
- Omit marketing language.
- Keep writing clear and concise.

FORMATTING TO CONSIDER

- Minimize download time by keeping pages simple.
- Make information easy to find and skim.
- Make dates stand out.
- Put contact information on every page.
- Provide other ways to view the information, including an RSS feed.

☑ **Examples**

E-Mail Story Print Story Bookmark Page

XML Subscribe to PR RSS

▷ Print vs. Web

Effective Web writing and design should take into consideration the differences between reading printed and online documents.

Tips

- In designing your Web page, consider the following differences:
    - Screens are landscape orientation and about 1/3 of a normal page size.
    - Monitors vary in size and in the number of colors they display. They shine light in readers' eyes, unlike print documents that use reflected light.
    - Web design is not WYSIWYG (What You See Is What You Get). A Web page will look different on every monitor and browser.
    - Fonts appear different sizes in different resolutions.
    - Text and graphics are grainy—about 50 to 100 dots per inch.
    - Readers must scroll or click to navigate.
    - Readers cannot carry the documents with them or annotate them without printing them out.
    - Print readers start at the upper left; on Web pages, readers see the entire screen as a whole.

- Readers want to scan Web pages quickly.
- Web pages load slowly.
- Web documents can use hyperlinks, color, animation, interactivity, and multimedia.

▷ Printing

A "printer-friendly" version of a Web page allows readers to print information. Several factors affect whether you should offer a printable version of your pages:

- The content of your pages. For example, this feature is especially useful for articles and stories that appear on several different pages.
- Whether readers will read information online or download and save the pages.
- Bandwidth restrictions.

Tips

- Use Cascading Style Sheets, which help support printing from the Web.
- Provide a link to a printer-friendly version of your Web pages.

> ☑ **Example**
>
> Print this page | Print this chapter

- Make the wording of the link clear, such as *Printer-Friendly Version, Printable Version, Print This Page.*

> ☑ **Example**
>
> PRINTER FRIENDLY

- Consider your page width and length to optimize for printing (535 pixels wide by 295 pixels long).
- Make the printable version simple: white background with black text.
- Combine all pages and sections (chunks) into one page.
- Do not use frames.
- Use no images or include only essential images. Make sure they are sized to print correctly.
- Write out the full Web addresses (URLs) for links. Also write out the full URL and article title/author/date for the page itself.
- Consider using a script to automate printing.

▷ Privacy/Privacy Policy

A privacy policy is a document that explains how your company uses the information it collects. On a business site, you may want to request private information from customers in the form of registration and online orders. Readers want to be assured that financial transactions and other communications are secure.

Tips

Protecting Your Own Privacy

- To protect your own privacy, be careful about putting private information (home address, home phone number, social security number, etc.) on a Web page or online resume.
- Be careful about posting specific dates, activities or other information on your Web site that reveals private details about your family.
- Provide contact information, such as e-mail and business addresses/telephone numbers.

Protecting Readers' privacy

- Develop a privacy policy for your Web site. It should state what information you collect from users and how you collect it. You may also include information about the security of transactions (often called a security policy).
- Put a link to your privacy policy on the home page and wherever you collect information from readers. Many sites put the link in the footer.

> ☑ **Example**
>
> If you're worried about what we're going to do with your e-mail address, review our Privacy Policy.

- Explain the following types of information:
  - ☐ *Collecting* and *using* personal information.

> ☑ **Example**
>
> Please sign in or click on REGISTER! below. WHY are you collecting this???

  - ☐ Collecting information from children.
  - ☐ Sharing and disclosing information.
  - ☐ Keeping information secure and confidential.
  - ☐ Using cookies.

- Make your privacy policy easy to read and understand.
- Update your policy regularly.

▷ Procedures

Procedures are step-by-step instructions that help readers understand a task or policy. They appear on the Web as a means of providing technical support and of making policies/procedures widely available to employees and customers.

Tips

Organizing Procedures

- Use the following organization:
    - Introduction
        - Purpose of document
        - Date
        - Revision history
        - Overview of document organization
        - Contact information
    - Procedural overview
        - Statement of purpose/objectives of the procedure
        - Difficulty level
        - Risk factor
    - Required equipment
    - Time to perform
    - Preparation/warnings
    - Conceptual information (the context in which the procedure fits, prerequisites, why the procedure is done, options, links to relevant documents).
    - Synopsis (advance organizer): list of all steps, a summary of each, and a link to details for each step.
    - Steps
        - Action
        - What step accomplishes
        - Picture (link to large graphics)
        - Results: how user can tell if procedure succeeded or failed
    - Troubleshooting

Chunking Procedures

- Find a logical way to chunk the procedures.
- Classify policies by type (service, warranty, etc.) and arrange in a logical order that can be quickly scanned.

☑ **Example**

Yale University's Division of Finance arranges Policies and Procedures alphabetically by category (Finance, Faculty, Facilities, Human Resources, ITS, Other, Procurement, University Properties) and alphabetically within each category.

Faculty

Faculty Appointments

Faculty Handbook

Faculty Related Policies

☒ **Examples**

The following is an excerpt from the table of contents for a library's circulation policies. The organization is not obvious.

Who May Use the Library
How to Obtain a Borrower's Card
Responsibilities of Library Patrons
Lending Periods
Audio Visual Material Lending Periods
Borrowing Limits Borrowing Journals
Recalls
Blocking Borrowing Privileges:
General Policy, Overdue Materials, Non-University Patrons,
Faculty Reserve Materials, Interlibrary Loan Materials

Using Links and Navigational Aids

- Provide a table of contents and an index. For several procedures, consider creating a main menu of procedure types. Then link to menus for each type.

☑ **Example**

TABLE OF CONTENTS
Introduction
What Is Arbitration?
What Disputes Are Eligible For Arbitration?
Who Are The Arbitrators?
Can I Be Represented By An Attorney?

- Break up long procedures (7-10 steps) into subtopics (chunking). Several ways to do this are as follows:
  - Put all steps on one page. Use a sub-level table of contents and internal links to short sections on the same page.
  - Outline the procedure on the first page. Make each step in the outline a link to a page describing just that step. Use *Next* and *Previous* links to connect the pages for each step.
  - Outline the entire procedure on every page but expand one step per page. For example, on the first page, outline the whole procedure but expand the first step to include the details about only that step. On the second page, outline the whole procedure but expand the second step.
  - Provide two versions: a summary version and a detailed version.

> ☑ **Example**
>
> *This library procedure is available in both a Quick List version and a detailed version.*
>
> Quick List
> Section B: Preliminary Procedures

- ☐ Use a frame, with the list of procedures on the left and steps on the right.

> ☑ **Example**
>
> System Case Preparation Procedure
> Floppy Disk Drive Physical Installation Procedure
> Hard Disk Drive Physical Installation Procedure
> CD-ROM Drive Physical Installation Procedure
> Processor Physical Installation Procedure
> Heat Sink Physical Installation Procedure
>
> [The PC Guide | Procedure Guide]
>
> Physical Installation Procedures
>
> This section of the Procedure Guide covers procedures that are related to the physical installation of hardware devices. You will most likely use these procedures when building a PC, performing upgrades, or replacing damaged hardware.
>
> **Note:** The procedures in this section deal primarily with the *physical* installation of devices. These are not comprehensive installation procedures that deal with all the facets of a component upgrade (such as selection, configuration, installing drivers, etc.)
>
> Next: System Case Preparation Procedure

- Link to more prerequisite or detailed steps.

> ☑ **Example**
>
> **Preparation / Warnings:**
> If you have not already done so, please read the section on general installation and assembly tips. Ensure you have already decided how you want to configure the CD-ROM drive, and that you have already set the appropriate jumpers. See this procedure if you have not already done this.

- Use links to help present decision points and options.

> ☑ **Examples**
>
> *This procedure contains links to more detail about options:*
>
> 3. Perform survey appropriate to the isotope(s) used.
> - For low-energy beta emitters (H-3, C-14, S-35, etc.), take swipe samples and count using liquid scintillation counting.
> - For high-energy beta emitters (P-32, etc.), take swipe samples and count using either liquid scintillation counting or a ratemeter + GM probe. Floors and surfaces can also be directly monitored using a ratemeter + GM probe.
>
> *This procedure lets readers decide which procedure is applicable:*
>
> Waste containing radioisotopes of half-lives less than 90 days are allowed to decay to background radiation levels before disposal. Waste with half-lives greater than 90 days are either stored or incinerated.
> **Less than 90 day half-life?** See Solid Radioactive Waste Management.
> **Greater than 90 day half-life?** See Preparing Solid Radioactive Waste for Pickup.
> **Radioactive carcasses?** See Preparing Radioactive Carcasses for Pickup.

- Link to definitions and a glossary.
- Link to forms, samples, checklists, charts, flowcharts, graphics, multimedia files, and downloadable files.
- Provide cross-references to other sections, Web sites, and printed documents.

☑ **Examples**

(3) Procedures for suspending and revoking approvals under section 3730.05 of the Revised Code.

Know the hazards of the materials you are working with. Material Safety Data Sheets, (MSDSs), are an important primary source of information on physical properties, health hazards, reactivity, and spill cleanup procedures.

- Provide contextual clues to readers about their location in the procedures.

☑ **Example**

This example makes the link to Application Procedures inactive.

Diagnostic Imaging Program
Application Procedures

Location: | NVCC Home | Health Tech Home | Diagnostic Imaging Program | Application Procedures

- Provide easy navigation to other sections and to the table of contents.

☑ **Example**

Jump to:
Cataloguing Issues
Processing Procedures   Disk Copying Procedures   Disk Labeling Procedures
Return to:
Bibliographic Services Procedures Manual
Bibliographic Services Resources
Technical Services Web Resources Home Page

- Consider using a flowchart format with links to steps.

☑ **Example**

Boxes in the flowchart links to details, as shown on the right..

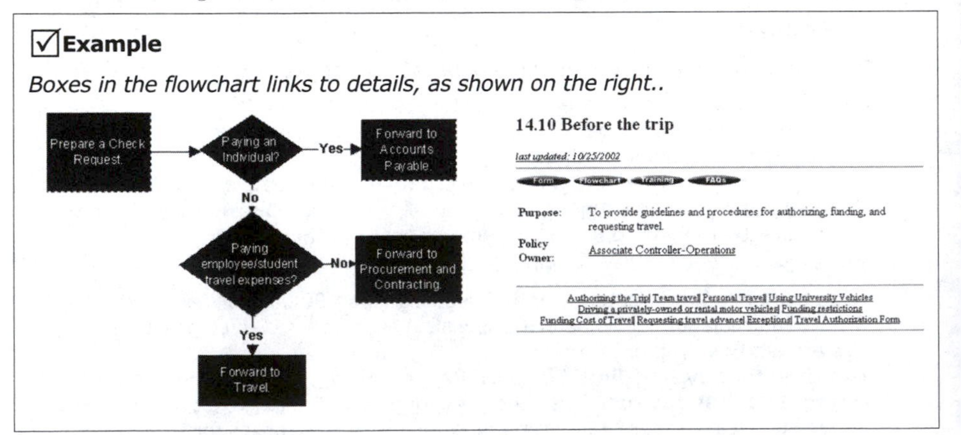

Writing Procedures

- Use active verbs.
- Use short sentences.
- Use parallel construction.

☒ **Example**

This list is not parallel and not easy to skim quickly.

■ Job Description ■ Binding Schedule Pulling Binding ■ Preparing Books for Binding
■ Replacing Missing Issues/Pages ■ Updating Binding Budget
■ Processing Returned Binding ■ Book Boxes

- Use positive language (*do* vs. *do not*).
- Use appropriate words for referencing (*refer to, see*) and branching (*go to, return to, proceed to, repeat*).
- State conditions first (*if, when*).
- Place cautions and warnings before the step.
- Place notes before or after the explanation, using a format that indicates they are not as important as cautions.

Formatting Procedures

- Use devices of emphasis (boxes, bold, caps, underline), especially for warning and cautions.
- Use standard outline or decimal format.

☑ **Example**

2 Emergency Procedures
  2.1 Fire
  2.2 Chemical Spills
  2.3 Radioactive Spills
  2.4 Biohazards
  2.5 First Aid
  2.6 Accident/Incident Reporting Procedure

- Use tables when appropriate to show information such as condition or contingency, step, result.
- Make policies/procedures available in other formats, such as Adobe Acrobat PDF or printer-friendly versions.

▷ Process/How Things Work

Process descriptions explain how things work or are done. They help readers understand a task rather than instruct them on how to do it. Process descriptions are used in a variety of Web pages, such as informational, training, and educational sites. They also are used in conceptual information and as part of online technical support.

Tips

- Provide process descriptions to help readers understand your product, procedures, and other conceptual information.

☑ **Example**

Epson uses process descriptions in online training for their products, such as this explanation of How 3LCD Works. The explanation includes a parts diagram and interactive video.

An LCD panel is made up of thousands of miniscule pixels. Together, these pixels can produce any pattern to create a blueprint for an image. As light passes through the LCD panel, it forms a recognizable picture from the configuration of pixels. This picture becomes what you see on the screen.

- Provide a link to "how it works" information, especially if it is important to your business.

☑ **Example**

Marriott Rewards Program:
How It Works
Earning Points
Earning Miles

Handspring VisorPhone:
Overview | **How it Works**
Details | **Reviews** | **FAQs**

- Provide an overview (advance organizer) of the entire process using text links or a graphic.

☑ **Example**

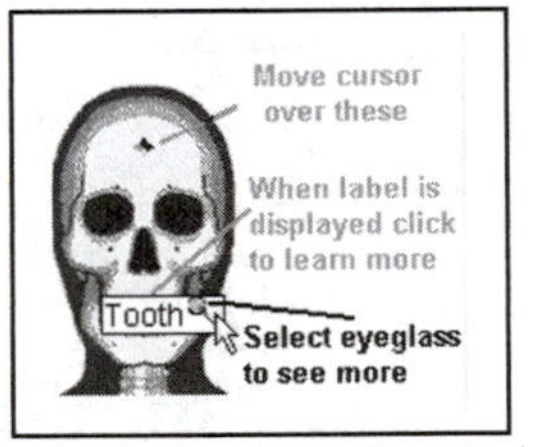

The Human Anatomy Online site provides hundreds of graphics. "Pick Points" displayed as diamonds on the image can be clicked to display an explanation of the part and how it works.

Etymology for Beginners contains a diagram of an insect. Clicking any one of the seven parts jumps to the appropriate explanation. The diagram within each section is shaded to indicate which part you have selected.

Insect body parts
The most visible parts of the body of an adult insect are: the head, the antennae, the mouth parts, the thorax, the wings, the legs, and the abdomen.

Head
The head is the anterior of the three body regions of an adult insect. It bears the eyes (usually a pair of compound eyes), the antennae and the mouth parts.

- Specify the intended audience and scope.

☑ **Example**

This page is intended for students, as a quick introduction. It explains the scientific principle behind electroplating, and offers two demos, but it does not discuss the real-world industrial situation.

- Use chunking to divide the process into stages.

☑ **Example**

Internal links are used to each section. Each section is one short paragraph.

Laser Basics: Operating principles

Lasers consist of a lasing medium, excitation mechanism, feedback mechanism, and an output coupler.

Lasing medium

The lasing medium of a laser is . . .

Excitation mechanism

The excitation mechanism of a laser is . . .

Feedback mechanism

A laser's feedback mechanism is used to . . .

Output coupler

The output coupler of a laser is . . .

- Break down the explanation into smaller sections.

☑ **Examples**

Heads & tails | Meteor showers

Made of dust, ice, carbon dioxide, ammonia and methane, comets resemble dirty snowballs. You may remember them as blurry smudges in the sky. Comets orbit the Sun, but most are believed to inhabit in an area known as the Oort Cloud, far beyond the orbit of Pluto. Occasionally a comet streaks through the inner solar system; some do so regularly, some only once every few centuries.

Your modem is a combination of hardware and software that work together to allow your computer to communicate over telephone lines. These sections give you a broad description of how your modem works.

What are modulation and demodulation?

How do I talk to my modem?

- Use transitions to show sequence and cause and effect. Be sure to use these transitions on the same Web page. Transitions on separate Web pages may cause confusion if readers do not read sequentially.
- Provide links to the following:
    - Definitions
    - Conceptual information and details

☑ **Example**

The microwaves are generated by the magnetron; they are transmitted down the waveguide; they are reflected off the fan stirrer and the walls of the microwave oven's cavity, and then they are absorbed by the food.

- Flowcharts, graphics, and multimedia
- Related topics in your site

☑ **Example**

What You'll Learn On This Page...
Electrons
Conductors & Insulators
Voltage & Current
See also...
Follow the Power Trail – How electricity gets to your home

- Related Web sites (external links)

☑ **Example**

HowStuffWorks articles contain tables of contents with links to subtopics for the article, related topics, external links, animations, and a printable version.

› Introduction to How Car Engines Work
› Internal Combustion
› Parts of an Engine
› What Can Go Wrong
› Engine Subsystems
> How to Help an Engine Produce More Power
> Q and A
> Lots More Information!
> What do you think?

Click on image to see animation
See How Camshafts Work for details.

▷ Product or Service Information

A variety of Web sites provide information about their products and/or services. This information may be the sole purpose of the site or pages available from the company home page. Product and service information are provided for promotional purposes, sales, and customer support, and to conduct business transactions over the Internet.

Tips

Providing Product/Service Information

- Use the following organization:
    - Introduction (name of product or service)
    - Summary: scanned image and concise description
    - Details (features and specifications)
    - Action
- Arrange information by user needs rather than require that users know your products or services.
- Provide the ability to search. This feature is key to your site's success.
- Help readers make purchasing decisions, and provide advice on online shopping. Keep technical specifications simple, and provide links to helpful information.

☑ **Example**

Circuit City provides links to information to help buyers make decisions.

Click & Learn

iPod specs

iPod basics

Get your new MP3 player up and running

- Reassure readers about security of transactions. Make shopping carts easy to use.

ITEMS TO INCLUDE

- Include the following types of information:
  - ☐ Catalog
  - ☐ Description of services
  - ☐ Fact sheets
  - ☐ FAQs
  - ☐ Features
  - ☐ Length of time in business
  - ☐ Links to contact information
  - ☐ Links to an informational page
  - ☐ List of services
  - ☐ New product announcements
  - ☐ Ordering information
  - ☐ Pictures (link to large images)
  - ☐ Portfolio
  - ☐ Product descriptions
  - ☐ Product reports
  - ☐ Product reviews
  - ☐ Promotions
  - ☐ Search feature
  - ☐ Specifications
  - ☐ Support information
  - ☐ Testimonials
  - ☐ Trade show appearances
  - ☐ Where to buy/dealers
- Provide clear links to the corporate or other important Web pages.

☑ **Example**

Our Company

Did you know that in addition to being the world's largest soup manufacturer, we're also a leading producer of juice beverages, sauces, biscuits, and confectionery products? We invite you to learn more.

We Recommend:

Investor Center

Media Relations

Career Center

Governance

Campbell Worldwide

▷ Promotional Language

Promotional language includes glowing adjectives and unsupported claims. It is commonly used in sales/persuasive writing. Readers prefer factual, objective writing rather than marketing hype. In addition, promotional language slows readers down and antagonizes them. The honesty and authenticity you convey also affect the way readers perceive your Web site's credibility.

Tips

- Avoid exaggeration, superlatives, and inflated adjectives (*most, best, perfect, greatest, hottest, premier, top-rated, world-class*).

☒ **Examples**

-One of the Largest Web Music and Movie Inventories. Out of This World Sales Everyday! Best Prices Around! Today is Your Day.
-Our goal is to provide you with the very best tools and services needed to operate a profitable business, or even your dream of running a global empire.
-The Best Selection, Bargains, Inventory and Discounts on Computer Products on The Planet
-These are simply the best tools available to improve your business.
. . . . a collection of some of the most rare, unusual and beautiful properties in stunning locations world-wide. All are exclusive, special and promise the most discerning of travellers an experience they will never forget.
-We offer complete web presence solutions incorporating unparalleled skills.
-This web site is to briefly show you the amazing technologies available to you, for your own amazing site.

- Avoid piling up adjectives.

☒ **Example**

ABC provides superior web design services, dazzling graphic designs, multi-tiered web hosting plans, innovative e-commerce solutions, powerful database driven applications, and Internet marketing expertise.

- Avoid promotional language in your page title and description for search engines.

☒ **Before**

Best and cheapest web design and hosting anywhere on the planet!!

☑ **After**

Offers design, hosting, search engine registration and e-commerce solutions.

- Focus on reader benefits.
- Link to support information.

 **Before**

Our staff has many years' experience in creating beautiful custom computer graphics, the newest features in website design, the best features in your electronic shopping cart, and more. We have numerous satisfied customers.

☑ **After**

You will work with our staff to plan a Web site with a design tailored to reach your customers.
Testimonials
Portfolio

- Be careful using logos, mission statements, and slogans that readers may associate with marketing hype.

▷ Pronoun Errors

Tips

Vague Referents

A referent is a word a pronoun refers to.

- Avoid vague referents, such as *it, this, that, this means that,* and *them.*

☒ **Examples**

-A clean, crisp, professional web site speaks volumes. **This** is where most of the work in web design is geared.
-After **that**, they click on the advertisement to get a full overview of your services.
-Most of the address blocks have been allocated to research, education, government, corporations, and Internet Service Providers, who in turn assign **them** to the individual computers under their control.

- Avoid vague referents in menus. Be clear what a link refers to.

☒ **Example**

Mailing Lists
How they were invented
How they work
How to use them

That vs. Which

Two types of adjective clauses are restrictive and non-restrictive.

TYPE	ESSENTIAL TO SENTENCE	RELATIVE PRONOUN	COMMA
Restrictive	Yes	That	No
Non-restrictive	No	Which	Yes

Tips

- Use *that* in a restrictive clause (needed for sentence to make sense).
- Use *which* in a non-restrictive clause (not needed for sentence to make sense). Separate the phrase from the main sentence with commas.

☑ **Example**

The W3C is an industry consortium that seeks to promote standards for the evolution of the Web and interoperability between WWW products by producing specifications and reference software.

☒ **Examples**

-This is the document which describes what HTML is and how it works.
-The Web Standards Project is a grassroots coalition fighting for standards which ensure simple, affordable access to web technologies for all.
-Home Sweet Home: Creating Web Pages Which Deliver

WHO/WHOM VS. THAT

Who and *whom* are commonly misused pronouns. One way to remember the rule is to substitute a different word in the sentence:

USE	PRONOUN TO USE	PRONOUN TO SUBSTITUTE
Subject	who	He, she, they, we
Direct object	whom	Him, her, them, us

Tips

- Use *who* and *whom* (rather than *that*) to refer to people.
- Use *who* as a subject.
- Use *whom* as a direct object.

☑ **Example**

For whom is the site designed?

☒ **Examples**

-Users that want real page layout should certainly demand Style-Sheets from their browser.
-You want to make sure that anyone that might be interested will be able to find YOUR site.
-Whomever visits those sites (home pages) can also access your page.
-Who to Call

▷ Proximity

Proximity refers to the closeness or spacing among objects on your page. The principle of proximity is that elements that are close are perceived as being related.

Tips

- Use proximity to show related items, such as headlines and text, images and text, and navigational items.
- Use the following visual methods to show proximity:
  - Boxes, lines, indentation, color, varying fonts, tables, and graphics
  - Grid
  - Methods of emphasis
- Use white space to show proximity.

☑ **Example**

Three groups of navigational items are indicated with white space and other formatting.

What's Hot Right Now

70% off 400 TC sheets

- AVON: Free ship & gift
- Kohl's: Everything on sale
- Get a free Fekkai sample
- Loaded PC under $280
- Tony Little trainer $95
- $1 extra laptop battery
- Save 10-30% at JCPenney

Today's Sales & Deals

Need contact lenses? Save up to 70% at VisionDirect.com.

- Avoid lone items separated with excess white space. Especially avoid too much space between headings and text, or they will not appear to be related.

☒ **Example**

Heading	**Heading**
Text text text text	Text text text text
Heading	**Heading**
Text text text	Text text text

▷ Publication Information

Publication information is used for an online newspaper, newsletter, journal, or magazine. This information helps readers determine the currency of information. The publication background provides readers with information about the source/history of the online version of a document.

Tips

- Provide the following publication information:
  - About the publication
  - Company or individual responsible
  - Contact information
  - Date
  - Editors
  - Editorial mission and sections
  - Editorial guidelines for contributed articles
  - Format and download information
  - Frequency of publication (e.g., quarterly)
  - History of the publication, including the original title and form
  - Issue
  - Number
  - Other locations: repositories, servers, mirror sites
  - Readership profile
  - Subscription information
  - Version
  - Volume
- Put a link to Publication Information on the home page.

▷ Pull Quote

A pull quote is a small text passage that is enlarged and set apart from the main text. A pull quote

- Adds visual interest to a Web page by breaking up the text.
- Captures attention.
- Interests readers in the text.
- Emphasizes key points made in the document.
- Summarizes your message.

HTML Code: One method of separating quotes is with the <BLOCKQUOTE> tag. You can also use a one-cell table and use the ALIGN attribute to specify text alignment.

Tips

- Keep the pull quote short.
- Pick an excerpt that will trigger interest.
- Avoid too many pull quotes on one page.
- Consider using large fonts, italics, boxes, rules above and below, and color for emphasis.

☑ **Example**

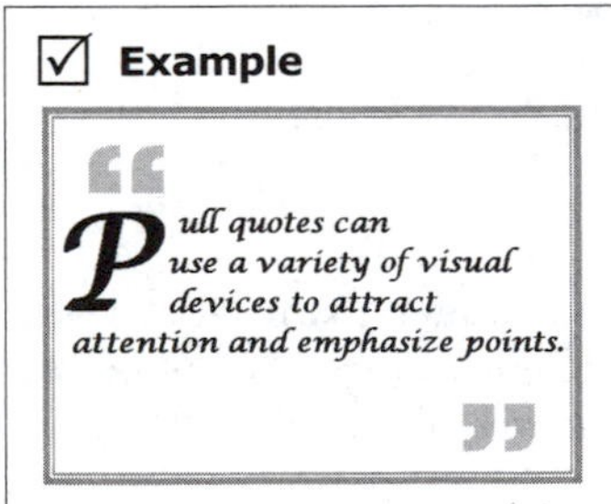

- Surround the text with quotation marks.
- For accessibility, consider labeling the quote as a pull quote.

☑ **Example**

Pull Quote: "IAQ is the type of thing that is an unseen comfort. It impacts people, but they don't always see it."

▷ Punctuation Marks

Punctuation is more difficult to read on the screen. However, it is important in helping people understand your meaning. It is also crucial in punctuating URLs (Web addresses), e-mail addresses, and code.

Tips

- Edit carefully for punctuation errors, which affect your site's professionalism and credibility.
- Avoid overusing punctuation such as commas, colons, semi-colons, hyphens, dashes, quotation marks, and parentheses.

- Avoid using punctuation for emphasis.

> ☒ **Example**
>
> The Mother of all award submission sites!!!

- Use one space after a period in online writing.
- For accessibility, end all sentences, headers, list items, initials, etc. with a period or other appropriate punctuation.
- Print out Web pages to check for punctuation errors.
- Be especially careful with punctuation used with computer documentation (addresses, code, etc.).
- To highlight important punctuation, use techniques such as bold, color, shaded boxes, and monospaced type.

> ☒ **Example**
>
> *The code and punctuation in this excerpt are difficult to see online.*
>
> To use these characters in your own HTML files, put the appropriate number into `&#__;` (e.g. `£` for the British pound (currency) sign), or, for the 8-bit alphabetic characters, use the alternative standard HTML 2.0 entity in parentheses on the right. (These are the only non-numeric character entities defined in HTML 2.0, except for `&`, `<`, and `>`, which should be used to escape the characters `&` < > in an HTML file, and `"` to escape a double-quote character in an attribute value.)

▷ Purpose/Objective

The purpose/goal of your site is the reason you created it: to motivate, inform, persuade, or entertain. The objective is what you want readers to think or do. The purpose and objective should be immediately clear and appear in the introduction. The purpose and objective help you plan your site and help readers know why the site exists. In addition, stating your purpose when you start your project lets you refer to it as you design it: *The purpose of this Web page (or Web site) is to . . .* The statement of purpose on your home page lets readers know the following information:

- Who the site is for
- What the site is about
- Who you are
- Why you created the site

Tips

- Tie your purpose and objectives with your audience analysis.
- Determine your site's goal/objectives:
    - ☐ Conduct research
    - ☐ Entertain
    - ☐ Establish corporate identity and visibility
    - ☐ Exchange information among members or organizations
    - ☐ Improve your corporate image and customer relations

- ☐ Inform
- ☐ Instruct
- ☐ Motivate: to read, to sense a need
- ☐ Persuade: to buy, to adopt a viewpoint
- ☐ Present information
- ☐ Promote your organization
- ☐ Promote, introduce, or sell a product or service
- ☐ Provide a collection of links
- ☐ Provide news
- ☐ Provide technical support
- ☐ Publish a newsletter or magazine
- ☐ Recruit volunteers or employees
- ☐ Request audience information
- ☐ Tell about yourself, your hobbies, and interests

▶ State the purpose of the site in the introduction or link to it.

☑ **Examples**

-The purpose of this site is to provide court information to the public.
-The purpose of this site is to provide a user friendly online resource,
for students or teachers, no matter what they are being challenged to write.

Usability.gov contains information about the purpose of the site on the About Us *page:*

What is the purpose of this site?
Usability.gov is the primary government source for information on usability and user-centered design. It provides guidance and tools on how to make Web sites and other communication systems more usable and useful. Topics include:

- a step-by-step guide on how to plan, analyze, design, and usability test a highly usable Web site;
- quick access to the latest usability methods;
- an introduction to usability, how to get started, and what it costs;
- the latest research-based Web design guidelines;
- templates and examples for assessing audience needs, writing a usability test report;
- and much more.

▶ State the purpose of each Web page in your site.

☑ **Examples**

-The purpose of this page is to provide answers to Frequently Asked Questions.
-The purpose of this page is to provide links to Web resources about . . .

▶ If your purpose is important, emphasize it by using headings, size, color, position, or animation.

☑ **Example**

The Mars Society places their purpose on the focal point of the home page using reverse text:

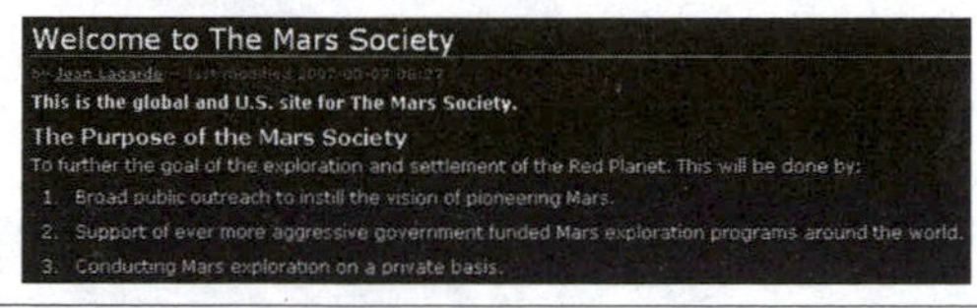

- Be concise and specific.

☒ **Examples**

-The purpose of this site is to present a small portion of the history of mathematics through an investigation of some of the great problems that have inspired mathematicians throughout the ages.
-The purpose of this page is to answer some questions that teachers often ask.
-The major purpose of this site is to begin an exchange of ideas between myself and other individuals who come across this site on the net. My personal goal is to explore my own ideas about the world, universe and existence.
-The purpose of this Website is NOT to make you into a Brain Surgeon . . .

▷ Questions/Question Marks

Question marks are used to punctuate the end of direct questions. Questions are an effective technique in writing for lower-level audiences, getting attention, and providing interactivity. They are often used in link text to interest readers in the site and in FAQs or as icons.

Tips

- Use a question mark after a direct question.
- Use questions for link text to attract attention.

☑ **Examples**

Selling Your Home?
Why Participate?
What's New?

- Use questions in your text to engage readers or direct them to a topic.

☑ **Examples**

Did you hear your modem dial? Can you access the Internet?
Need Help?

- Do not use a question mark after an indirect question.

☒ **Examples**

-How to Create HTML?
-Instructions & examples?
-Perhaps your present web site needs a facelift?

- Use parallel construction when using questions in headings and lists.

☒ **Example**

The following list of links is not parallel. One way to correct the error is to word all as questions.

This page will show you -
What is a table?
Why you should use tables
How to construct a simple table

▷ Quick Reference

A quick reference is a concise summary of frequently used material, such as commands, part numbers, and instructions. Quick references are available on Web sites as part of technical support, reference information, user guides, and electronic books (online).

Tips

- Consider including a quick reference when you are providing technical documentation, technical information or reference material.

☑ **Example**

Webopedia provides a Quick Reference section for information on common Internet and computer facts.

- Make the quick reference available in several formats, including a downloadable version.

☑ **Example**

Python quick reference HTML - one large HTML page.
Python quick reference Zip - Zip file for easy download.
Python quick reference Help - Windows compiled help.
Python quick reference Text – Plain text.
Python quick reference PDF – PDF.

- Organize information into categories.

☑ **Example**

HTML Tag Quick Reference Guide
This table summarizes the basic HTML tags. Click on a particular category of tags to jump directly to it, or browse the full reference list.

Document	Images	Forms
Basic Text	Imagemaps	META Tags
Lists	Tables	Style Sheets
Links	Frames	Special Characters

- Provide links to specific reference material.

☑ **Example**

Quick Reference
This section groups together quick reference guides.
Contents contains a clickable summary of the contents of the Developer Zone in alphabetical order.
Tag Summary contains a summary of HTML tags and their switches.
Color Chart is a list—complete with sample colors and hex numbers—of the standard Netscape colors.
Search Services is a summary—complete with appropriate links—to each of the major search services, how to submit your site, and how each service indexes sites and supports meta tags.
JavaScript Operators contains a summary of JavaScript-supported operators along with an example of each.

- Annotate contents of the quick reference sections.

☑ **Example**

The 7 Layers of the OSI Model: Use this handy guide to compare the different layers of the OSI model and understand how they interact with each other.

- Arrange quick reference information in a logical order: alphabetical, most common to least common, chronological, etc.

☑ **Examples**

Alphabetic List	Functional List
<a>	Fonts
<address>	Forms
<area>	Frames
<b>	Images
<base>	Links

- Provide appropriate types of information: reference (what is) and tasks (how do I).

☑ **Example**

Getting Started
Overview
Getting Help on WilsonWeb
Searching with WilsonWeb
Using Your Search Results

- Use tables and lists to summarize information.

☑ **Example**

This table (partially shown) provides a list of search engines.

Do you want to...?	...Then try these tools!		
Browse a broad topic?	Yahoo www.yahoo.com/	Librarians' Index to the Internet lii.org/	About.com www.miningco.com/
Search for a narrow topic?	All theWeb www.alltheweb.com/	Google www.google.com/	WiseNut www.wisenut.com/

- Use icons and screen captures to identify interface elements.

☑ **Example**

Clicking any element on this toolbar opens a small window that describes what the button does.

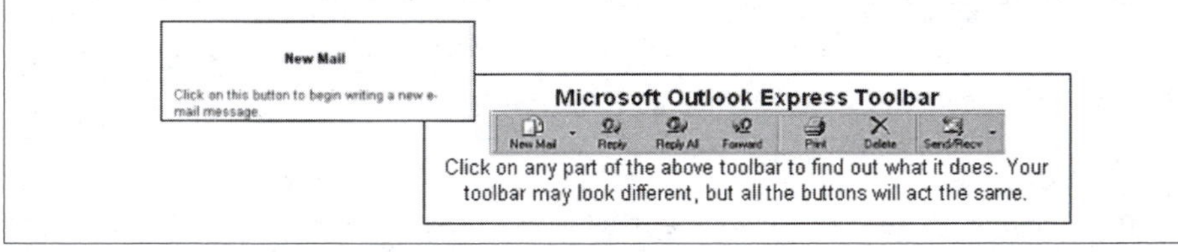

▷ Quotation Marks

Quotation marks enclose direct quotes, titles, and letters or words being highlighted.

Tips

- Use quotation marks when you are quoting verbatim and with pull quotes.
- Use direct quotes as a "hook" to interest readers in your page.
- Weigh the pros and cons of directly quoting the original or linking to it. By linking, you lead readers to the original but encourage them to leave your site.
- Use quotation marks to indicate that a technical term is being used for a special purpose.

☑ **Example**

The commands are almost never case sensitive and are either "container" or "separator" commands.

- Avoid using quotation marks to highlight a word or phrase that is the subject of discussion or for emphasis. Instead, use italics.

☑ **Example**

Virtually every computer operating system includes a piece of software called a *text editor*.

- Use quotation marks around single letters.

☑ **Example**

For users on a text-only browser, these items would appear as just "o" instead of something like [IMAGE].

- Do *not* use quotation marks around commands, filenames, code, and other user input. Readers may mistakenly believe they must type the quotation marks.

☑ **Example**

<IMG SRC= "URL">
Don't forget to put quotation marks around the URL.

☒ **Example**

At the top of the file, type "<title>"

- Avoid unnecessary quotation marks when referring to links or clicking highlighted text or icons. Instead, use the context to help readers understand the words.

☒ **Example**

When you want to go to our home page, click on the "Home" option.

- Put commas and periods *inside* quotation marks.

☒ **Example**

One color in the color table may be denoted as "transparent".

- Put semi-colons and colons *outside* quotation marks.

☒ **Examples**

A bright RED page would be FF0000;"

In the following example, the black background of the image has been made "transparent:"

▷ Readability

Readability refers to the difficulty or ease with which a text is read. Factors that affect readability include vocabulary, sentence structure and length, paragraph length, and formatting. Because your message is the most important part of your Web page, the text should be readable. In addition, Web readers tend to skim pages. Online documents look grainier than printed documents and fit less information (less than 1/3) on a page. Also, users will have different browsers, so your document will look different on different systems.

Tips

- Use a vocabulary and sentence style appropriate for your audience. The average reading level in North America is eighth or ninth grade.

> ☑ **Example**
>
> *The National Institutes of Health provides* ★**Easy-to-read**★ *versions of some health information. The following examples illustrate the difference in sentence length and word choice.*
> Lactose intolerance is the inability to digest significant amounts of lactose, the predominant sugar of milk. This inability results from a shortage of the enzyme lactase, which is normally produced by the cells that line the small intestine.
>
> *(Easy-to-read)* Lactose intolerance means that you cannot digest foods with lactose in them. Lactose is the sugar found in milk and foods made with milk.

- Be concise to keep word count low.
- Keep pages, paragraphs, and sentences short.
- Make readability a priority over color and fancy effects.
- Use large, clear, legible fonts.
- Use a background that contrasts well with text (preferably black text on a white or light background). Keep backgrounds simple on pages with lots of text.
- Keep line length short.
- Use lists to help break up long sentences and paragraphs.
- For long articles and stories, give readers the option to print the page.

▷ Redirect Page

A redirect page notifies readers that your Web page has moved. They are then transparently redirected to the new URL (address).

Tips

- Include the following information:
    - ☐ Notification that your page has moved
    - ☐ Notification that they will be redirected in ___ seconds
    - ☐ A link to the new address
    - ☐ Code that will send users to the new page automatically
      <META HTTP-EQUIV=refresh CONTENT=time;URL=url>
      time=number of seconds the page is displayed
      url=new address
    - ☐ An e-mail address to report any problems
    - ☐ Date
    - ☐ Suggestions to inform the referring site of the new address
    - ☐ Reminder to readers to reset their bookmark
- Avoid having readers see error messages such as *HTTP 404 - File not found.*

▷ Reference Information

Online reference material on a Web site provides support information. Web reference sites, such as Research-It and Bartleby, provide tools such as encyclopedias, almanacs, dictionaries, thesauri, quotations, style guides, and atlases or specialized reference material.

Tips

- Divide information into numerous short topics.
- Use a logical organization, such as alphabetical or by category.

> ☒ **Example**
>
> *This list of links (partially shown) is difficult to skim:*
> Fundamental Physical Constants
> most commonly used constants in physics and engineering
> Astro-physical Constants
> solar system data, current moon phase, and more ...
> Nuclear & Particle Data
> electron, proton and neutron data, table of nuclides and more ...
> Air Composition
> gas composition of air
> Coefficients of Friction
> static and kinetic coefficients of friction for various materials

- Put reference material in a separate Web document rather than embed it in pages. It will then be available from other documents.
- Provide a variety of ways to access the information, such as browsing, searching, table of contents, index, and cross references. Also provide links to special materials, such as illustrations, charts and tables, and bibliographic record.
- Use layering; take advantages of links to details.

> ☑ **Example**
>
> *An online periodic table (partially shown) lets readers "click on the element symbol for more info."*

Periodic Table of Elements

Group	1	2	3	4	5	6	7	8	9	10	11	12	13	14	15	16	17	18
	IA	IIA	IIIB	IVB	VB	VIB	VIIB	VIII			IB	IIB	IIIA	IVA	VA	VIA	VIIA	VIIIA
Period																		
1	1 H							Click on the element symbol for more info.										2 He
2	3 Li	4 Be											5 B	6 C	7 N	8 O	9 F	10 Ne
3	11 Na	12 Mg											13 Al	14 Si	15 P	16 S	17 Cl	18 Ar
4	19 K	20 Ca	21 Sc	22 Ti	23 V	24 Cr	25 Mn	26 Fe	27 Co	28 Ni	29 Cu	30 Zn	31 Ga	32 Ge	33 As	34 Se	35 Br	36 Kr

- Use patterned information to present information consistently for each item. Use headings, tables, and list format.

> ☑ **Example**
>
> **Font Style**
>
> | | |
> |---|---|
> | **Syntax:** | font-style: <value> |
> | **Possible Values:** | normal \| *italic* \| *oblique* |
> | **Initial Value:** | normal |
> | **Applies to:** | All elements |
> | **Inherited:** | Yes |
>
> The **font-style** property defines that the font be displayed in one of three ways: **normal**, ***italic*** or ***oblique*** (slanted). A sample style sheet with **font-style** declarations might look like this:
>
> ```
> H1 { font-style: oblique }
> P  { font-style: normal }
> ```

- Check reference material for accuracy.
- Keep reference material updated regularly. Indicate to readers what information is new or changed.

▷ Repetition

Repetition is repeated use of design, navigational, and textual elements. It has the following functions:

- Provides unity and consistency.
- Gives a visual identity.
- Helps readers learn how to use the site, navigate, and find information.
- Helps readers know what to expect.

Tips

- Repeat the following elements from page to page:
  - Background
  - Banner
  - Bullets
  - Colors
  - Fonts
  - Grid/page layout
  - Header and footer
  - Headings
  - Icons
  - Lines (rules)
  - Logo
  - Metaphor
  - Navigation
  - Textual elements (such as titles and headings)

▷ Research

A key step in planning a Web page is conducting research. Research lets you gather content and determine what others have written.

Tips

- Research what others have written on the same subject. Check both print and online documents. Contribute something new.
- Check other Web sites. This step allows you to do the following:
  - Refer to other sites as examples and use later as links.
  - Determine if your subject is unique.
  - Get ideas about good page design and writing style.

- Research what your competition (if appropriate) has done.
- Identify where the information you will use is located and its format, such as multimedia, graphics, or text.
- Interview experts.
- Review the rules of your ISP (Internet Service Provider), or company, institution, organization, to determine if your content is acceptable. Also find out any technical issues and limitations, such as file size.

▷ Resume

A Web resume is similar to a traditional resume but uses hyperlinks and possibly multimedia. It may be a Web page or a link from a home page. Electronic resumes have several advantages. One is low cost and wide distribution. An electronic resume can link to samples and multimedia files (electronic portfolio). Electronic resumes also support biographical information on company or organizational sites. Finally, an electronic resume can indicate to employers that you are able to create a Web page and use the Internet.

There are other formats for resumes: attached to e-mail or submitted through electronic forms. The format for attachments is usually plain text (ASCII), RTF (Rich Text Format), Microsoft Word, or Adobe Acrobat (PDF). Plain text resumes are useful for submitting via forms or when companies scan your resume.

Tips

Items to Include

- Include the following:
  - Cover letter
  - Name, address, phone, e-mail address/*mailto:* link
  - Career objective/type of position sought
  - List of keywords
  - Technical skills (e.g., hardware, software, languages)
  - Non-technical skills (e.g., supervisory, leadership)
  - Educational: degrees, licensures, certificates
  - Experience: positions, job titles, dates
  - Honors
  - Hobbies, interests, extracurricular activities
  - References
  - Relocation, salary, full/part-time/contract information
  - Date stamp
  - Links to support information: (e.g., samples of artwork & publications, photo, home pages of companies you worked for)

Writing Your Electronic Resume

- Pick the best organization: reverse chronological order or qualifications in order of strongest to least important.
- Use specific details rather than adjectives.

- Use strong action verbs.
- Be concise.
- Omit irrelevant information and non-essential words.
- Be careful when including personal information.
- Proofread for errors.

Formatting Your Electronic Resume

- Chunk information into categories with links to specific sections. These links can be internal or links to separate pages.
- Provide either a table of contents on a home page or a table of contents with internal links to sections. The table of contents can also be located in a frame.
- Use lists to summarize your knowledge of specific areas.
- Use layering. Highlight and summarize information; then allow readers to link to more detail.

> ☑ **Example**
>
> *Begin with links to detail pages. Begin each detail page with a summary. At the bottom of each page, provide navigational aids such as*
>
> My next job was at . . . . where I was . . .
> My previous job was . . .
> Return to the top page of my résumé.
> Look at my home page.

- Provide navigational aids.

> ☑ **Example**
>
> [Cover Letter]
> [Photo] [Professional Goal] [Geographic Area(s) Considered] [Experience]
> [Present Position] [Prior Position] [Education] [Outside Activities; References]
> [My Resume Files For Download]

- At the top (and bottom) of each page, provide a reminder of your name, location in site, and contact information.
- Consider providing your resume in downloadable form in a variety of file formats, such as plain text, Microsoft Word, and Adobe Acrobat PDF.
- Keep backgrounds simple and fonts legible.

Creating a Plain Text Version

- Use simple, common fonts, such as Times Roman or Courier.
- Avoid tables, columns, and tabs and other spacing adjustments
- Eliminate devices of emphasis and graphical elements, such as bold, italics, underlining, bullets, rules, and borders.
- Use a plain background.

▷ RSS Feed

An RSS feed is a method of delivering and publishing frequently updated syndicated digital content and sharing content between sites. RSS stands for Really Simple Syndication (or Rich Site Summary). Each RSS text file contains two types of information: static information about the site and dynamic information about new information. This information is surrounded by matching start and end tags in XML format. You can provide a partial feed or full feed. A partial feed includes a headline, content summary, and link to the content on your Web site. A full feed provides the entire content. Subscribers can view the feeds by installing a dedicated news reader or a Web-based news reader.

Tips

- Provide RSS feeds for news, events, headlines, updates, announcements.
- Provide a link to the feed.
- Give readers instructions about how to obtain the feed.

> ☑ **Example**
>
> XML RSS To add the feed of Breaking News items, right-click on the XML icon and click Copy Shortcut. Paste that shortcut into your news reader where it allows you to add a feed. Or, you may click the Add to My Yahoo! button.

- Determine whether to provide a partial or full feed. For a partial feed, provide a descriptive title and concise summary.

▷ Run-On Sentence

A run-on sentence contains two or more sentences without the necessary punctuation or conjunctions. Correct the error by making two sentences, inserting a semi-colon, inserting a coordinating conjunction and comma, or making one sentence into a subordinate clause.

> ☒ **Before**
>
> -There are too many links for one page each index is a separate page.
> -I suggest visiting the site it's worth it.
>
> ☑ **After**
>
> -There are too many links for one page; each index is a separate page.
> -I suggest visiting the site; it's worth it.

▷ Sales & Persuasive Sites

Sales writing on the Web persuades readers to act (buy, volunteer, contribute) or adopt a viewpoint. Persuasive writing can also inspire confidence in your company or organization.

Tips

Items to Include

- Include the following information:
  - ☐ Contact information
  - ☐ Corporate information

- ☐ Examples and case studies
- ☐ FAQs
- ☐ Forms
- ☐ Informational page
- ☐ Office locations
- ☐ Philosophy or mission statement
- ☐ Product/technical support information
- ☐ RSS feed
- ☐ Staff credentials
- ☐ Support information (e.g., testimonials, resumes)

Organizing Persuasive Pages

- Use the following organization:
  - ☐ **Attention:** Get readers' attention.
  - ☐ **Interest:** Help readers visualize the most important benefit and gain interest.
  - ☐ **Desire**: Create desire for the product, service, or cause. Give facts to prove your point.
  - ☐ **Action**: Provide links to ordering or action information.

Getting Attention

- Get readers' attention in the opening:
  - ☐ Offer something free
  - ☐ Emphasize reader benefits
  - ☐ Ask a question
  - ☐ Begin with a quote or testimonial
  - ☐ Give a statistic

Creating Interest and Desire

- Be specific.

> ☒ **Examples**
>
> *These companies give no specifics.*
>
> Our aim is to please you. Whether you want to develop an extensive on-line catalog or a simple web site to advertise your company we are there for you.
>
> **Web Design:** Okay, folks. Here's the deal. In today's society, nothing is more impressive than a business that has a quality web site. You're in luck. I make those. Drop me an e-mail if you're interested in expanding your business, band, or personal ideas to the internet.
>
> *Many Web sites use statements like the following:*Our prices are much lower than our competitors who offer the same level of service.

- Use statistics.

☑ **Examples**

As of July 21st, two of our Realtor clients have closed on over a combined $13,000,000 worth of sales made solely from their respective web sites.

Making Action Clear

- Make it easy to act.
- Make it clear what readers should do.

☑ **Example**

Many organizations have a separate How You Can Help *page.*

- Provide contact information.

Using Persuasive Wording

- Emphasize service and reader benefits rather than features or yourself.

☑ **Example**

why join STC?

You may call yourself a technical writer, editor, illustrator, Web designer, or any of several other job titles... but if your job involves communicating technical information, the Society for Technical Communication has the resources you need to help you do your job better... and to move your career forward.

- How Can I Learn More?
- What Do Members Get?
- Who Belongs to STC?
- 10 Reasons to Join (PDF)

☒ **Examples**

We can enable your organization to take advantage of all that the Web offers. We accomplish this through years of Internet wisdom and experience, continuous improvement, unquestionable integrity, cutting edge technology, and hard work.

What I Bring to Your Business: You will find me a kind, bright, creative, hard-working, enthusiastic, and highly-organized marketing professional. Problem solving and "thinking outside the box" are what I do best!

- Use concise, descriptive headlines and subject lines.
- Use concrete words, positive wording, and action verbs.

☑ **Example**

Now you can:
- Create your own online resume
- Search available jobs
- Use career alerts

☒ **Example**

The following does not use concrete benefits, such as shopping at home conveniently or saving money. It also does not end with a reader benefit; instead, it asks readers to support the site. The final sentence does not show much confidence in the site.

Advantages of ordering books from this site: You will not waste your time going to local bookstores, which may not have the books recommended here. You will get the books at a discount price, which you may not get from your local bookstore. (The discounts vary around 10%-40%.) You will help support this site. If you don't want to buy from this site, you're still welcome to browse around and (maybe) check them out at your local bookstores.

- Avoid promotional language, such as superlatives (*most, best, perfect, greatest*).
- Avoid clichés.

☑ **Example**
No job is too big or too small for us.

- Avoid criticizing the competition. Instead, focus on what you can offer.

☒ **Example**
To be frank, I am appalled when I surf the web and see the rates that other web design services charge.

- Use a confident style.

☒ **Examples**
-But if you're not interested in having a Web page written, I've put some other things that I hope you'll find interesting on the site. Please have a look around.
-Give us a try, we are sure you will not be disappointed!

- Use short, concise paragraphs, words, and sentences.

Using Links

- Use layering. List the benefits and let readers jump to details.

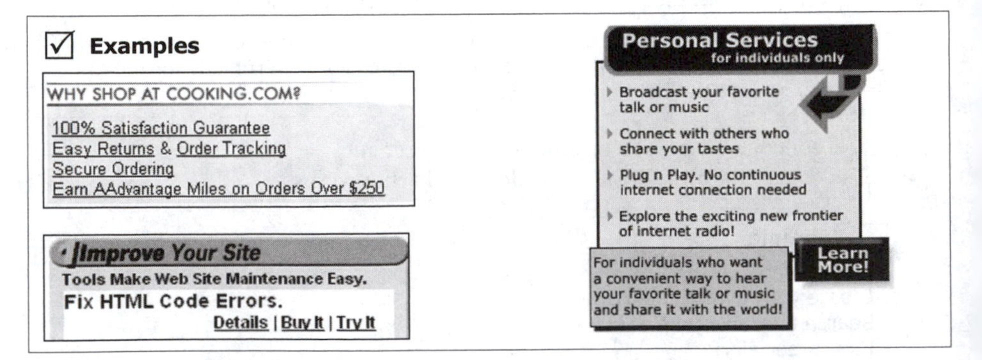

The Society of Professional Journalists lists benefits of joining.
Three reasons you should be a member:
1. Continuing Professional Education
2. Career Services & Support
3. Journalism Advocacy

Each item links to a summary of benefits, which, in turn, contains links to even more detail.
Professional development has been a long-time cornerstone of the Society's mission. SPJ programs the **SPJ National Convention**, **specialized workshops**, **regional conferences** . . .

- Link to support information (testimonials, reviews, portfolio, resumes, demos, samples), third-party research, and detailed facts/features on separate pages.

☑ **Examples**

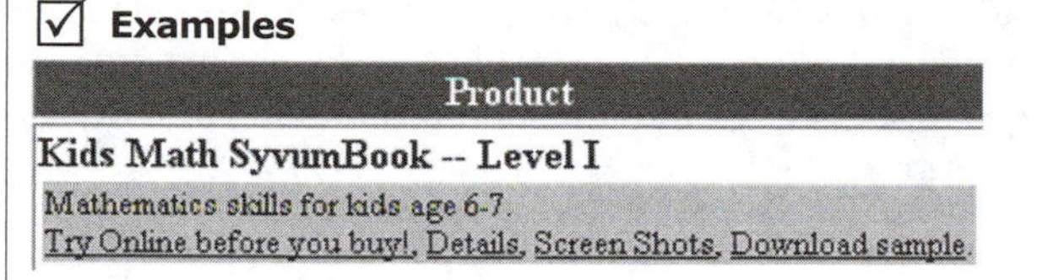

Product
Kids Math SyvumBook -- Level I
Mathematics skills for kids age 6-7.
Try Online before you buy!, Details, Screen Shots, Download sample.

Testimonials - Read some comments from our happy customers.
A writer's workshop offered on CD-ROM is advertised on a promotional Web site. It contains links to the instructor's biography, kudos (reviews from users), demo, samples critiques, course details, and an order form:

It's convenient!
It's effective! Work at your own pace on your own schedule in your own home and get feedback on your writing fast--while it's still fresh in your mind. (*link to e-critique service details)*

- Use links that will attract readers.

☒ **Examples**
Sales Pitch
Hype

- Think twice about offering links to competitors' sites.
- Link to action information. Make the link obvious and available on every page.

Formatting

- Use simplicity in your page design. Many sales sites put too much detail on the home page.
- Use graphics.
- Emphasize benefits by using tables, bold, bullet lists, short paragraphs, boxes, color, and headings.

☑ **Example**

When you join STC, you receive these and other fine benefits.

- Award-winning publications to keep you up-to-date on current trends, tools, and practices vital to the industry.
- Networking opportunities through STC's 140 local chapters.
- Educational conferences and seminars at reduced member rates.
- The opportunity to join STC's 20 Special Interest Groups (SIGs).
- Forums for sharing your expertise, expressing your professional opinion, and building a reputation within the industry.
- Full access to STC's Technical Communication Career Center - containing current job openings specifically geared toward the technical communication field.
- STC's Annual Salary Survey, which provides vital statistics on pay and benefits for technical communicators in the U.S. and Canada.

▷ Scannability

A scannable page is one that can be looked at quickly. It emphasizes the important information and aids selective reading. You create a scannable page by positioning visual cues. Readers of Web pages tend to scan or skim information. You should thus design your page so readers can find information quickly.

Tips

- Use elements that enhance scanning, including the following:
    - ☐ Advance organizers
    - ☐ Chunking information
    - ☐ Emphasis
    - ☐ Graphics
    - ☐ Grid
    - ☐ Headings
    - ☐ Introductions
    - ☐ Lists
    - ☐ Tables of contents/menu
    - ☐ White space

☑ **Example**

What is the World Wide Web? The Web is about communication. The Web gives a chance for new kinds of **roles to exist among people** who provide and consume information, communicate, and interact. The Web is a unique kind of medium for communication and is another step in centuries of change in the way people use media.	The Web isn't just technology, although its operation depends on it--the Web does have **a technical basis as a client/server system** for hypermedia communication. But not all the jazzy-interactive software in the world will convince you or anyone to use the Web and abide its curious unpredictableness and chaos--there's probably not a single compelling reason to use the Web, but rather **a collective set of personal opportunities that each person decides are important**.
Read more	• The World Wide Web Consortium. (2000). "A Little History of the World Wide Web (1945-1995)." http://www.w3.org/History.html

▷ School Web Site

Many schools, school districts, and colleges/universities create their own home pages and Web sites. School Web pages introduce readers to the school and community and point to useful Web resources.

Tips

Items to Include

- Include the following types of information:
  - ☐ Activities
  - ☐ Alumni directory
  - ☐ Building information
  - ☐ Calendar and schedules
  - ☐ Catalog link
  - ☐ Clubs
  - ☐ Community information
  - ☐ Contacts (e-mail addresses, phone numbers)
  - ☐ Curriculum
  - ☐ Departments
  - ☐ District links
  - ☐ Events
  - ☐ Extracurricular activities (clubs, organizations, sports)
  - ☐ Guidance
  - ☐ History
  - ☐ Job postings
  - ☐ Map and directions
  - ☐ Mission
  - ☐ News
  - ☐ Newsletter
  - ☐ Personnel/directory
  - ☐ Phone directory
  - ☐ Policies
  - ☐ Position on technology
  - ☐ Profile and information about the school (size, location, type school)
  - ☐ Programs
  - ☐ Projects
  - ☐ Publications
  - ☐ Recognitions/awards/honors
  - ☐ School store
  - ☐ Staff information
  - ☐ Student work (art, writing, etc.)
  - ☐ Teacher and class home pages; home pages for schools/departments
  - ☐ Technology news
  - ☐ Virtual tour
  - ☐ Web resources (List Page)

▷ Scope

A scope statement in the site's introduction defines what is and is not included in the site. A scope statement may also be used on a page within the site. The scope guides readers' expectations about the type of material included in the site or page.

Tips

- Clearly state what is and is not covered in the Web page or site.

> ☑ **Examples**
>
> SearchHelp is a simple guide to search engines and directories. It is not a searcher, but it will help you use search engines more effectively. SearchHelp will focus on the major search engines and search directories.
>
> The David Rumsey Historical Map Collection contains to date over 7,180 maps online and focuses on rare 18th and 19th century North and South America cartographic history materials. Historic maps of the World, Europe, Asia and Africa are also represented.
>
> This list is more concerned with the issues of designing pages than the technical problems of HTML or PhotoShop.

> This site does **not** include resource collections, lists of links, commercial products, or online schools.
>
> This site does not cover P2P projects (which are more about file sharing than about distributed computing), private distributed computing projects (projects in which the general public can't participate), or projects that research ways to do distributed computing.

- Be specific.

> ☑ **Examples**
>
> 10,972 entries (more than one myriad), 97,970 cross-references, 4,986 figures, 189 animated graphics, 984 live Java applets, and counting...
>
> The IPL Online Literary Criticism Collection contains 4745 critical and biographical websites about authors and their works that can be browsed by author, by title, or by nationality and literary period. The collection is not inclusive of all the work on the web, nor does it plan to be. The sites are selected with some thought to their overall usefulness.

- If applicable, state your criteria for including or excluding content.

> ☑ **Example**
>
> This page maintains links with on-line grammars of as many languages as can be found on the Web. It includes all types of grammars: reference grammars, learning grammars, and historical grammars. Grammars are selected for their accuracy and effectiveness for learning the language they describe.

- If your scope statement is long, link to it or include it on another page, such as About This Site.
- In determining your scope, consider the number of writers and designers you will need and your ability to keep the site updated.

▷ Scrolling

Scrolling is moving vertically or horizontally on the computer screen. Users use scroll bar arrows, scroll bar slides, Page Up and Page Down keys, or scroll mice. Because scrolling is necessary in long documents, it should be a consideration in determining the page length. Scrolling is much faster than clicking links and waiting for pages to load. However, studies show that Internet users disliking scrolling.

Tips

- Keep pages short to avoid scrolling. If you have long pages of text, split it up into separate pages.
- Avoid *horizontal* scrolling.
- Minimize *vertical* scrolling. Avoid pages longer than about 1½ to 3 screens of information because readers lose the context. Repeat navigational information at the top and bottom of such pages.
- Avoid requiring readers to scroll for the following:

- ☐ The home page
- ☐ Important information
- ☐ The introduction
- ☐ Navigational aids
- ☐ Warnings and cautions
- ☐ Information that must be compared

- ▶ Don't tell readers they must scroll.

> ☒ **Examples**
>
> Scroll (e.g., using the Page Down button) to see the full page content!
>
> Scroll down for information about our classes & links to more information.
>
> **the <u>site search tool</u> and some other helpful page links are <u>at the bottom</u>**

▷ Search Engine

A search engine is a program that lets readers search for keywords in your site or in database files or documents. Users expect this feature on large sites. A search for a keyword or phrase requires that the word(s) entered matches designated keywords. Both free and commercial search engines are available from a number of Web sites. Providing keyword search capabilities allows readers to quickly find information in a large Web site without having to learn how you have organized it. Search engines are also useful for content that changes often.

Tips

- ▶ Provide a search engine if you have a large Web site or one that includes a database or a collection.
- ▶ Find a search engine compatible with your server.
- ▶ Use a search engine that is appropriate for your readers' experience using search techniques.
- ▶ Inform readers on the home page that the search engine is available and provide a link to it, or put it in a prominent location.
- ▶ Allow for flexible searching (e.g., using advanced searches).
- ▶ Allow for misspellings.
- ▶ Give instructions for using the search engine and search tips.
- ▶ Give users control over the range of their search (e.g., global, within a section or sub-section).

☑ **Example**

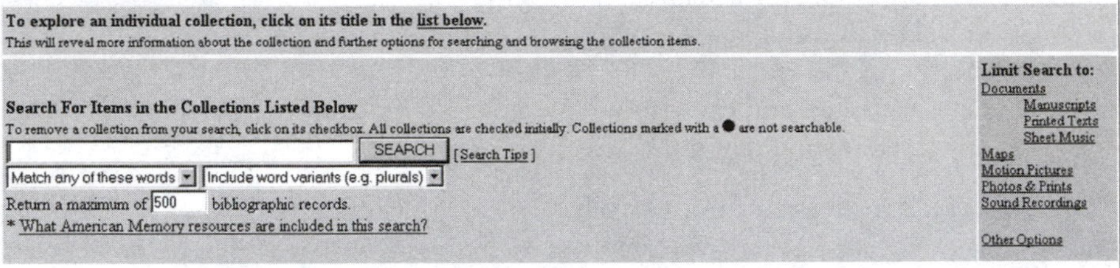

To explore an individual collection, click on its title in the list below.
This will reveal more information about the collection and further options for searching and browsing the collection items.

Search For Items in the Collections Listed Below
To remove a collection from your search, click on its checkbox. All collections are checked initially. Collections marked with a ● are not searchable.
SEARCH [Search Tips]
Match any of these words | Include word variants (e.g. plurals)
Return a maximum of 500 bibliographic records.
* What American Memory resources are included in this search?

Limit Search to:
Documents
Manuscripts
Printed Texts
Sheet Music
Maps
Motion Pictures
Photos & Prints
Sound Recordings
Other Options

▷ Search Engines: Writing for

Search engines allow readers to find Web sites based on search queries using keywords and phrases. Sites are listed in order of relevancy. This ranking is based on the location and number of keywords on your page.

Tips

- Use keywords in the following locations to help search engines find your site:
    - ALT tags
    - Description
    - Links to pages within your site
    - META tags
    - Text (headings, large fonts, bold)
    - Text links
    - Title
    - URL (domain name)

▷ Semi-Colons

A semi-colon connects closely-related independent clauses and phrases containing commas.

Tips

- Use a semi-colon to connect two independent clauses that are closely related.
- Use a semi-colon between two independent clauses connected by a transitional word or expression (*consequently, furthermore, however, indeed, in fact, moreover, nevertheless, then, therefore, thus*).

☑ **Example**

The HTML is very simple; however, there are important decisions that go into adding video clips and sounds to a web page.

☒ **Example**

Using these is a good and aesthetically pleasing means of navigation however, the page shouldn't be solely dependent on these for navigation. *(add semi-colon before "however")*

- Use a semi-colon to separate items in a series containing commas.

- Do not use a semi-colon after a dependent clause.

☒ **Example**

So if you want to find the perfect book for you; all you need to do is click.

- Avoid semi-colons because they are difficult to see online and may look like commas.

▷ Sentence Fragment

A sentence fragment is incomplete grammatically but punctuated as a sentence.

Tips

- Check for phrases or dependent clauses without a main clause.
- Check that every sentence has both a subject and verb.
- Check for the following causes of sentence fragments:
    - ☐ Gerunds (-ing words) without a helping verb and subject
    - ☐ Infinitive verbs (to . . .)
    - ☐ Subordinating words (e.g., *if, then*)
    - ☐ Explanatory phrases

☒ **Examples**

-As well as discuss with you site design and materials that you want included on your site.
-We will design a site that works! Fast loading, focused on the contents instead of pretty gizmos, and ready to do the job for you. Because we know that your prospects are busy people.
-The basics of color theory as it pertains to web design.

☒ **Before**

Always remember how your clients will be accessing the web page. Through network connection, modem or both.

☑ **After**

Always remember how your clients will be accessing the Web page: through a network connection, modem or both.

▷ Sentences

Sentences on the Web should be *short* and *simple* for the following reasons:

- Because the size and resolution of the screen typically represents half of a written page, sentences seem longer when they're online.
- Because the resolution is not as fine as on paper, reading is more difficult.
- Readers prefer short, simple sentences.
- Readers read Web pages by scanning quickly for information.

Tips

SENTENCE LENGTH

- Use short sentences (about 20 words or less).

> ☒ **Example**
> There are a number of other services besides web site design that your web site will need to give you a well rounded design service, so you will not have to spend extra money on hiring others, such as graphic design, banner design, creating interactivity between your surfers and your web site.

- Eliminate wordiness.

> ☒ **Examples**
> Another problem with using headers incorrectly is that document indexing software will not be able to properly weight the text (that is, text in headers is generally given a higher precedence than other text when you are doing searches—using headers for font control could make it difficult for you or others to search your documents and find the relevant information they are looking for.
>
> *This excerpt from an introduction provides readers with immediate links but is too long.*
> Such design essentials as color, fonts, shapes, textures, the use of photography and the importance of structure in design, plus some more specific issues such as web site navigation are treated in detail here and even illustrated by case web design projects.

SENTENCE STRUCTURE

Simple sentences: SUBJECT + VERB

Our rates are low. We want to offer Web site services to everyone.

Compound sentences: SENTENCE + (CONJUNCTION) + SENTENCE

Our rates are low, and we want to offer Web site services to everyone.

Complex sentence: DEPENDENT CLAUSE + SENTENCE

Because we want to offer Web site services to everyone, our rates are low.

- Avoid long, complex, convoluted sentences. (This is one of the most common problems with many Web documents. Authors try to include too many ideas in one sentence.)

> ☒ **Example**
> It will evolve from a conventional hotlist of cool sites into a page with ad hoc categories of cool designed to help me think through the sociological and formal reasons for the online cool phenomenon (ultimately for a chapter in my book-in-progress on the relation between literature and information, the well-read and the well-informed).

- Use simple, direct sentences with one subject and one verb.

☒ **Before**

URLs will become more familiar as you spend time on the Internet, so the most important thing to remember as you begin using URLs to locate sources on the Internet is that URLs must be entered accurately in order to locate a site.

☑ **After**

URLs will become more familiar as you spend time on the Internet. Remember to enter URLs accurately to locate a site.

- Limit sentences to one clause. Avoid compound and complex sentences and embedded clauses.
- Avoid long introductory clauses.

Before

Before actually starting your tour of the Internet via the World Wide Web, it is necessary to start off your new learning curve by reading recommended online resources.

☑ **After**

Read recommended online resources before beginning your Internet tour.

- Use a dependent clause when it helps clarify your meaning.

☑ **Example**

To ensure faster and more reliable downloads, please choose a site within the continent or region closest to you.

- Put the main point at the beginning of the sentence. Use the first part of the sentence to
  - Repeat an important part of the previous sentence.
  - Lead into next topic.
  - State an important fact.

☒ **Before**

Matching the words in the user's query to words in the title and meta keywords and meta description tags is how most of the major search engines determine placement or ranking of the results. Unlimited forwarding email addresses to any number of desired email-boxes also comes with your account.

☑ **After**

Most major search engines rank results by matching words in the query to words in the title, meta keywords, and meta description tags. Your account also includes unlimited . . .

- Avoid incomplete sentences or using phrases to be concise. The pace may become too quick, and readers must work to fill in missing words.

> ☒ **Example**
> A Complete Website Solution. Everything required to design, develop and maintain a professional/commercial website. Professional expertise to provide customized solutions specific to your requirements and personalized customer service.

Formatting Sentences

- Use bullet lists when possible.
- Use white space to make sentences stand out.
- Use bold to emphasize keywords in sentences.
- Use bold to highlight the topic sentence.
- Use indentation to show the importance of sentences.

▷ Seven Plus or Minus Two Rule

A rule of thumb is to limit items to seven plus or minus two. Experiments have shown that people can remember seven separate items (plus or minus two). This rule helps prevent information overload.

Tips

- For a Web document, never have more than seven (plus or minus two) of any of the following:
  - Choices on the screen
  - Headings
  - Menu items
  - Items in bulleted or numbered lists
  - Items of information in a chart or diagram
  - Subtopics
  - Steps in procedures or instructions
- Be careful applying this rule to navigation. Use the number of links appropriate for your audience and content.
- If you have more than seven items, break them up into categories (chunking).

▷ Shifts in Tense, Person, or Number

- Keep the following consistent:
  - Tense

> ☒ **Example**
> The mail server authenticates you, and then will allow you to receive mail.

  - Person

☒ **Examples**

The following shifts from we to I:
How are we doing? Send me some notes so that I can verify my feedback is working!

- ☐ Number

☒ **Examples**

-These pages feature essays and articles written about the net and such issues as copyright, democracy, access, politics and culture. It contains guides for using Internet protocols and software.
-Hot spots are areas on the picture that act as a hyperlink when clicked with a mouse.

▷ Shortcuts

Shortcuts include a variety of navigational aids to help readers go quickly to information. They are also called *Quick Links, Fast Find, Fast Track,* or *Quick Summary.* Shortcuts are used as a method of layering information: to link to a site map, search engine, directory, bottom line summaries, important data, and frequently-accessed information. They are useful for return visitors or expert readers.

Tips

- Provide shortcuts as part of your navigation system to allow experienced readers who know what they want go to information quickly.
- Use the following techniques to help readers get to information quickly:
  - ☐ Links to frequently-accessed information
  - ☐ Drop-down lists

☑ **Examples**

Shortcuts

GO DIRECTLY TO CNET'S NEW REVIEWS:

Links on this list are alphabetical and action verbs.

I Would Like To...
Access My Student Account
Apply For Admission
Apply For a Job
Find An E-mail Address
Find A Phone Number
Get Directions to Campus
Learn More About Mercer
Make A Gift
Request a Transcript
Take A Virtual Tour

- ☐ Breadcrumbs

Examples
quick links: Engineers · Purchasing · Company · News · Jobs

- List of links (some sites use their site map or index to provide shortcuts). These can be alphabetical or categorical.

☑ **Example**

Site Shortcuts
Academic Access Program
ActiveTest
ActiveTest SecureCheck
ActiveTune

- List of URLs (Web addresses)

☑ **Example**

site shortcuts

DESCRIPTION	SHORTCUT
About National Statistics	www.statistics.gov.uk/aboutns
Agriculture, Fishing & Forestry theme home page	www.statistics.gov.uk/agriculture

▷ Signature

A signature is often used to describe information at the bottom of each Web page (signature footer). The signature allows readers to identify the site's sponsors and defines the end of the page. A signature is also used at the end of e-mail messages.

HTML Code: <ADDRESS> </ADDRESS>

Tips

- Include any of the following:
  - ☐ Author's name
  - ☐ Address
  - ☐ E-mail address
  - ☐ URL (Uniform Resource Locator)
  - ☐ Date stamp
  - ☐ Contact information
  - ☐ Copyright information
  - ☐ Motto or quote (used in e-mail messages)
- Include a signature at the bottom of each Web page.
- Clearly identify the people responsible for the Web site.
- Limit a signature to about six lines or less.

▷ Simplicity

A simple Web page or site has a clean design without complex features. Simple pages look easy to use and help avoid information overload.

Tips

- Keep pages—especially the home page—clean and simple.
- Use ample white space.
- Keep HTML code and special effects simple.
- Keep navigation clear and simple. Use layering to let readers click to more detail.

☑ **Example**

General Mills' home page keeps navigation simple.

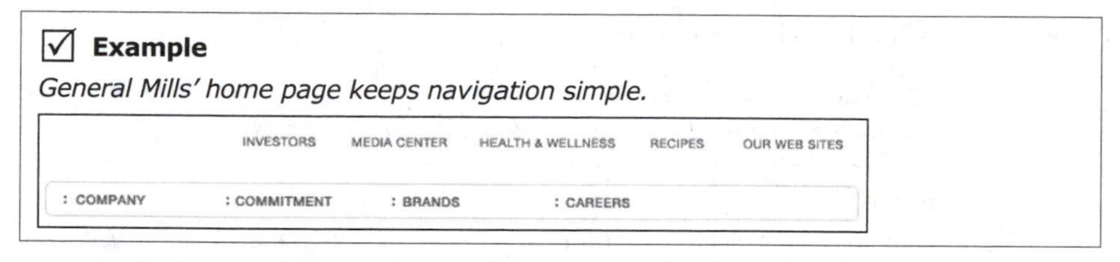

- Use graphics sparingly.
- Include only the most important information, and word text concisely.

▷ Site Guide/Site Help

A site guide explains how to navigate and use a Web site. This feature is useful for readers who are new to the Internet or who need more information about using your site. A site guide accommodates both new and experienced readers because advice is available but does not clutter the home page. However, most readers know how to use a site based on experience from using other Web pages. In general, most readers are too impatient to read instructions about using a site. If you have carefully designed your Web site and provided simple, clear navigation and a well-organized structure, your site may not need instructions.

Site help and tips provide instructions for using features in your site.

Tips

- Determine if readers need basic instructions on the home page on how to use links.

☑ **Example**

SITE TIP: Use Watch Finder for product searching, or enter text in the Search box for watches and other site info. Browse the whole site from the dropdown menu.

☒ **Examples**

Determine if your readers need the following types of information on the home page:

-Simply click on your favorite category, select a link and you're on your way!
-Please use the toolbar on the left-hand side of the screen to navigate through the link sections.
-It's suggested that you begin your visit by clicking the "About" button and reading the information provided in that area, then move along to the Product and Service areas. You may wish to bookmark this page now for future reference.
-You are currently on the Home page. The Home page provides an easy point of reference to the rest of the site. If you ever get lost or simply want a quick way back to the Home page, just click on Home.

- In general, avoid instructions that most readers who use the Web would know.

☒ **Example**

You will need to use either your back button or use the back button placed at the bottom of each page to return to this page.

- Customize the site guide for your audience.

- Use one of the following options:
  - ☐ Link to separate instructions. Give this section a title such as *How to Use This Site, About This Site, Guided Tour, Help,* or *Hints.*

> ☑ **Example**
>
> *This site provides links at the top of the home page to information for new users and help.*
>
> First time here? | About us | Help

  - ☐ Explain how to navigate within the text itself (e.g., *Click the items below*). Once users have learned how to navigate, however, this information is distracting.

- Include information such as the following:
  - ☐ How the site is organized.

> ☑ **Example**
>
> **Using The Site**
> 1stHeadlines is divided into 6 categories:
> U.S. & World, Business, Health, Lifestyles, Sports and Technology.
> Clicking on the appropriate category in the navigation bars located at the top and bottom of each page will display headlines for that category sorted by Source. Each category can be viewed with headlines sorted by Source or by Time.

  - ☐ How to navigate and an explanation of navigational buttons or other navigational aids.

> ☑ **Example**
>
> How to Book **How to Book** This button takes you to a booking information page with an easy-to-use e-form and Terms and Conditions of Booking.

  - ☐ A list of symbols, icons, and colors used and their meaning.
  - ☐ Topics covered in each section, annotated descriptions, and suggested reading order.
  - ☐ Who should read each section.

> ☑ **Example**

Section	Designed For	Description
Search Engine Submission Tips	Webmasters, site owners and marketing types	This section explains how search engines find and rank web pages, with an emphasis on what webmasters can do to improve how search engines list their web sites.
Members-Only Area	Webmasters, site owners and marketing types	This section is available to Search Engine Watch members and provides extensive details about how particular search engines work and in-depth content on submission issues. It may also be of interest to search engine users.

- ☐ Browser basics, such as changing font sizes and printing.
- ☐ Required browser settings and plug-ins.
- ☐ Instructions for using special features of the site, such as the search engine and forms.

▶ Provide extensive instructions or tips on sites that are difficult to use.

> ☑ **Example**
>
> **Tips:** Click on a state to see the cities in that state where job opportunities exist. If you are interested in a particular city, select it by clicking on it. To select or deselect multiple cities, hold down the Ctrl key while making selections. The number of jobs found will change to let you know how many results will be returned when you click the get results button.

▶ Place tips near to where the user is performing the task.

> ☑ **Examples**
>
> **Search By WebID:**
> Go
> What's a WebID?
>
> All Companies
> ABS
> Acer
> Select multiple companies by pressing the CONTROL or COMMAND key

▶ Organize help topics.

> ☑ **Examples**
>
> **Search Help** · Basics of Search · Interpreting Results · Customize Results · Advanced Search Tips
>
> **Web Search Features** · Cached Links · Dictionary Definitions · File Types · I'm Feeling Lucky · News Headlines · Phone Book · Similar Pages · Site Search · Spell Checker · Interpreting Results · Street Maps · Web Page Translation · Who links to you?
>
> *Yahoo! organizes help by*
> Top 5 Questions
> Helpful Links
> Help for Popular Yahoo! Services

▷ Site Map

A site map is a representation of your site's organization. It lists all the pages in your site and their contents. It can be graphical (table or image map) or textual. Some sites use an index as a site map. You can use a site map to

- ▶ Show the structure and organization of your entire Web site.
- ▶ Help readers find information.
- ▶ Provide another way for readers to navigate.
- ▶ Display all your topics.
- ▶ Show readers where they are located (contextual clues) and what topics they can go to.
- ▶ Accompany your site guide.

Tips

- Use a map when your site is large.
- Make the site map available from each page.
- Make the map easy to read. Keep it simple with adequate space between links.
- Link items on the map to the topics.
- Provide a text-based map for users without graphics.
- Use any of the following visual devices:
    - Topic list

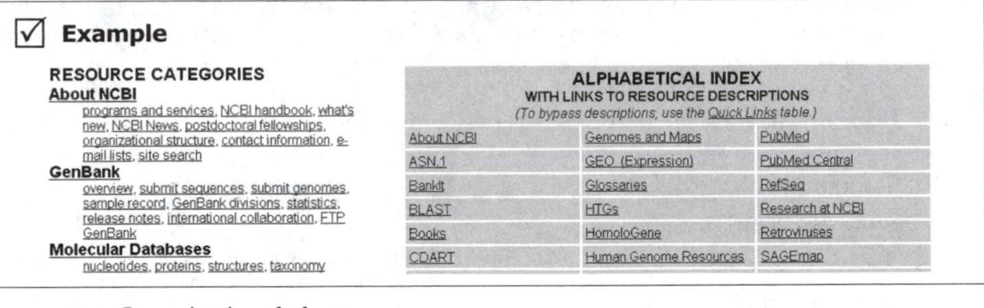

- Organizational chart
- Expandable/collapsible menu

Examples

Organizational chart (left) and expandable/collapsible menu(right).

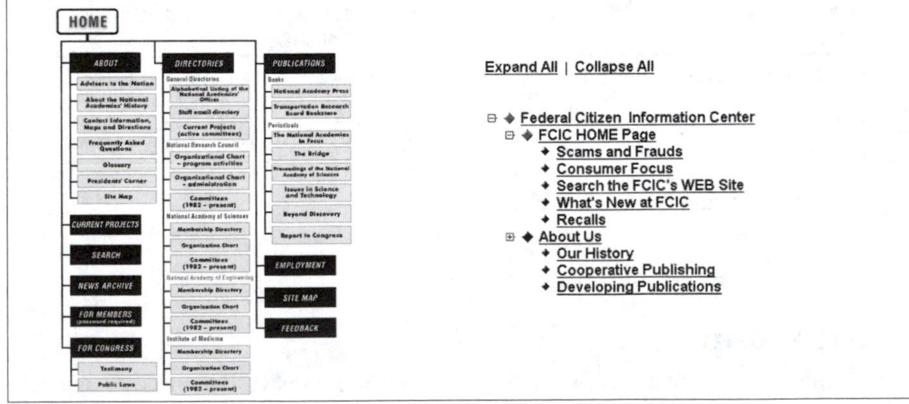

- Graphical diagram
- Nested tables

☑ **Example**

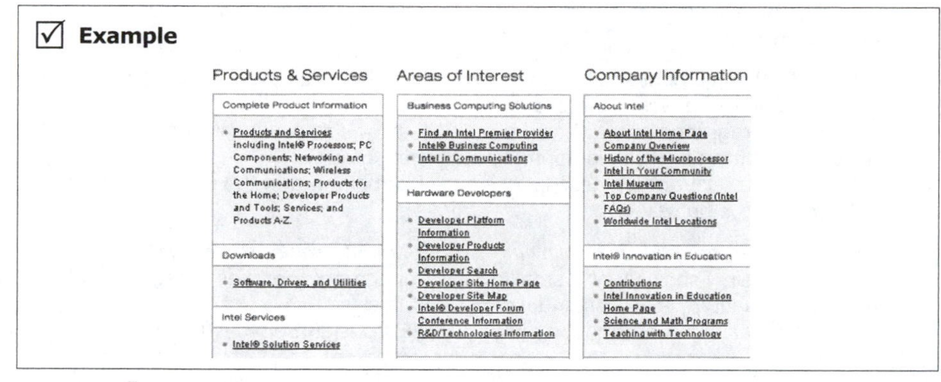

- ☐ Image map

▷ Spamming

Spamming is inundating a Web page, e-mail, or online forum with unwanted information. It includes sending junk posts to many Usenet recipients. Spamming also includes a variety of ways to attract search engine robots and spiders to your site. The term allegedly comes from a Monty Python skit in which Spam was used to represent too much of an unwanted product. Spamming is considered poor netiquette.

Tips

- ▸ Avoid the following spamming methods:
  - ☐ Overusing keywords so search engines will find the site.
  - ☐ Using invisible text to increase the keyword frequency. Some search engines can detect this practice.
  - ☐ Over-advertising your site to other sites even when it has no relevance to their audience.
  - ☐ Posting unsolicited messages to numerous newsgroups.
  - ☐ Sending junk e-mail.

▷ Spatial/Temporal References

Spatial and temporal references are phrases such as *above, below, previous, latter,* and *aforementioned.* Not only are these phrases vague, but readers may have to scroll to find the location. Every browser formats a page differently, and page length varies. So you cannot assume where items will be placed. Also, phrases like *see below* are meaningless when a document is read aloud. By using words like "previously" and "earlier," you assume readers have read separate Web pages sequentially.

Tips

- ▸ Avoid spatial and temporal references.

☒ **Examples**

-Every HTML document should have a title. As noted *previously*, titles appear at the top and center of a web browser's open window.
-A couple paragraphs *above*, I said that HTML allows multiple computers to interpret a document written in HTML in approximately the same way.
- We will post relevant job descriptions here normally within the next two weeks. *(date of page is not given)*

In these excerpts, the word "following" would be preferable to "below."
-The chart below shows that a small amount of compression has a negligible effect on the image quality, yet a substantial effect on file size.
-See below for more information about our WEB Design Services!!

- Provide a link to the item you are referencing. Also provide easy return navigation.

☒ **Example**

The link below jumps to another section on the same page. However, readers must scroll to return to the referring sentence and may forget to return altogether.

Authors who want their documents to be accessible to all would do well to avoid the `<FONT>` element, which in any case will soon be unnecessary as Style Sheets are more widely implemented (see below).

- Use figure numbers and refer to numbers rather than position.

▷ Special Characters

You use special codes to include symbols (%, *, °,®, ©) and accented characters on your Web page. (Characters such as ` - = [ ] \ ; * , . / ~ ! @ # $ % ^ ( ) _ + { } | : ? | ' do not require special code).

HTML Code: Most special symbols use the ISO-8859-1 character set code. The format is the ampersand symbol (&) plus the character code. For example, < is the < (less than) symbol and > is the > (greater than) symbol. (Most Web authoring software lets you insert a symbol without knowing the code.)

Tips

- Avoid over-using symbols and special characters because they are difficult to read online.

▷ Spelling

Misspellings are one common way of judging a Web site's credibility.

Tips

- Check your spelling because spelling errors detract from your credibility and perceived accuracy.

☒ **Examples**

These excerpts from real Web pages show how spelling errors affect your opinion of the Web page. Would you want to read further or trust the accuracy of content?

-One Web page: HMTL for beginers
-A Really Great Exapmle

- Use the spell checker in your authoring software or word processor. If your HTML editing software does not have a spell checker, save documents as ASCII text, then use the spell checker in a word processor. Spell checkers are also available on the Internet.
- Check for words that your spell checker will miss. These words are spelled correctly but misused.

☒ **Example**

This is **were** you will enter your HTML code.
Don't use **to** many huge pages on your Web site.

- Pay special attention to spelling of page titles, keywords, and headings.
- Watch for commonly misspelled words.

☒ **Examples**

While there are many spelling errors on Web pages, these words are examples of some of the most commonly misspelled words, especially on home pages, business, and sales sites.

A lot: I have seen alot of Websites.
Acknowledgement: If you would like an email acknowlegement, please enter your email address.
Advertisement: Advertisment & Communications
Already: I'll assume you have all ready read the Getting Started Page.
Appointment: Get an apointment for a first visit.
Accessible: Making Accessable Web sites
Accommodate: How does it recognize and accomodate different audiences?
Acquire: Many hosting companies will aquire a domain name for you to use.
Affordable: Every business needs an effective, afordable e-business solution.
Assistance: We provide custom development asistance with . . .
Becoming: Becomming part of the team
Beginning: Recent Developments Week begining 12/01/02
Business: Want to start a small buisness?
Calendar: Advertising Rates and Editorial Calender
Categories: We break web pages down into three separate catagories.
Convenient: . . . offers the most convienent location and the best rates.
Definitely: Definately check this out now.
Develop: You may also let us develope a Flash presentation for your site.
Dissatisfied: If you are ever disatisfied, you may cancel the contract.
Excellent: Excellant Condition.
Existent: The multiplayer features are non-existant in this release.
Familiar: Become familar with HTML.
Independent: How to become an independant travel consultant

Interrupt: interrupt
Knowledgeable: Our customers know they will get a knowlegable person to help them.
Maintenance: Web Site Maintainance or Redesign
Miscellaneous: Hardware, Miscelaneous
Noticeable: The only noticable change is our new domain name.
Occurred: Write down the line where the error occured.
Omitted: Because of the nature of this site, these things are omited.
Permitted: No copying of any part of this site is permited without express written permission.
Questionnaire: If you could just take a moment, please fill out the questionaire.
Quantity: Quanity: ☐
Receipt: Allow two weeks from reciept of payment on international orders.
Receive: Fill out the form and recieve a free issue of Yahoo Internet Life.
Recommend: Recomend this site to a friend.
Separate: A seperate link is provided here.
Site: This sight is being renovated.
Similar: Links to Similiar Sites
Transferring: The Internet is far from being the most secure way of transfering information.
Truly: Your web site is truely low-cost advertising
Undoubtedly: The most popular service will undoubtably be high speed Internet access.
Valuable: Valuble links for writers & publishers.
Yield: High rankings in the search engines and other internet promotions yeild the most important element of a profitable web site . . . visitors!

- Watch *ie/ei* words and words with double letters.
- In determining keywords for your site to submit to search engines, use common misspellings.

▷ Splash Page

A splash page is an entry page before the home page. It contains only a graphic, logo, quote, animation, or brief text (*Click Here to Enter*) that links to or launches the real home page. Some sites use the splash page to have readers select which browser or language version of the page they wish to view. Splash pages have pros and cons:

PROS

- Provide an attention-getting entry to the site.
- Allow you to create a look, theme, or branding for your site.

CONS

- Keep readers from getting to the home page.
- Often require plug-ins.
- May load slowly and cause readers to leave.
- Require repeat visitors to navigate the splash page to get to the real home page.

Tips

- Use a splash page only if you believe it is an appropriate entry into your site.

- Use a splash page for a purpose, such as to convey an important message.
- Avoid large graphics and special effects that take long to load.
- Include text such as a welcome and link to ENTER HERE. With many splash pages, it is often unclear what to click.
- Automatically load the home page after a few seconds by using the META REFRESH tag.
- Allow readers to bypass the page.
- Submit the home page rather than the splash page to search engines.

▷ Stickiness

Stickiness is anything about a site that encourages readers to stay on your site and return.

Tips

- Provide useful, relevant, updated content that is targeted to your audience.
- Present up-to-date information, such as breaking news, new jobs, special deals, etc.
- Make the site easy to use.
- Allow readers to personalize the site.
- Provide community features and feedback.
- Encourage interactivity.

▷ Storyboard

A storyboard is a graphical representation of your site. It is created by sketches on paper or by using a computer; some organizations use storyboard templates. The purpose of a storyboard is to help your plan your site's layout, structure, and flow. A Web site storyboard has two components: a mockup of each page in your site, and a map of all your pages and their interrelationships.

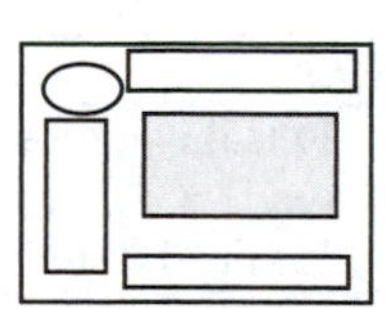

Page Storyboard: A sheet of paper, sticky note, or a box drawn on a larger sheet can represent one Web page. A storyboard can take the form of rough sketches of the layout of each page and its contents (text and graphics).

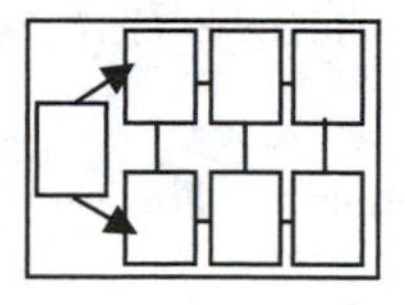

Navigation Storyboard: Arrows drawn between pages (similar to flowchart format) show navigation and links among pages.

Tips

- For the storyboard for each Web page, include any of the following types of information:
  - ☐ Type of page (home page, topic page, etc.)
  - ☐ Purpose
  - ☐ Summary of page contents
  - ☐ Graphics to be used
  - ☐ Sketch of the page layout and grid
  - ☐ Navigational links

- Show how all the Web pages are linked with each other and to the home page.

▷ Structure

The structure is the organization of your topics and their relationships. The structure affects your navigational system.

Tips

- Pick the organization that best fits your purpose and content. Consider using a combination of organizational structures.
- Provide readers with a mental model of your site. A site map is often useful.

Types of Structure

Types of structure include *linear/sequential*, *linear with alternative paths*, *hierarchical*, *web/network*, and *grid*.

Linear/Sequential Structure

Definition: A sequential structure forces readers to go though a sequence of screens. You can allow users to navigate one or both directions. This organization is similar to flipping pages in a book.

Uses: This structure is best for information that should be read in a specific order, or when you want readers to read all or most of the pages. Sequential structure is recommended for instructions, procedures, tutorials, arguments, tours, or presentations.

Navigation to provide: Sequential (next, previous) and links to the home page, first page, and last page.

PROS	CONS
▸ Lets the author control how readers go through topics. ▸ Limits the site's complexity.	▸ Restrictive.

Linear with Alternative Paths

Definition: This variation on sequential structure gives readers some branches from the linear path.

Uses: This type of organization is useful for accommodating both novice and experienced readers.

PROS	CONS
▸ Gives readers more flexibility. ▸ Allows incorporation of supplementary materials.	▸ Somewhat restrictive.

Hierarchical Structure

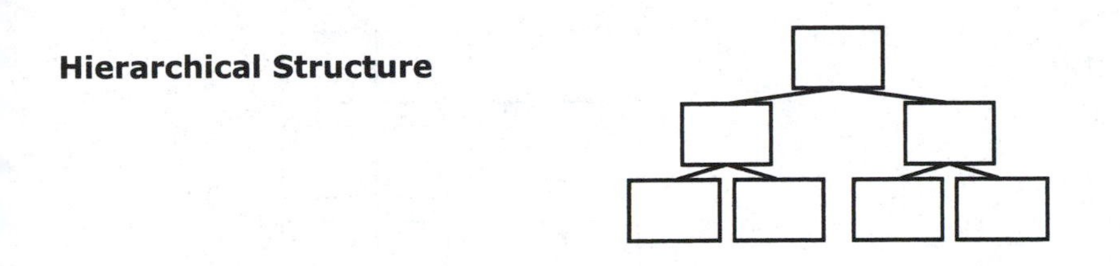

Definition: A hierarchical structure uses a branching organization with a series of levels (parent-child). Lower levels contain more detailed information. You can combine linear and hierarchical organizations.

Uses: This structure is best for subjects that lend themselves to an outline with general (major topics) to specific organization.

Navigation to provide: Up/Down, links to the home page, lateral links (Previous, Next).

PROS	CONS
▶ Flexible; gives readers more control and choices. ▶ Supports a common organization: general to specific. ▶ Familiar to people. ▶ Easy to form a mental model of this structure.	▶ Because it requires more navigation, users may get lost. ▶ Continuity of the material may become lost. ▶ May have too many branches and levels. ▶ More limiting than a web structure.

Web or Network Structure

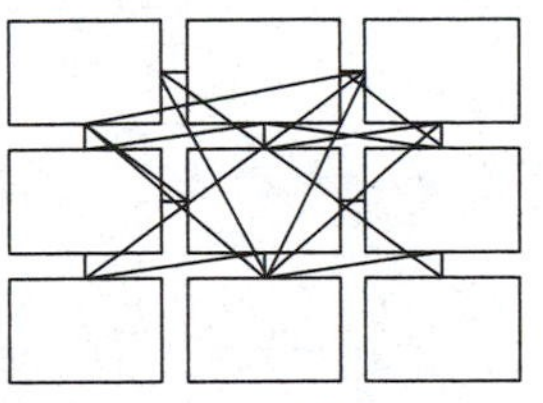

Definition: A web (network) structure contains interconnected Web pages.

Uses: A web is best for unstructured information with many cross-references.

Navigation to provide: Links to home page and to every topic.

PROS	CONS
▶ Encourages browsing.	▶ Often confusing.

Grid Structure

Definition: A grid structure is similar to a table or database model.

Uses: A grid is best for highly-structured technical information and data.

Navigation to provide: Up, down, laterally, and possibly diagonally.

STRUCTURES TO AVOID

- Avoid making your organization too *flat* or too *deep*.

Flat Structure

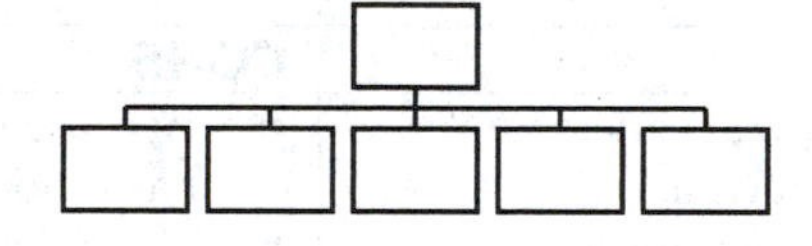

Definition: A flat structure contains most topics at the same level.

PROS	CONS
▶ Easy to navigate quickly.	▶ The menu becomes a large list of unstructured items.

Deep Structure

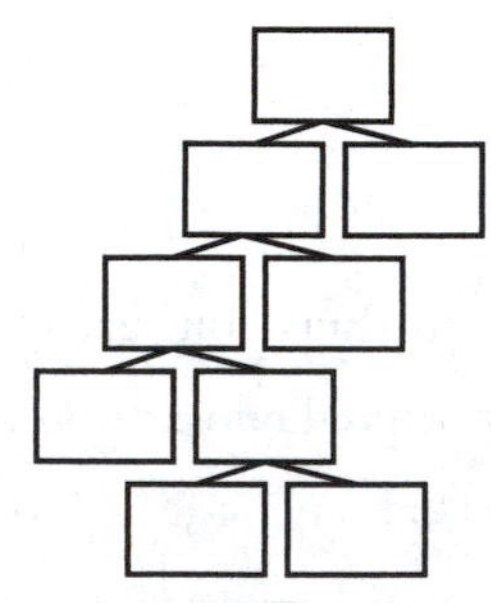

Definition: A deep structure has few main topics and many levels.

PROS	CONS
▶ Contains few initial options.	▶ Requires too many clicks to get to the topic. ▶ Buries information at deeper layers. ▶ May require too many nested menus.

▷ Style Guide

A style guide provides standards for content, format, and usage. Guidelines provide consistency, particularly when several authors contribute to a Web site.

Tips

- Create and follow an online style guide.
- Set a schedule for updating the style guide.
- Tailor the guide to your organization or publication.
- Include the following types of information:

Background Goals of site Intended audience	**References to Use** Stylebook/manual of style Dictionaries Online style guides	**Links** Link policies Use, wording, and format of links Use of external links
Content Acceptable content and subject matter Procedures for submissions Procedures for reviewing	**Design** Templates Page specifications (size, colors, fonts) Required components Position of page elements Use of graphics	**Graphics and Multimedia** Specifications Placement Captions File sizes File formats ALT tags **Identity Standards** Logos Trademarks
Style Standards Usage **Grammar and Punctuation** Grammar Spelling Punctuation	**Tags** Cascading Style Sheets HTML conventions META tag information Heading sizes Font tags Use of emphasis Special characters Table sizes Use of frames **Accessibility Guidelines**	**Glossary** Terminology/usage Company names Product names
Technical Issues File storage File naming conventions Location of templates, fonts, and graphics, required code	**Legal Issues** Copyright notices Permission policies Privacy policy for posting personal information	**Contacts** Webmaster Graphic designers Other

▷ Style Sheets (Editorial)

An editorial style sheet for a Web site is a document (usually in chart or alphabetical list format) listing the usage, spelling, and formatting of words. An editorial sheet helps with consistency and proofreading.

Tips

- Use an alphabetical list or a table format.

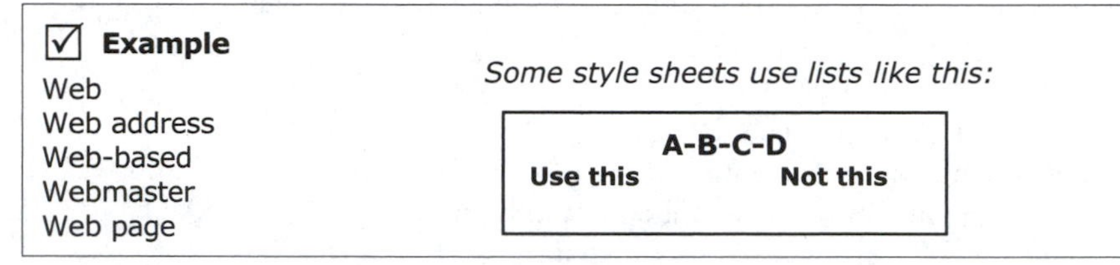

Example

Web
Web address
Web-based
Webmaster
Web page

Some style sheets use lists like this:

A-B-C-D	
Use this	**Not this**

- Include the following types of guidelines. For most of these items, list only variations from traditional rules or situations with acceptable variations.
  - ☐ Abbreviations and acronyms
  - ☐ Bibliography and references
  - ☐ Emphasis: use of bold, italics, underlining
  - ☐ Formality (contractions, pronouns)
  - ☐ Graphics: captions (spelling, punctuation, source credits)
  - ☐ Grammar
  - ☐ Numbers
  - ☐ Photos: credits
  - ☐ Punctuation
  - ☐ Spelling
  - ☐ Symbols
  - ☐ Quotations (italics or quotation marks)
  - ☐ Spelling and capitalization
  - ☐ Terms: trade, business, proper, place
  - ☐ Units of measurement
- Customize the style sheet for each project or client.

▷ Summary or Abstract

A summary is a condensed version of the main points. It may be used in various locations, including after a headline (with a link to the complete article), at the end of an article, or at the end of a topic. It is more complete than an annotation used to explain links. A summary or abstract condenses the main points of an article, document, or bottom line information. Because users skim sites, they read summaries to get the key information or determine what information they want to go to.

Tips

General

- Use a summary to help readers decide whether they want to jump to the complete article or detail page.

☑ **Example**

SUMMARY

Nutrition in emergencies

Malnutrition is rampant among refugees and displaced populations, representing 21.5 million people in 1999. Many are at risk of malnutrition and mortality. The risk depends on factors such as the state of civil insecurity, food unavailability and inaccessibility, and inadequate delivery of assistance. **Full text**

- Use a summary at the beginning of a page for readers who want to skim.

☑ **Example**

Monster Interview Coach

Summary

- This question requires quick thinking and lots of rehearsing.
- Think about new ways of selling yourself.
- Spice up your answers -- learn from these examples.

DRAFTING A SUMMARY

- Underline key words and points, especially noting the topic sentences.
- Ask, "What is the main point?"
- Ask yourself questions a reader might ask.
- Write an outline of all the points made in the document.
- Write a rough draft based on the information underlined.
- Keep the following from the original: main points, order, proportion, emphasis, key data, statistics, and facts.
- Omit the following: introductory material, background, common knowledge, examples, explanation, description, and extra wording.
- Depending on the desired length, use either one sentence or one paragraph to summarize each paragraph or section in the original.
- Add transitions between points. Make the summary about 6-20% of the length of the original, depending on how you will use it.
- Rearrange sentences.
- Edit for conciseness.

FORMATTING THE SUMMARY

- Highlight the key points on each Web page.
- Link to the summary, full article, details, topics and documents mentioned in the summary, and to *next* or related topics.

☑ **Examples**

REV-ving Up the Clock

J.D. Alvarez and Amita Sehgal

[Summary] [Full Text] [PDF]

This page summarizes the points and links to the pages mentioned.

Summary of Options

Page 4 of 4

1 - 2 - 3 - Summary

Several points were raised on preceding pages, so here's a summary of the general design options that I covered. Again, these are**choices**, not rules.

Make the site navigable by every browser on the face of the Web.(Page 1)

Make the look, feel and function consistent.(Page 2)

Make links to every section from every single page.(Page 3)

- For long summaries, break the text into several pages. Provide links between pages and downloadable and printable versions.
- Use lists and tabular format to summarize factual information.
- Use devices of emphasis, such as boxes or bold, to highlight a summary.

☑ **Examples**

1. eBook Conversion

Summary: Get your manuscript converted into an eBook. Choose from many formats, and get a cover professionally designed if you like.

HIGHLIGHTS
If content is short, mostly text, and timely, put it all in the e-mail message.
If content is long, includes images and embedded links, and has long shelf life, post it on the Web and provide links to it in the e-newsletter.
Tips on writing e-newsletter blurbs (article summaries)
About the author

▷ Support Information

Support information is detailed information used to enhance material presented on your Web page. Support information gives your site credibility and a reason for readers to visit.

Tips

- Provide links to relevant support information:
    - Advice
    - Articles
    - Background information
    - Biographies
    - Case studies
    - Customer profiles
    - Demos
    - Documentation
    - Downloads
    - Endorsements
    - External validation
    - Fact sheets (FAQs)
    - Gallery
    - How-to articles

- ☐ Hypermedia
- ☐ Information about writers
- ☐ Interviews
- ☐ Past company performance
- ☐ Press releases
- ☐ Recommendations from others
- ☐ Research
- ☐ Resumes
- ☐ Reviews
- ☐ Samples
- ☐ Statistics/data
- ☐ Success stories
- ☐ Test results
- ☐ Testimonials
- ☐ Tips
- ☐ Tours
- ☐ Web sites
- ☐ White papers

▷ Support: Technical

Technical support includes documentation and help available on a Web site. It can include a wide range of material, including instructions, procedures and policies, FAQs, reference information, troubleshooting information, user guides, and links to a form or e-mail request for specific help. Technical support is provided on company, organization, and service Web sites to promote good customer relations and provide assistance. Online support material has the advantages of wide distribution, low cost, and quick updates. It also helps reduce technical support calls.

Tips

Items to Include

- ▶ Include the following types of information:
  - ☐ Advisories
  - ☐ Announcements
  - ☐ Application notes
  - ☐ Bug reports
  - ☐ Bulletins
  - ☐ Catalogs
  - ☐ Contact information
  - ☐ Customer support offerings and levels
  - ☐ Disclaimers
  - ☐ Downloads
  - ☐ FAQs
  - ☐ File library
  - ☐ How to use the support site
  - ☐ Instructions
  - ☐ Knowledge base
  - ☐ Links to FTP site
  - ☐ Manuals/user guides
  - ☐ News
  - ☐ Newsgroups
  - ☐ Parts information
  - ☐ Policies
  - ☐ Procedures
  - ☐ Registration information
  - ☐ Quick references
  - ☐ Reference information
  - ☐ Resource library
  - ☐ Service center locator
  - ☐ Services offered/programs
  - ☐ Service partners
  - ☐ Service publications
  - ☐ Solutions
  - ☐ Specifications
  - ☐ Technical notes
  - ☐ Tips and tricks
  - ☐ Top questions
  - ☐ Training and tutorials
  - ☐ Troubleshooting information
  - ☐ Updates
  - ☐ Upgrade information
  - ☐ User to user forums
  - ☐ Vendor support sites
  - ☐ Warranty information
  - ☐ White papers

ORGANIZING SUPPORT

- Use an appropriate organization, such as by product type or model.
- Classify the product categories, documentation, and other types of support available.

> ☑ **Example**
> Product Support
> Service and Warranty
> Product Registration
> Training
> Downloads

- Provide detailed document information, such as product, language, modification date, and annotation of topics included.

☑ **Example**

Title	Date posted	Format	Size (MB)	Estimated download time
HP LaserJet Printers - Microsoft Windows XP and Windows Vista Printing Comparsion	23 Mar 2007	PDF	1.33	56.6K: <1m 512K: <1m

- Arrange specific support information in a logical order.

> ☒ **Example**
> *This information (partially shown) is not easy to find.*
> **Initial Setup**
> **Correctly installing the Printer**
> What is the "Station ID" and how do I set it
> How do I send a fax to more than one fax number?
> How do I clear "NO RESPONSE/BUSY" error?
> The machine does not print anything?
> I cannot enter Special Characters and Symbols when setting the Station ID.

USING LINKS

- Make it clear how to find support information. Make the link prominent on all pages.
- Provide a variety of ways to access the support information:
    - Search engine
    - Browse option
    - Quick find (type in product name and model)
    - Shortcuts to drivers and other commonly-accessed information

☑ **Example**

Quick Links

- Getting Started
- News & Highlights
- Frequently Asked Questions
- Microsoft Solutions
- Order Status

- Let readers branch to the appropriate information (e.g., drop-down or cascading menus) by product and by type of support needed.
- Use layering to provide links to details.

☑ **Examples**

Imagine 128 PCI Video Card
Contents
Specifications
Jumpers/Switches/Controls
Tech Notes
Graphic Pictures

Express Service
This responsive support package provides advance hardware replacement, telephone technical support, and software updates.

VIEW SERVICE DETAILS   WHERE TO BUY

- Provide download menus with any necessary instructions. Make these files available in a variety of file formats. When linking to the full document, indicate the file format, date, size, and whether access is restricted.
- Provide a link to request information or print documents.
- Provide contact information for e-mail and phone support.

▷ Table of Contents

A table of contents is a list of topics arranged into categories, similar to a table of contents in printed documents. It can be at the top of a page or on a page by itself. It is usually more detailed than a menu. It is different from an index because it is an outline based on the site structure rather than alphabetical. (NOTE: Some Web sites have alphabetical lists of topics that they call a Table of Contents rather than an index.)

A table of contents has the following uses:

- Serves as home page for navigation.
- Serves as an overview or map for your site.
- Shows the hierarchy and organization of topics.
- Shows the topics and type of information contained in the site.
- Gives readers an idea of the site's scope.
- Allows users to jump to topics of interest.
- Lets users know if they have visited all the pages.

- Accommodates different levels of users.
- Is used for navigation in electronic books, magazines, and journals.

Types of Tables of Contents

- **Abbreviated/abridged** table of contents with links to sub-tables of contents. This "contents at a glance" shows only the first two levels of heading. Readers can then choose to expand the contents to more levels. This format is useful for different levels of readers or for very long lists. The expanded outline can be on the same page (using internal links) or to a new page.

> ☑ **Examples**
>
> *This site lets readers choose the number of levels to view.*
>
> 1-level TOC · 2-level TOC · 3-level TOC
>
> *This site provides a Brief and Detailed Table of Contents (partially shown).*
>
> **56K Modem Troubleshooting Guide**
> **Brief Table of Contents**
> Initial Connect Speed Issues
> Interoperability: V.34, x2, K56flex 1.0 and 1.1, and V.90
> Known Problems with Specific Modems
> Disconnects
> Troubleshooting Links
>
> **Detailed Table of Contents**
> Initial Connect Speed Issues
>   Help! My 56K modem connects at 33.6 or less!
>   Even if I don't connect at 56K, I'm guaranteed 33.6, right?

- **Short annotated** table of contents. It lists the main headings and describes each in a few sentences.

> ☑ **Example**
>
> SECTION 2: The Ad Styles and the Survey
> Different Ad Styles
> Standard Ads: Common and simple.
> Rotating Ads: Interesting and efficient.
> Rotating Ads With Effects: Eye catching.

- **Detailed** table of contents. It contains all levels in a hierarchical tree that may be up to four or five levels.
- **Table of contents page:** A separate page rather than listed on the home page. This technique allows the home page to be a less overwhelming welcome page. The contents page can then be longer and more detailed.
- **Customized:** Separate tables of contents for different audiences.

Tips

Organizing a Table of Contents

- When listing headings and subheadings, put them in the same order that they appear on the Web pages.
- Pick a logical organization: simple to complex, alphabetical, etc.

> ☑ **Example**
>
> **Today's Boston Globe**
> The following is an alphabetical list of all the stories that ran in today's Boston Globe that appear in the Globe Online. You can also search today's Globe or the archives, or choose a section at right.

- Keep the structure fairly shallow (no more than three links deep).
- Have no more than about fifteen choices for any one group of topics.
- Consider providing several tables of contents, each organized in a different way.

> ☑ **Example**
>
> *Education Virtual Library*
> Alphabetically by Site (Complete Listing)
> Listed by Education Level
> Listed by Resources Provided
> Listed by Type of Site
> Listed by Country

Formatting a Table of Contents

- Use a design that will be familiar to readers: menus, lists, buttons, and toolbars.
- Consider using icons. However, make it clear what they represent and provide text as well.
- Use headings to break up long lists.
- Avoid long one-column lists that require scrolling.
- Avoid horizontal lists; they are difficult to scan.
- To show hierarchical organization, use indentation (nesting) or numbers with decimals. Distinguish main sections from sub-sections by using different font sizes and bullet styles.

> ☑ **Example**
>
> - Epistemology
>   - Epistemology, introduction
>   - Reflection-correspondence theory
>   - Knowledge
>   - Model
>   - Homomorphism

- When reading order is important, use an ordered (numbered) list.

- If you use more than one list, make their layout consistent.
- Use links rather than page numbers.
- Provide a link to the table of contents on each Web page.

WRITING A TABLE OF CONTENTS

- Consider annotating entries. A brief annotation of each topic allows readers to determine which topics to visit.
- Use informative and specific titles.
- Make sure the items in the contents correspond exactly to the wording of the titles they refer to.
- Use gerunds *(-ing* form of the verb*)* for tasks.
- Use noun strings for reference information.
- Put *Table of Contents* in the page title to help readers when they refer to their bookmarks (____ Table of Contents Page or Table of Contents: ______). This name will also appear at the top of the browser.

▷ Tables/Grids

A table aligns text or numbers in rows or columns.

HTML Code: <TABLE> </TABLE>

Some of the table features you can specify include the following:

Table data <TD>
Table heading <TH>
Table row <TR>
Border thickness <BORDER=>
Size of line between cells <CELLSPACING>
Space between text and cell borders <CELLPADDING>
Horizontal text alignment (left, middle, right) <TR HALIGN>
Vertical text alignment (top, middle, bottom) <TR VALIGN>

You can also create nested tables (tables within tables).
Tables can be fixed width or a percent of the screen.

Tips

USES FOR TABLES

- Use tables to do the following:
  - Arrange text and graphics in a grid layout.
  - Control the position of text and graphics.
  - Shorten the line length.
  - Create sideheads.
  - Condense and summarize key points and options.

☑ **Example**

The table summarizes main points. Each **>>More** *link goes to a separate page.*

At a Glance	
Vision	Sun's vision is to provide a tightly integrated system with Sun Linux tuned to our hardware platform giving customers more stability, manageability, and usability. » More
Compatibility	Sun's commitment to the Linux operating system brings additional value to customers of its Solaris[tm]/SPARC[tm] architecture. » More
Systems	Linux is consistent with Sun's computing vision of employing open standards and nonproprietary interfaces to develop products and services that address the needs of a variety of environments. » More
Operating Environment	Keeping to our true systems approach, Sun Linux is tuned to our hardware platforms. » More
Sun[tm] ONE	Sun will support Linux with the entire Sun ONE portfolio; Java, middleware and tools. » More
Support & Services	With total care support for software and hardware, Sun is providing an unmatched investment protection solution for customers and developers alike across all tiers of the data center. » More
Community	With the development of Sun Linux, Sun plans to increase contributions to the Linux community through participation with open source organizations. » More

☒ **Example**

This partial list of options would be easier to read in a table.

If you just want to know what a technology does, you just need to go down one level to **What Is It?** where we give you a plain-English, non-technical explanation.
If you want to know where to get it and what the system requirements are, go down another level to **What You Need.** If you've got everything you need to experience the technology, head down another level to **Trying It Out**, where we've selected sites for you where you can experience the technology firsthand.

- ☐ Categorize and organize information.
- ☐ Compare and contrast data and show relationships.
- ☐ Layer information.

☑ **Example**

A grid allows you to compare items and link to more detail.

Search Engine	Google www.google.com	Yahoo! Search search.yahoo.com	Ask.com www.ask.com
Links to help	Google help pages	Yahoo! help pages	Ask help pages
Size, type Size varies frequently and widely. See tests and more charts.	HUGE. Size not disclosed in any way that allows comparison. Probably the biggest. Biggest in tests.	HUGE. Claims over 20 billion total "web objects."	LARGE. Claims to have 2 billion fully indexed, searchable pages. Strives to become #1 in size.
Noteworthy features and limitations	Popularity ranking using PageRank™. Indexes the first 101KB of a Web page, and 120KB of PDF's. ~ before a word finds synonyms sometimes (~help > FAQ, tutorial, etc.)	Shortcuts give quick access to dictionary, synonyms, patents, traffic, stocks, encyclopedia, and more.	Subject-Specific Popularity™ ranking. Suggests broader and narrower terms.
Phrase searching (term definition)	Yes. Use " ". Searches common "stop words" if in phrases in quotes.	Yes. Use " "	Yes. Use " ". Searches common "stop words" if in phrases in quotes.

- ☐ Help readers skim and find information quickly.

Formatting Tables

- ▶ Minimize the number of columns (2 to 4) and rows.
- ▶ Keep tables short by chunking large tables.
- ▶ Try to avoid vertical scrolling.

- Always avoid horizontal scrolling. To keep the table from being too wide with higher screen resolutions, avoid fixed-width tables. Instead, use absolute table widths.
- Use lines to separate items but avoid dark lines, which impair reading.
- Use shading, color-coding, and white space to distinguish groups and sections.
- Use checkmarks, symbols, and other graphics.

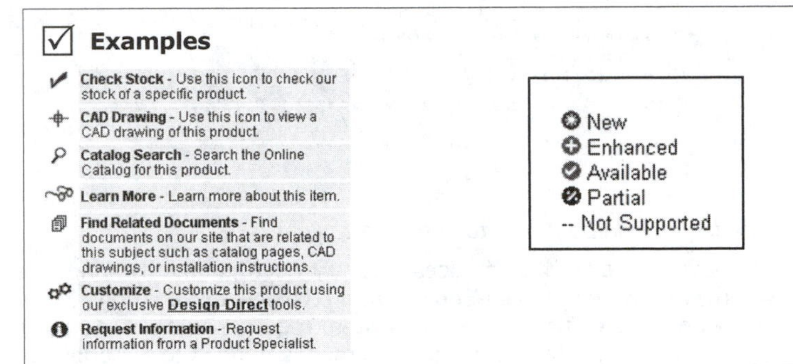

- Size tables on the same page consistently.

Using Links

- Provide links to large tables.

> **Examples**
>
> View Larger Table
> The BEST Subject Directories to use- TABLE of features

- Use links for layering complex tables or explanations.

> **Example**
>
> *Links lead to other tables or definitions.*

Manufacturer	Russound	Russound	Russound	Xantech	Xantech
Model	DPA-1.2	DPA-4.8	DPA-6.12	PA640	PA1235
Number of Channels	2	8	12	6	12
Continuous Power Output	2 x 35W @ 8 Ohms 2 x 50W @ 4 Ohms	8 x 50W @ 8 Ohms 8 x 75W @ 4 Ohms	12 x 50W @ 8 Ohms 8 x 75W @ 4 Ohms	6 x 40W @ 8 Ohms	12 x 35W @ 8 Ohms
Bridgeable	Yes, to 1 x 70W @ 8 Ohms	Yes, to 4 x 100W @ 8 Ohms	Yes, to 6 x 100W @ 8 Ohms	Yes, to 3 x 140W @ 8 Ohms	Yes, to 6 x 150W @ 8 Ohms
Individual Gain Adjustments	Yes	Yes	Yes	Yes	Yes (Via IR)

Writing and Organizing Table Elements

- Use a logical organization for table items.
- Label rows and columns by using descriptive headings and subheadings.
- Use parallel construction within the table.

☒ **Example**

A table comparing Common Internet File Formats uses the following non-parallel headings:
Suffix
Format Description and Type
How to Deal with It
Comparison of GIF and JPG

- Provide instructions for reading complex tables.

▷ Tagline

A tagline describes what a company or organization does and what makes it unique or different from competitors.

Tips

- Make it clear what you do and how you differ from competitors.

☑ **Examples**

-What We Do:
Novell is the leading provider of Net Business Solutions.
-Cisco Systems is the worldwide leader in networking for the Internet.
-Evite is the world's No.1 online invitation service and event planning destination.
-Official Website of the Olympic Movement

☒ **Example**

Specializing in website design services for the modern business.

- Make the tagline short.

☑ **Example**

WebCT is the leading provider of e-learning solutions to the global higher education market.

☒ **Example**

ABC provides comprehensive eLearning and knowledge management solutions for corporate and higher education partners. From our learning management system and knowledge base solutions, to a full range of planning, hosting and implementation services, we offer the support necessary to assure access to information, education and training.

- Be specific.

☑ **Example**

AIESEC, the world's largest student organization, is the international platform for young people to discover and develop their potential so as to have a positive impact on society. In addition to providing over 5,000 leadership positions and delivering over 350 conferences to our membership of over 22,000 students, AIESEC also runs an exchange program that enables over 4,000 students and recent graduates the opportunity to live and work in another country .

- Make it clear what your site does.

> ☑ **Examples**
>
> Spacelink: An Aeronautics and Space Resource for Education Since 1988
>
> Internet marketing reference and index of the best sites & articles
>
> Boating Courses, Boating Tips, Boating Safety, Boating Contests
>
> ☒ **Example**
>
> It's information is about the Internet compiled from the Internet to provide actionable information at your fingertips.

- Link to details.

> ☑ **Examples**
>
> The Web Standards Project is a grassroots coalition fighting for standards that ensure simple, affordable access to web technologies for all.
>
> **Who we are**
> HFI offers consulting, training and products in software usability.
> We use The Schaffer Method™ of user-centered design to make Web sites, Intranets, and applications usable for our commercial and government clients.

▷ Telegraphic Style

A telegraphic style omits key articles, pronouns, conjunctions, and explanations. This writing style forces readers to work hard to fill in missing words. The tone may also appear too curt.

Tips

- Be concise, but don't omit key information or create an impersonal, robot-like tone.
- Avoid deleting articles, pronouns, and conjunctions.

> ☒ **Examples**
>
> -Mission and purpose should be clear up front.
> -What is Java applet?
> -Host system environment is continuously professionally managed to assure optimal performance.

- Provide explanations.
- Use lists and tables to help readers skim. However, don't omit lead-in sentences that provide the context for information.
- Use a telegraphic style when appropriate, such as online resumes, annotations, reference entries, procedures, footnotes, and specifications.

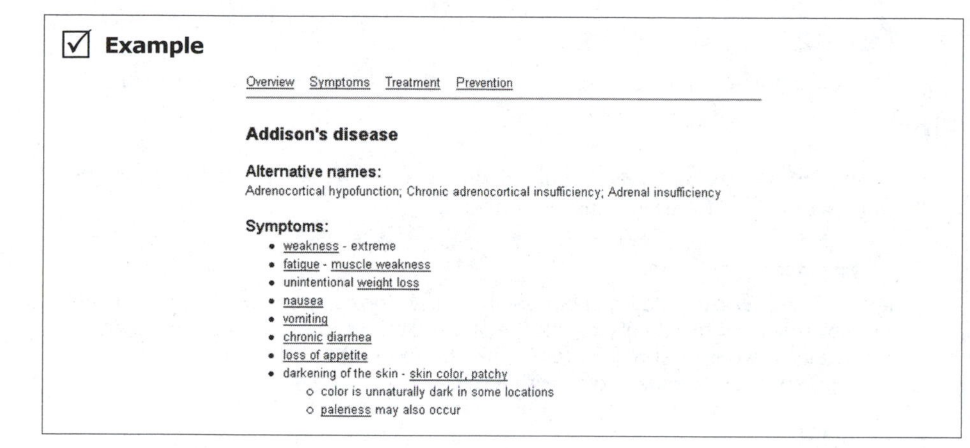

Example
Overview Symptoms Treatment Prevention
Addison's disease
Alternative names:
Adrenocortical hypofunction; Chronic adrenocortical insufficiency; Adrenal insufficiency
Symptoms:
weakness - extreme
fatigue - muscle weakness
unintentional weight loss
nausea
vomiting
chronic diarrhea
loss of appetite
darkening of the skin - skin color, patchy
color is unnaturally dark in some locations
paleness may also occur

▷ Template

A Web template is a pre-designed Web page with placeholders for text, graphics, and navigation, and with the HTML tags already inserted. Templates can also use Cascading Style Sheets.

You can use a template in the following ways:

- Substitute your own text and graphics.
- Modify the page to suit your needs.
- Obtain templates of Web pages from many sources, including HTML software, CD-ROM discs that accompany books, and the Internet. Templates are usually categorized by the type of page you want to create (e.g., Corporate Web site).
- Create your own templates.

Using a template has the following advantages:

- Helps you get started. Templates are particularly useful for those new to Web authoring.
- Encourages consistency within your own site and with multiple authors.
- Creates a unified look for pages created by several Web authors.
- Saves time.
- Simplifies changes.
- Lets you focus on content.

Tips

- Plan your site's structure and write your content before you begin putting information in the template.
- Try to vary a commercial template to create your own unique look. Pick a "look" or theme for your site.
- Design and distribute templates to maintain consistency in pages created by multiple authors in an organization.

▷ Tense

Verb tense is the form of the verb that indicates time—past, present, and future action.

Tips

- Use present tense because it helps readers scan information quickly. Use other tenses only when present tense is confusing.

> ☒ **Example**
> Next, you **will want** to learn about adding online forms, graphics, sound and video to take advantage of the Web's interactive and multimedia capabilities. And once you understand the basics and start coding pages, you **will want** to find some development tools to make your work easier.

- Avoid shifts in tense.

> ☒ **Example**
> The server *will send* the copy of the file back to the browser. An HTTP response *is sent* back to the client.

▷ Testing

Test your site by creating a prototype and checking content, design, accessibility, and links.

Tips

- Test the site
  - Using different browsers.
  - On different screen sizes and resolutions.
  - On different platforms.
  - With graphics switched off.
  - When printed out.
- Validate your pages to check code and Cascading Style Sheets. HTML validators are available either as downloadable programs or Web site services.
- Test your links. Link checking programs are available on the Web and in some editing programs.
- Check for conformance to W3C Recommendations and other standards.
- Test downloading time on different types of connections.
- Use log files to gather statistics about your pages, such as most visited.
- If you are writing a company Web site, have your legal department check the content.

▷ Theme

The theme is a dominant look and feel for your site. Themes are also used in online newsletters and magazines in the form of feature articles on one subject. Using a theme

- Provides consistency.
- Creates a mood.
- Supports your site's focus.
- Attracts readers to your Web site.

Tips

- Select a theme that matches your content and the type of organization, company, or group you represent.
- Convey your theme with the following:
    - ☐ A metaphor
    - ☐ Color scheme
    - ☐ Design elements and layout
    - ☐ Graphics (including your logo)
    - ☐ Site title
    - ☐ Headings and headlines
    - ☐ Feature articles
    - ☐ Writing style

> ☑ **Example**
>
> **CoffeeHouse For Writers**
> *Slogans*: A Mind-Brewing Experience! Good to the Last Word!
> *Banner graphic*: Coffee grinder
> *Colors*: Shades of brown
> *Newsletter*: The Daily Grind
> *Link to workshop information*: Stir up your Passion for Writing
> *Writing sample:*
> What's a writing perc? Well, in this case "perc" is short for "percolate." You might wonder how the Writing Perc will help you brew ideas.
>
> *Link icon*: Coffee cup

▷ Thumbnail Graphics

A thumbnail graphic is a small, low-resolution graphic that links to a larger version. A thumbnail allows

- The page to download more quickly.
- You to display multiple images on one page.
- Readers to decide if they want to view the image or other destination.

Tips

- Avoid thumbnails that are too small or too detailed.
- Provide text in addition to the thumbnail.
- Provide the following information:
    - ☐ Name of the image or other destination

- ☐ Description
- ☐ File type
- ☐ File size
- ☐ Any necessary instructions

☑ **Example**

The thumbnail and title both link to an enlarged graphic.
The annotation explains what readers will see.
SPIDER CONCEPT MAP: The "spider" concept map is organized by placing the central theme or unifying factor in the center of the map. Outwardly radiating sub-themes surround the center of the map.

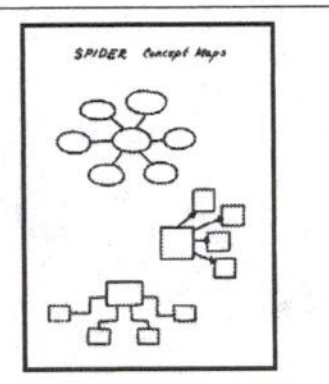

▷ Title

There are two types of titles: those that appear in the document head (HMTL code) and those that appear at the top of the page.

The title in the HTML code (<TITLE>) is important because it

- Appears in the top of the browser's window when the page is displayed. If your page does not have a title, the browser will display *Untitled*, *No title*, the page's URL, or the filename.
- Is saved in a reader's bookmarks.
- Will probably be used when people create links to your site on their page.
- Appears when a search engine indexes pages and submits search results.
- Helps you keep track of the documents in your Web site.
- Is often the first impression readers have of your site. If the title presents an unprofessional image, readers may not jump to the page. If your page is called "Untitled," it may appear that you are unable to create a Web page.

HTML Code: <TITLE> </TITLE>
The document title appears in the document <HEAD>

Tips

Using Keywords

- Use titles that summarize or describe the topic by using keywords.

☑ **Example**

Secrets of Web Advertising, Advertising on the Web, Web Advertising *(best)*

☒ **Example**

Yet Another HTCYOHP Home Page

- Use keyword phrases over single words.

☑ **Example**

Web Page Design vs. Web, Page, or Design

- Put keywords as close to the beginning as possible. This technique is especially important for helping readers find the title again in a bookmark list.

> ☑ **Examples**
>
> -Hypertext: How it Started *(best)* vs. How Hypertext Started
> -Frames Tutorial *(best)* vs. Guide to Frames Usage
>
> ☒ **Examples**
>
> *These titles are difficult to find in a long list of bookmarks.*
>
> Welcome to Ringling Brothers
> Find a job at LexisNexis

- Because titles appear in search engine lists, write a title that does not contain keywords that may have another meaning.

> ☒ **Examples**
>
> Cookies *may call up recipe sites or pages about the Web metric tool.*
> Construction *may call up home improvement sites or Web site contruction: (e.g.* Home Page Construction Kit).

- Don't repeat keywords (spamming).

WRITING TITLES

- Use a title on every Web page.
- Make the title match the title of the page itself; use a unique title for each page.
- Make the title as specific as possible.

> ☑ **Examples**
>
> Paging Yourself *vs.* How to Publish on the World Wide Web *or* Web Publishing Tips *(better)*
>
> <title>This tax tip explains the conditions workers must meet to deduct job-related education expenses. </title>

- Use a title that describes the type of page.

> ☑ **Example**
>
> *Use words like* Main, Home Page, Contents, Index, Resources, or Reference *in the title. Also consider using your domain name as the title, such as* REALLYBIG.COM Web Builder Network

- Select words that will attract readers. Emphasize reader benefits and make the page sound worth visiting.

☑ **Examples**

-Employment Resources for Technical Communicators
-Getting a Job in Technical Communication
-12 Web Page Design Decisions Your Organization or Business Will Need to Make.

☒ **Example**

Random Tips and Hints on Constructing WWW Pages

- Keep the title succinct, simple, and easy to remember.
- Titles range from 60-130 characters. The first 65 characters are most important. If your title is too long, readers may get an error message when they try to bookmark it. Long titles also take up too much space in the browser address bar.

☒ **Examples**

-Tips, Tricks, How To, and Beyond: Web Design Resources
-All the basics tools needed to do good Web page design

- Make sure the title makes sense out of context.

☑ **Example**

Introduction to HTML Elements

☒ **Example**

Recommended Reading Resources

- If you have a business site, you may want to put the name of your business (or an abbreviation) in the title of every page on your site.
- For documents relating to the same topic, use the title to help users remember their location (contextual clues).

☑ **Example**

Page Design: Styles and Templates
Page Design: Devices of Emphasis

- Check other Web sites before selecting a title, or you may pick a title that others have used.

☒ **Example**

Resources for Writers *is the title of at least four Web sites.*

- Use initial capital letters as in book titles.

Titles to Avoid

- Avoid creating negative expectations about your page.

☑ **Example**

HTML for the Conceptually Challenged *may not attract some readers.*

- Avoid beginning with *The* or *A*.
- Avoid titles that sound sales oriented, especially those that overuse punctuation such as !!! and $$$.

☒ **Example**

!!Mammoth Cave, Kentucky, Historic Diamond Caverns!!
. : Knox Dining Guide : .

- Avoid jargon, acronyms, and technical terms so novices can understand the title.
- Avoid titles that are difficult to remember or spell.

☒ **Examples**

<TITLE>Mmm-EAT it's about food</TITLE>
<title>ahhh-allegra.com - Welcome</title>

- Avoid cute and clever titles.
- Avoid misspellings and typos.

☒ **Example**

<title>The Succesful Web Site</title>
<title>Dining Out with THe Kids</title>

▷ Tone

The tone of a Web site is its "personality"—how it comes across to a reader. You convey the tone through the theme, design, and graphics, as well as through your word choice and arrangement of words. Types of tones include professional, business-like, factual, energetic, knowledgeable, motivational, persuasive, condescending, informal, satirical, offbeat, upbeat, folksy, bland, fun, and personal. In general, the tone of a Web page should be friendly, engaging, and lively so that readers are drawn to reading the information.

☑ **Example**

Note the differences in tone conveyed in the introductions to two literary sites:

BookTalk: Welcome to the Publishing insider's page where you'll find out who's hot and what's up. We've got the Buzz, the Chatter, the Rumble. We know the Scene. And you've clicked to the right place.

Publishers Weekly is the international news magazine of book publishing and bookselling. Industry professionals depend on **PW** for in-depth interviews with top authors, publishing industry news, bestseller lists, and early reviews of adult and children's books.

Note the differences in tone conveyed in the introductions to two Web page design services.

You want a web site that blows the competition away? Then *don't* give the project to your MIS code cruncher, or the college freshman next door.

In order to provide our clients with very fast time-to-market with a market proven product set, we have established solutions consulting offerings in specialized markets.

Tips

- Use a tone appropriate for your audience.

☒ **Example**

We can provide you with all of the following services to meet your needs, so you don't have to learn a bunch of technical stuff to have a presence on the Web.

- Use conversational language and informal writing to create a personal tone. Talk directly to readers (you orientation).

☑ **Example**

This beginning of a Web page uses a conversational, informal tone:
Now that you've learned to set your margins and use tables, here is a collection of little hints and tricks that should help you put these things into practice. This collection is likely to grow, so come back often.

- Use a consistent tone within a Web page. When several authors contribute to a Web site, also be sure to unify their style.
- Use a design theme that fits your tone.
- Use graphics that match the tone of your text. For example, don't use cartoons on a page that attempts to convey a professional tone.
- Beware of using a formal, bureaucratic, or sarcastic tone.

☒ **Examples**

The policies established and adhered to at CS are very important with respect to all business conducted by CS. It is recommended that you familiarize yourself with these policies in order to assure you fully understand and appreciate how you should anticipate business to be transacted when working in conjunction with CS.

This excerpt comes from an author who is trying to promote his seminar. Will the tone win customers?
If you want, you can spend hundreds of hours... you can go all over the web and see hundreds of pages of what various so-called experts describe as design for the web. You'll be amazed at how many never even touch on design! They're all wrapped up in html and gifs and frames and shockwaves and lots of mind polluting clutter. Evidently they either don't care, or didn't know that html and gifs have nothing to do with designing for the web... or any monitor for that matter.

- Avoid a condescending or parental tone with novices.

> ☒ **Example**
>
> If you are a newbie and these terms confuse you, don't worry. Everything you need to know is explained here, though it may take some time to understand it all.

▷ Topic Sentences

A topic sentence is the first sentence in a paragraph. A topic sentence is important because it does the following:

- Lets busy readers skim the material.
- Summarizes the main point of the paragraph.
- Previews the content and organization.
- Helps readers decide whether to read the paragraph.
- Provides the context for the information.
- Explains why the information is important.

Tips

- Write strong topic sentences.

> ☒ **Before**
>
> One of the Web's more powerful features is that it allows users to interact with your Web pages. Forms are a great way to get information from visitors. Forms enable you to gather feedback, conduct surveys, and receive orders for products. This information can then be sent to your e-mail account, or faxed to you.
>
> ☑ **After**
>
> Forms let you get information from visitors and make your site more interactive.
> For example, you can gather feedback, conduct surveys, and receive product orders.
> This information can then be sent to your e-mail account or faxed to you.

- Use a topic sentence in combination with specific headings to help readers skim.

> ☑ **Example**
>
> **Define your purpose**
> The first step in producing an effective website is to define your purpose. What do you want your site to accomplish? For instance, assume your company sells exotic fruits. Your initial purpose may be to attract new customers and enable them to purchase fruit online. Although you may redefine your purpose after you have received input from representative users, your statement of purpose will guide you throughout the process of defining your audience, developing your strategy, and creating the content of your site.

- Use topic sentences to provide advance organizers.

> ☑ **Example**
>
> HTML has two types of styles for individual words or sentences: logical and physical. *Logical styles* tag text according to its meaning, while *physical styles* indicate the specific appearance of a section.

- Consider emphasizing the topic sentence by using bold or indenting the paragraph text beneath it.

> ☑ **Example**
>
> **A hypertext link is a special tag that links one page to another page or resource. If you click the link, the browser jumps to the link's destination.**
>
> **There are two parts to a link:**
>
> - One part tells the human what to do.
> - The other part tells the browser what to do.
>
> Here is an example:
>
> | HTML CODE | BROWSER DISPLAY |
> | --- | --- |
> | <A HREF="http://www.microsoft.com">Go to Microsoft</A>Microsoft</A> | Go to Microsoft |

▷ Training/Tutorial

Tutorials and training courses delivered over the Internet are often referred to as WBT (Web-Based Training). Online training allows you to

- Deliver information to a wide audience using a variety of platforms.
- Update information quickly.
- Decrease costs and time required for traditional training, such as travel and facilities.
- Provide flexible, convenient access to training.
- Accommodate multiple languages and audience levels.

Tips

Organization

- Chunk the material into modules. Each module should focus on a specific skill or concept.
- Consider the best way to structure the information: sequential, hierarchical, etc.

Introduction

- State the objectives.

> ☑ **Example**
>
> This course will enable you to create Web pages with text, images, and navigational features.

- Identify the intended audience for the course.

> ☑ **Example**
>
> **<u>Perl</u>** This course is intended for System Administrators, Web Administrators, Computer Programmers, and power users. Knowledge of another computer programming language recommended

- Consider providing versions for different audiences.

> ☑ **Example**
>
> *The Kodak tutorial on The Language of Light is available in two versions:*
> We have provided two versions of this Language of Light photo program.
> Photo enthusiasts use the selections from the provided menu.
> Photo educators use our special section designed for classroom discussion.

- Describe the prerequisites.

> ☑ **Example**
>
> You should be familiar with editing HTML files and testing them in a web browser.

- Describe any required materials and requirements.
- Estimate the amount of time required to complete the course.

> ☑ **Example**
>
> **About this course:**
> - There are six lessons in this course that present the information you need to know to complete the final exam.
> - Each lesson contains information and practice exercises. You must answer each practice exercise correctly to continue through the lesson. If you answer a practice exercise incorrectly, you will be given the opportunity to enter an answer until it is correct.
> - You will be able to complete the entire lesson in approximately 15-25 minutes (depending on your modem connection).

- Tell readers why the material is important.

> ☑ **Example**
>
> This lesson is important because it demonstrates how "event objects" can be different from "effect objects."

- List the tasks readers will have accomplished by the end of the course.

> ☑ **Example**
>
> Once you've gone through the different sections of this workshop, you will be able to:
> - Locate web sites for downloading pre-made Shockwave files and ready-to-go JavaScript code.
> - Write the HTML code that inserts a Shockwave file into your own web pages.
> - Copy, paste, and modify functional JavaScript code into your own HTML files.

- Suggest the order readers should go through the materials or materials they may skip.

> ☑ **Example**
>
> There is not a strict order for going through this workshop. The pull-down menu at the top of every page allows you to quickly jump to another section.
> If you already know how HTML pages are transferred, how web pages make their way around the Internet, and how to identify and edit HTML tags, you can skip this chapter.

LESSONS

- Provide an overview/introduction to each lesson (advance organizer). Review the last lesson and lead into the current lesson.

> ☑ **Example**
>
> In the previous lesson you learned how to write intervals, and how to identify given intervals. In this lesson, you will learn how to invert them.

- For each lesson, state the objectives.

> ☑ **Example**
>
> **Objectives of this lesson:** At the completion of this lesson, you will:
> 1. Given a list, identify Web sites used as resources for weather, geographic information, and other travel-related data.
> 2. Distinguish the difference between Web sites having high or low performance features.
> 3. Identify the standard operational features of Web sites.

- Provide exercises and tests with links to answers.
- Provide interactivity.
- Provide a review/summary of what was learned in each lesson.

> ☑ **Example**
>
> **Introduction to Seattle**
> In this lesson, you learned to identify the following Seattle landmarks:
> - Mount Ranier
> - Pike Place Market
> - The Space Needle
>
> Now you can test your knowledge of Seattle landmarks. [Try the Test]

NAVIGATION

- Provide a table of contents or menu of lessons.
- Annotate the links.

> ☑ **Example**
>
> Lesson 5: Tables
> This lesson addresses how to create "regular" data tables (information in columns and rows), how to modify table properties, and how to use tables for layouts. It also talks briefly about screen resolution issues.

- Include entry and exit points for every lesson.
- Allow flexible navigation: to main menu, lessons, and forward/backward.
- Provide standard navigational aids on every screen.
- Provide contextual clues so readers know where they are at all times.

> ☑ **Example**
>
> UNIT #, LESSON #, page___ of ___

- Provide a link to instructions for using the course.
- Provide links to conceptual information, cross references, reference material, resources, and a glossary.

FORMATTING

- Keep the screen design simple.
- Put a small amount of information on each screen. Consider one topic per screen
- Consider using a metaphor (e.g., workbook or office).
- Use multimedia elements (sound, video, graphics, animation) only if they contribute to the course and do not create great download delays.
- Make the tutorial available in other versions, such as a printable file.

▷ Transitions

Transitions are connecting words used in the following locations:

- At the beginning of a section that introduces a new topic.
- Between topics.
- At the end of a page.
- Between paragraphs.
- Between sentences.
- In annotated tables of contents to provide continuity.

Transitions provide continuity and flow. However, they are often not appropriate in Web documents. Because readers can choose their own paths through the Web, you can never assume they have read another page first. Because readers skim for important ideas, they often read only topic sentences and summaries.

The table shows some formatting techniques that can substitute for transitions. They are *not* recommended for every type of situation but can make text easier to skim.

TYPE OF TRANSITION	EXAMPLES	SUBSTITUTE
Adding	Also, in addition, first, second, next	Lists
Comparing	Similarly, just as	Table
Contrasting	Although, however, on the other hand	Table
Illustrating	For example	*Example* link Label before the text: **Example**
Referring back	As mentioned	(avoid or use link to page mentioned)
Showing cause and effect	Because, therefore, consequently	Diagram
Summarizing	In conclusion, to summarize	Highlighting device (box, bold) Beginning with summary (inverted pyramid) Lists
Showing location	Above, behind, next to	Graphic
Showing time	After, before, during	Diagram

Tips

Transitions in the Table of Contents

- Use transitions to annotate menu or table of contents items when the sequence is important.

> ☑ **Example**
>
> *This annotation looks back to previous topics.*
>
> **Six Steps for Designing Your Webpage**
> Now that you've researched copyright issues and have determined the kind of webpage that you want to make, follow these six steps for creating a content-rich, visually appealing webpage.

Transitions in Opening Sections

- Limit use of traditional transitions and spatial references in Web pages. Examples include *as shown above, later we will, next.*
- Be sure each group of text is context independent. Assume this is the first screen for readers and provide the appropriate context. Ask yourself, is the page independent? Does it make sense without reading any of your other Web pages?

Transitions Between Topics

- At the beginning of a Web page related to previous pages, provide a transition.

☑ **Example**

Now that you've written—and rewritten—you want to design your page so your points are easier to comprehend.

☒ **Example**

Relativity Example - More Details
If you're reading this page on its own, it won't make much sense. This page is designed to give more detail to the twins paradox example shown on this page.

- Link keywords to necessary background information or previous topics.

☑ **Example**

Now that you understand the theory behind HTML, the next thing to learn is the practical application of HTML.

- Provide transitions between topics on the same page.

☑ **Example**

Now that you have a basic understanding of the differences between search engines and search directories and how they work, it is important to know what you can expect from individual search sites. Let's take a look at some of the major search sites:

Transitions at End of the Web Page

- Sum up the topics discussed on the page and lead into the next topic, with necessary links.

☑ **Examples**

WebDesign Clinic#4: Creating a Storyboard for Your Site
Now that you've decided on the goals and audience (Clinic#1), tone (#2), and basic design elements (#3) for your site, you're ready for hands-on work planning its overall structure.

Let's move to the next step, choosing a domain name.

Paragraph Transitions

- Use transitional devices within a paragraph:
  - Transitional words such as *next, in addition, for example, therefore.*
  - Words such as *this, these,* and *it* that refer back to the previous sentence.
  - Repetition of synonyms and keywords.

☑ **Examples**

Transitions are underlined in the following:

Fractional T1 means some fraction less than a full T1. These fractions are known as channels and each channel is capable of speeds up to 64Kbps. With this type of connection, we are now talking about speeds anywhere from 128Kbps to 1.5Mbps. For example, a 256Kbps fractional T1, the server will be capable of serving up to 3000 users per hour.

If your site features time-sensitive information, provide the date on which the information was last updated. This technique helps people know what's new on your site since the last time they visited. On the other hand, if you don't update your site regularly, don't include a date since that makes it look like you aren't maintaining your site.

☒ **Example**

The following example is choppy because it lacks transitions:

Make your home page small and simple. It should never be longer than two screens. Get it all onto one of you can. There's nothing wrong with offering a ton of good information. Just don't attempt to offer it all on the home page. In fact—offer none of it. Use the home page as a map to all the great information. Multiple pages are much easier to surf than one big page that melds into a useless block of text the more you attempt to read. Think of your home page as the body of a spider. Everything comes out from and is attached to it.

- Variations in sentence structure. Using dependent clauses can help show the relationships between ideas.

☑ **Example**

This sentence uses varied sentence patterns (dependent clauses) and transitional words for flow. These techniques are underlined.

Comments: Users read faster when line lengths are long, although they tend to prefer shorter line lengths. When designing, first determine if performance or preference is important. If user performance is critical, use longer line lengths to increase reading speed. However, if user preference is critical, use shorter line lengths.

- Make paragraphs easy to skim quickly. In general, use layout devices such as bold topic sentences and keywords, headings, indentation, tables, spaces between sentences, and lists. These techniques allow you to use fewer transitions.

☑ **Example**

In developing **audience information**, you seek to answer the question "who will use this web?"

One method to generate this information is to list as much as you know about the audience's background, interests, proclivities. All this information may not be complete at any time during the web-development process, but you can develop a store of information over time. Also, this audience information may change over time as different users access the web.

Key Audience Practices
- Don't try to reach too broad an audience.
- Ask representative members of your audience for feedback on your web.
- Keep gathering as much information as you can on your audience's needs, interests, and abilities.

▷ Transparent Writing

Transparent writing does not draw attention to the writing but rather the content. For example,

- *The following list* calls attention to listing.
- *For more information* calls attention to the links to contact information.
- *Click here* calls attention to linking.

Transparent writing lets readers focus on your content rather than your style and mechanics.

Tips

- Don't draw attention to your writing and how you are presenting the information.

☒ **Examples**

-This screen is for those who are interested in the educational system; its problems, possible solutions, and how advances in computer technology can provide new opportunities in teaching and learning.
-I will keep this site fairly simple, for example not overloading you with every link possible known to man that has to do with learning HTML. I'll link the most efficient sites, and provide specific information about each link.
-Please read and continue to scroll down. We work hard to make your time with us productive.
-This is the Introduction to the Web version of the Frequently Asked Questions/ Frequently Posted Legends list

- Don't draw attention to links. Avoid phrases such as Click Here.

☒ **Example**

Please see our Services section to get an idea of the costs involved.

▷ Troubleshooting

A troubleshooting Web page helps readers solve problems with a product.

Tips

Organization

- Organize by general categories and sub-categories.
- Include the following items:
    - □ Introduction
    - □ Table of contents or list of topics with links

- ☐ Problem/Symptom
- ☐ Explanation
- ☐ Diagnosis
- ☐ Solution
- ☐ Contact information
- ☐ Date

FORMATTING

▶ Use layering to allow users to branch to their particular product or problem.

☑ **Example**

Ion Exchange Troubleshooting Guide
The guide consists of a set of tables, which describe the problem, possible causes and offer Solutions. Where appropriate, there are links to additional information in the Resource Center, such as cleaning procedures, etc. The tables are organized according to the common water treatment applications. Select an application area below for a drop-list of problem areas, which link to the appropriate table:

Step 1: Select an application.

Choose:

Step 2: Select a category.

Make A Selection:

Step 3: Select your problem area.

Make A Selection:

▶ Consider using any of the following formats:

- ☐ List of problems at the top of the page with internal links to solutions

☑ **Example**

Sound Card
Jump to the self-diagnostic flowchart, or select the specific problem you're having with your sound card:
- I'm getting no sound.
- I'm having problems installing the sound card driver.
- The volume on my sound card is too low.

I'M GETTING NO SOUND.
1. Make sure your speakers are plugged into the sound card correctly. Most sound cards come with more than one connector in the back. You should verify the speaker is connected to the sound card's speaker or line-out connector.

- ☐ List of problems with links to separate pages for each solution
- ☐ List of abbreviated solutions with links to separate pages for more detail

☑ **Example**

The first step toward resolving these is to isolate the configuration where the problem occurs. Windows 95 offers a few ways to do this:
-Restart in Safe Mode.
-Start with an alternate hardware profile (i.e., a test configuration).
-Use the Step-by-step confirmation start up option.

- ☐ Frame with problems on the left and detail on the right
- ☐ Menus and sub-menus
- ☐ Question with YES and NO linking to different Web pages
- ☐ Flowchart

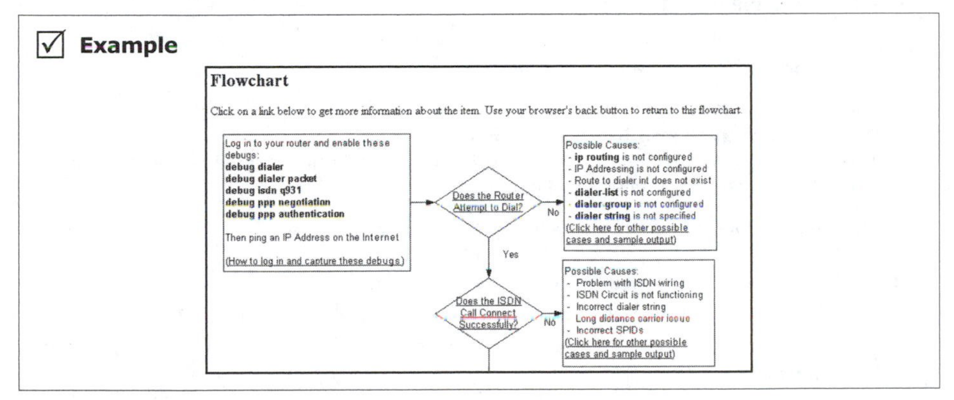

- ☐ Table. Avoid putting large tables online.

☑ **Example**

Link to solutions or other components to avoid large tables.

Light Indicators:	**Problem/recommended solution**
1. Error solid, Pass flashing	BLOW SOFTER ABORT: SUGGESTED CURE

- ☐ If/then table

▶ Make lists of problems easy to skim quickly. Arrange items in a logical order, or use branching instead.

☒ **Example**

This is only a partial view of a longer list that is difficult to skim quickly.

Troubleshooting Repair Advice | Troubleshooting
Engine hesitates
The engine surges or misfires while moving
A hissing sound is heard from the engine
Whirring from the engine that gets worse as engine speed increases
Loud exhaust
Engine backfires when you press on the gas pedal
Engine hesitates, and a popping is heard from the engine
Engine makes a clicking noise when idling

Popping noise from exhaust
Engine makes a ticking noise

- Use numbered lists for steps.
- Bold the problems.
- Offer a printer-friendly version.
- Solicit feedback on the value of the support.

ERROR MESSAGES

- Arrange error messages in a logical order.
- Use layering for error messages.

☑ **Example**

Each error message links to detail.
RealAudio Player and RealPlayer Errors
The following are descriptions and solutions for the RealAudio Player, RealAudio Player Plus, RealPlayer and RealPlayer Plus errors.

Error #1 --- General error. An error occurred.

- Use language that is clear and easy to understand.
- Tell readers what to do to solve the problem and link to information.

☑ **Example**

"Server busy" message
This may occur right after a large update is released. Try again later, or try a standalone installer from Apple Downloads.

▷ Underlining

Underlining is a horizontal line beneath text. Underlining is used for emphasis, such as highlighting terms. However, underlining interferes with letters and can be mistaken for a hyperlink.

HTML Code: <U> </U>

Tips

- Because hyperlink words are underlined, avoid underlining words that are not links.

☒ **Example**

The underlined words in this Web page excerpt are not links but terms. Unfortunately, they look like links. Use other methods for emphasis.
Electrocardiography is a testing method commonly used to determine if the heart has been damaged. Also called E.C.G. or E.K.G., an electrocardiogram is a graphic record of the heart's electrical impulses. A Holter monitor is a 24-hour portable monitor of the electrocardiogram, used to detect silent ischemia or other problems.

- Use bold, italics, or color for emphasis.
- Italicize book titles.

▷ Updated Information

Providing updated information means keeping both the content and links on your site current. Many Web pages contain information that changes often, such as news and weekly or monthly articles. Old information affects the accuracy, credibility, and "stickiness" of your site. It can even be dangerous to provide old information about crucial topics. Updated information also ensures repeat visits.

Tips

KEEPING YOUR SITE FRESH

- Keep the content of your Web site up-to-date, fresh, timely, and relevant.

> ☑ **Example**
>
> This directory is updated and monitored constantly to make sure its content remains fresh and informative.

- Provide revisions and updates to material in your site so readers have a reason to return.
- Check and repair broken links.
- Request feedback from readers about improvements and broken links.
- Be careful about changing the interface too often. Readers will have to relearn how to use your site.
- Keep up with changes in HTML and Web design techniques.
- Include a disclaimer about any information that may become quickly outdated and leave you open to lawsuits.
- For major site updates, inform readers. Provide a schedule and maintain links to frequently-accessed information. Consider providing a site history.

TYPES OF UPDATES TO PROVIDE

- Provide the following types of updates:
    - Articles
    - Blogs
    - Editorials
    - Limited-time offers
    - New links to resources
    - New products and services
    - News and announcements
    - RSS feed
    - Tip of the day
    - Upcoming events

Formatting Updates

- Tell readers how often you update the content by using the following elements:
    - ☐ Date created
    - ☐ Date last modified

> ☒ **Example**
>
> Updated two days ago.
>
> ☑ **Examples**
>
> This page last revised Sunday, October 20, 2002.
> This site last revised Sunday, November 3, 2002.

- ☐ Article dates
- ☐ What's New or Updates link
- ☐ *New* NEW or *Updated* UPDATED labels
- ☐ A message stating what will be coming to your site

- Make it easy for a visitor to determine what is new and when things were changed.

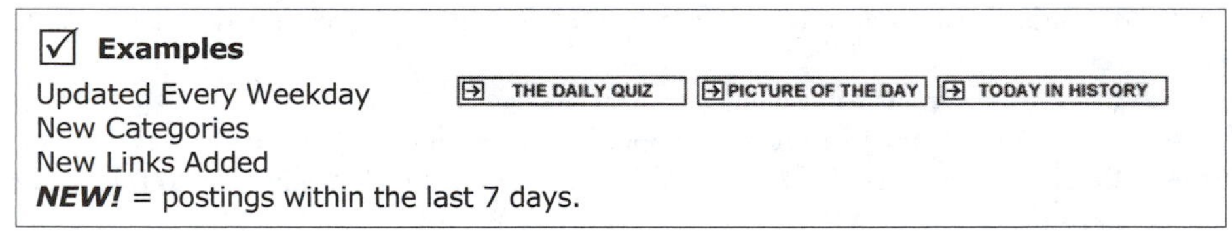

- Consider using a table, lists, or separate update page for explaining updates and providing links to the new pages.

> ☑ **Examples**
>
> **Summary of Alterations Made in 2006**
> *Click on the desired link for more details*
> Dec 13 - new bibliography created, the Storysource
> Nov 15 - Slan Hunter news, one new link added
>
> **Update History**
> This page chronicles monthly updates and revisions to this Web site.

- Choose an organization and layout that makes changes and updates easy. For example, place new information at the top or bottom of a list. Tell readers where new items are located.
- Put changeable information in a fixed location or separate page to make updates easier. Also use a grid so that constant (static) and changeable information (dynamic) are consistently located. Readers will then always know where to find the new information.
- Investigate scripts (e.g., CGI and JavaScript) that allow you to deliver daily changing content.

AVOID THE FOLLOWING

- Avoid a statement in the beginning that the page is out-of-date and no longer maintained. If the page is an archive, warn readers that information is outdated.

> ☒ **Example**
>
> The information in this document may be outdated or incorrect.
>
> ☑ **Example**
>
> **Note:** This is an archive page; and is no longer maintained. Some of this information may be out of date.

- Avoid *Under Construction* messages. 
- Avoid using phrases such as "in the next two weeks" when your page is not dated.
- Avoid using an automatically updated date stamp when it is obvious that the content is outdated.
- Avoid a page that contains nothing but a notice that it is being created or improved.

> ☒ **Example**
>
> I am working on a new design! Check back again!

▷ URL (Uniform Resource Locator) Format

A URL (Uniform Resource Locator) is a unique address for every item on the Internet. The URL includes the protocol (transfer method), domain (server) name, directory path, filename, and type of site. You write out your URL on a Web page in the signature or footer for people who print the page. You also give the complete URL in announcing your site and in telling readers how to cite or link to your page.

Tips

The following are tips for using URLs within text:

- Do not use bold or italics for a URL, although these are acceptable formats.
- If the URL is at the end of the sentence, it is acceptable (but not necessary) to place a period after it.
- Do not hyphenate a URL.
- To break up a long URL onto several lines, use the following breaks:
  - Before or after the separate units
  - After the protocol (e.g., *http*)
  - After the slash /
  - Before a period
- Use brackets < > or parentheses ( ) to show where a URL begins and ends.
- If necessary, omit the *http* in the address unless it is a different protocol (e.g., *ftp*).
- Rearrange words in a sentence so that the URL appears at the end.
- Include the URL of your Web page at the bottom of the page for readers who print it.

▷ Usability Testing

Usability testing involves a variety of methods to identify how users interact with your site. It usually involves observing a group use your site. Notes about their actions and comments help you gather information about the ability to complete tasks, time required, and navigational issues. Usability is one of the most important steps in the writing process to identify problems and ideas for improvement.

Tips

- Make usability testing part of your site design process.
- Consider developing a prototype, a mock-up of your site that contains only some of the content but the essential page design and navigation. This draft allows you to test scenarios before you invest time and money developing the entire site.
- Create a plan that identifies the following:
    - What you will test (navigation, icon recognition, formatting, graphics, download time, content, etc.)
    - Number and types of participants
    - Location
- Develop a checklist and scenarios of tasks you want participants to perform.
- Use the data you collect from problems and reactions to revise your site.

▷ Usage: Computer and Internet Words to Watch for

A number of words associated with the Internet are commonly misspelled. In addition, there is not common agreement about proper spelling for many of them.

Tips

- Be consistent within your Web site. Develop a style guide and editorial style sheets for commonly misspelled words.
- In general, spell generic names with a lowercase letter (e.g., home page) and capitalize words that refer to a specific entity (e.g., World Wide Web).
- Watch for the following words:

Backup/back up: *Back up* is a verb; *backup* is a noun.

Disk/disc: *Disc* refers to an optical storage medium, such as CD-ROMs. *Disk* refers to a magnetic storage medium, such as a floppy disk.

Email vs. e-mail vs. E-mail: All three are common spellings for both the modifier and noun. On many Web sites, the word is not hyphenated. In other media, it is hyphenated. Capitalize only at the beginning of sentences, headlines, and headings.

E-Zine vs. Ezine: Both are common spellings for electronic magazine.

Home page vs. homepage: Both are common spellings, but *home page* is preferred.

Input/output: Use as nouns rather than verbs.

Interface: Use as a noun rather than a verb.

Internet/Net

- In general, always use an initial capital: *Internet.* However, some style guides recommend using lower case when you are referring to a collection of networks that form one large virtual network.
- Always precede by *the* unless it is being used as a modifier (Internet service provider).
- Capitalize the shortened form, *Net.*

Lay out vs. layout: *Lay out* is a verb; *layout* is a noun.

Log on and log out vs. logon and logout: *Log on* and *log out* are verbs; *logon* and *logout* are nouns.

Multi-media vs. multimedia: *Multimedia* is correct.

Offline vs. off line: *Offline* is an adjective; *off line* is an adverb.

On line vs. online vs. on-line: *On-line* and *online* are both acceptable.

Print out vs. printout: *Print out* is a verb; *printout* is a noun.

Log on/logon: *Log on* is a verb; *logon* is a noun or adjective.

Set up/setup: *Set up* is a verb; *setup* is a noun or adjective. *Set-up* is incorrect.

Site/sight/cite: *Site* is a collection of Web pages; *sight* is vision; *cite* is a verb meaning to quote.

Start up/startup: *Start up* is a verb; *startup* is a noun or adjective.

Web site vs. Website vs. web site vs. website vs. web-site: *Web site* is preferred. Web-site is used as an adjective. *Web site* refers to a site with more than one *Web page.*

World Wide Web: Refer to it as the **Web** (if the context is clear), the World Wide Web, or W3.

When Web is used as an adjective (Web page), capitalize the initial letter: Web address, Web browser, Web page, Web site.

(A small letter is often used for generic references (web surfers, web sites, web server, webmaster).

The acronym is WWW.

▷ Usenet/Newsgroup Posting

A Usenet posting is a message sent to a newsgroup or other discussion list. A posting is a community feature used to communicate with others in a discussion group.

Tips

ORGANIZATION

- If your posting is part of a threaded discussion, identify the thread (ongoing discussion); then add new information.
- Begin with the point of your posting.
- Summarize the context of the posting. Do not repeat the entire original.
- Include the following, depending on the purpose of your message:
    - ☐ Ask a specific question or series of questions.
    - ☐ Don't ask for information you could obtain elsewhere.
    - ☐ State an opinion/present an argument.

WRITING

- Don't post a message until you are familiar with previous messages, threads, and FAQs. (*Lurking* is reading online postings without contributing.)
- Avoid *flaming* (publicly criticizing someone) or *trolling* (posting to get negative reactions).
- Do not make remarks that defame others or may cause financial harm to individuals or companies.
- Avoid subtlety because it is not communicated well over a computer.

FORMATTING

- Make the headline short and to-the-point.
- Use brackets to surround excerpts from the original post.
- Keep your comments short.
- Number any questions.

▷ User Guide

A Web user guide is a manual made available online. An online manual provides reference information and instructions.

Tips

- Make it easy to find user guides. Place them in a *Support* section and allow users to navigate to the appropriate product.
- Make the guide available in different formats, such as Adobe Acrobat Portable Document Format (.pdf) and a printable version.

> ☑ **Example**
>
> →PDF →HTML HP 9000 Model A-180 User's Manual

- Include a detailed table of contents with links to sections.

☑ **Example**

Basics

To begin, click the chapter heading in the left panel and then click the topic that you want to read.

Using Your Computer Outside Your Home Country

Changing the Date and Time

Power Cord Requirements

Selecting the Country for the Internal ThinkPad Modem

Using Audio and Modem Features

- Provide navigation to the entire document and to topics within chapters.

☑ **Example**

Prev | Next | QuickStart Guide
Related Pictures

- Provide cross-references.

☑ **Example**

To load envelopes, banners and other print media, see Chapter 2.

- Chunk the information so that text is no more than 1 to 3 screens long.
- For longer pages, list topics at the top with internal links to headings. Avoid simply putting an existing manual online without links.

☑ **Example**

Overview of the Hub
Cisco Micro Hub Overview
Networking Terms
Cisco Micro Hub Features
Hub Connectors
Hub LEDs
Hub Connectors
Hub LEDs

- Use screen captures or other graphics.
- Provide full document information, including version number and date.

▷ User Options/Personalization

Providing user options lets readers make choices about how to view your site. Allowing users to make choices has the following advantages:

- Makes your Web site reader-centered. Readers who feel comfortable with a site are more likely to return.
- Makes your site independent of browser, hardware, platform, and connection speeds.
- Accommodates different audience levels.
- Accommodates reader needs to download a page quickly or print your pages.
- Is a technique of "branding." Personalization is a method of creating a relationship with readers and giving them a good impression of your site.

Tips

- Provide options such as the following:
  - ☐ Browser type.
  - ☐ Complete or abbreviated instructions.
  - ☐ Abbreviated or detailed versions (e.g., table of contents or menu).
  - ☐ Text only or graphical page.
  - ☐ Link to large graphics or multimedia files.

> ☑ **Example**
>
> Click here to see a picture of a typical HTML Help file.

  - ☐ Site without special devices, such as Flash, Java, or frames.

> ☑ **Example**
>
> Text-only site/ADA compliant
>
> ENTER SITE WITH FLASH INTRO   ENTER SITE WITHOUT FLASH INTRO

  - ☐ Printer-friendly version.

> ☑ **Example**
>
> You can choose to access this material in one of three ways:
> A. PRINTABLE: white background, printable version
> B. STANDARD: standard web version
> C. LECTURE VERSION: flash, music, and full graphic version--(requires a flash plug-in, sound card, Pentium-speed processor, and a fast internet connection)

  - ☐ Different method of organization.

> ☑ **Example**
>
> The appendix is available as either a tabular summary of checkpoints or as a simple list of checkpoints.

  - ☐ Downloadable files in various formats, including handheld versions.

☑ **Example**

Get Wired News Your Way
Newsletters, handheld versions, alerts ...

- ☐ Sections geared to their interests.

☑ **Examples**

Internet Guides for
Teachers | Students | Librarians | Internet Beginners | Trainers
Non-Programmer's Guide to JavaScript   Programmer's Guide to JavaScript

- ☐ Customized interface and environment.

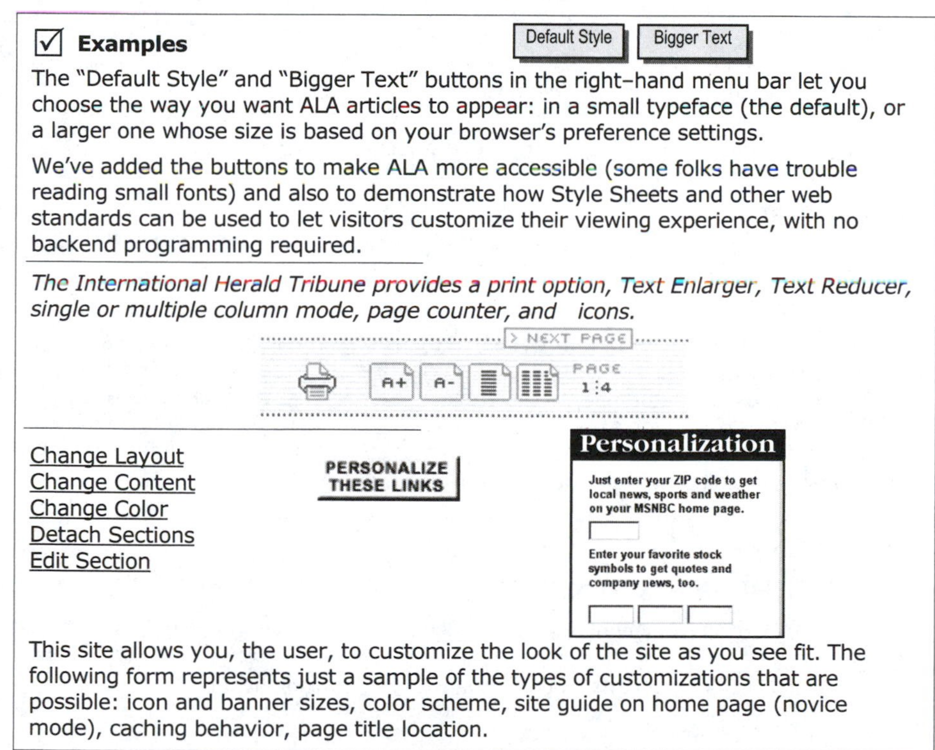

Examples
Default Style
Bigger Text
The "Default Style" and "Bigger Text" buttons in the right-hand menu bar let you choose the way you want ALA articles to appear: in a small typeface (the default), or a larger one whose size is based on your browser's preference settings.
We've added the buttons to make ALA more accessible (some folks have trouble reading small fonts) and also to demonstrate how Style Sheets and other web standards can be used to let visitors customize their viewing experience, with no backend programming required.
The International Herald Tribune provides a print option, Text Enlarger, Text Reducer, single or multiple column mode, page counter, and icons.
> NEXT PAGE
A+
A-
PAGE
1 :4
Change Layout
Change Content
Change Color
Detach Sections
Edit Section
PERSONALIZE THESE LINKS
Personalization
Just enter your ZIP code to get local news, sports and weather on your MSNBC home page.
Enter your favorite stock symbols to get quotes and company news, too.
This site allows you, the user, to customize the look of the site as you see fit. The following form represents just a sample of the types of customizations that are possible: icon and banner sizes, color scheme, site guide on home page (novice mode), caching behavior, page title location.

▷ Verbs

Verbs describe action. Using strong verbs makes your writing more direct, you-oriented, and concise.

Tips

Verbs to Use

- ▸ Use strong, active verbs, especially for navigational links.

☑ **Examples**

CONNECT PURCHASE INVESTIGATE PREPARE

I am here to... Find software Buy an Iomega product Get support

How can I... Back up my important files? Transport my data quickly and easily?

- Put the action of a sentence in the verb rather than create nouns from verbs.

☒ **Before**

The creation and maintenance of the Web documents and resources is the most time consuming aspect of managing a World Wide Web library service.

☑ **After**

The most time-consuming aspect of managing World Wide Web library service is creating and maintaining the Web documents and resources.

- Replace nouns ending with -tion, -sion, -ance, -ence, -ment, -ing.

☒ **Before**

This page provides links to tutorials and reference material related to the development of interactive Web sites.

☑ **After**

This page provides links to tutorials and reference material about developing interactive Web sites.

or This page links to tutorials and reference materials that will teach you how to develop interactive Web sites.

- When possible, use task-oriented verbs for links.

☑ **Example**

Many home pages emphasize reader action by using links such as buy, find, read, get.

Find your job now Announce your availability Use our job tools

- Use active voice to clarify who or what is performing an action.

▶ Use the following verbs when referring to user input:

TASK	VERB TO USE
▶ Use a key	▶ Press. Do not use *hit* or *strike*.
▶ Use a mouse	▶ Click or double-click
▶ Use either the mouse or keyboard	▶ Select
▶ Enter information with the keyboard	▶ Type

Verbs to Avoid

▶ Avoid weak, stretched-out verbs.

☒ **Before**

This page is an educational resource intended to **provide assistance with the implementation of** image maps. Several sites on the Web **can assist with the creation of** graphics.

☑ **After**

This page will help you create image maps. Several Web sites can help you create graphics.

▶ Avoid creating verbs from nouns (nominalization).

☒ **Examples**

-To commerce-enable your web site . . .
-In order to proceed with the translation of your website, we generally adopt one of the following two approaches.

▶ Avoid verbs such as *be, has, make,* and *do.*

☒ **Before**

You also have to consider the operations aspects of the site, such as guaranteeing response time and availability.

☑ **After**

You must consider . . . (or) Consider . . .

▷ Visual Cues

Visual cues are devices that show readers how information is organized. They also serve as decision-making aids, which help readers make choices about

▶ Where to go.
▶ What to read.
▶ What is important.

Readers want to quickly find important information on a Web page. Visual cues create easily scanned information. Visual cues also show readers the organizational structure of topics and subtopics. These aids help readers make quick decisions, thus reducing information overload.

Tips

- Use the following techniques:
  - ☐ Alignment and indentation
  - ☐ Boxes
  - ☐ Chunking
  - ☐ Consistency
  - ☐ Devices of emphasis
  - ☐ Headings at different levels
  - ☐ Icons
  - ☐ Labeling
  - ☐ Lists
  - ☐ Position of elements
  - ☐ Hierarchical arrangement of items
  - ☐ Horizontal lines to separate information
  - ☐ Indentation
  - ☐ Patterned information
  - ☐ Site map
  - ☐ White space

☑ **Example**

Every entry on this page is formatted the same way. Visual cues include alignment/indentation, chunking, consistency, headings, lists, position, lines, patterned information, and white space.

Consider Monitor Size

Guideline: Design for computers with 17-inch monitors with screen resolutions of 800 x 600 pixels.

Comments: About 40% of users use 17-inch monitors; 26% use smaller monitors (including laptops); and 34% use larger monitors.

Strength of the evidence: ●●●○○

·How to interpret "strength of evidence" scale

Sources:
- themes.org
- seir.sei.cmu.edu
- arstechnica.com

▷ White Paper

A white paper is an overview document designed to educate and inform readers on a technical topic, standard, policy, issue, service, or product.

Tips

Writing White Papers

- Use the following organization:
  - ☐ Summary of the bottom line
  - ☐ Overview of the problem
  - ☐ Available solutions
  - ☐ Best solution and why
  - ☐ Summary: review of problem and solution
- Use specific, factual, unbiased information.
- Support claims with third-party testimonials and other evidence.
- Write for your target audience. If necessary, gear the information to a wide audience, including business, technical, and novice readers.

- Include background, definitions, examples, and simple visuals.
- Use clear, simple, informal language; avoid technical jargon and marketing language.

Formatting White Papers

- Keep each document short and concise.
- Use an attractive, skimmable layout.
- Break into sections or chapters.
- Provide a table of contents with links.

Organizing White Papers

- Place white papers in a logical location, such as Technical Support.
- For numerous papers, provide browse and search capabilities.
- Group white papers into categories.
- Indicate if white papers are sponsored by any company or group.

> ☑ **Example**
>
> **White Papers**
> COM Papers DCOM Papers COM+ Papers MTS Papers ActiveX Papers

- Provide annotations.

> ☑ **Example**
>
> **ActiveX Papers**
> How to Write and Use ActiveX Controls for Microsoft Windows CE
> This paper introduces ActiveX controls, discusses how to build and distribute ActiveX controls for Windows CE, and describes how to use ActiveX controls in Windows CE–based applications.

- Provide clear titles.
- Specify dates, file types, and file sizes.
- Offer different file formats, including Adobe Acrobat and a printable version.

▷ White Space

White space (negative space) is

- The space between elements on the page.
- Margins and indentation.
- Space between lines and columns of text.
- Space between words and characters.

You use white space to

- Aid scannability and readability.
- Break up dense text.
- Create balance.
- Direct the eye.

- Emphasize headings and show where sections begin.
- Group and separate chunks of information.
- Keep the page uncluttered and reduce information overload.
- Show the document's organization and structure.
- Surround important information.

HTML Code: Use any of the following techniques to control spacing:

- Cascading Style Sheets use properties to control letter spacing, word spacing, line height, and white space.
- The break tag <BR> forces a line break.
- Preformatted text (<PRE>) controls text spacing.
- Small transparent GIF graphics can adjust indentation, leading, and graphical borders. You specify the IMG tag's WIDTH and HEIGHT.
- Tables can control line width and add space on either side.

Tips

- Use white space as a design element rather than trapped space.
- Look at just the white space on your Web page. Note where it is located and what functions it serves.

> ☑ **Example**
>
> *The following screen from Dell has been exaggerated to illustrate the location of white space. Note how it draws the eye from top left, down the vertical list, and across the bottom. It also separates the screen into distinct information zones.*
>
> 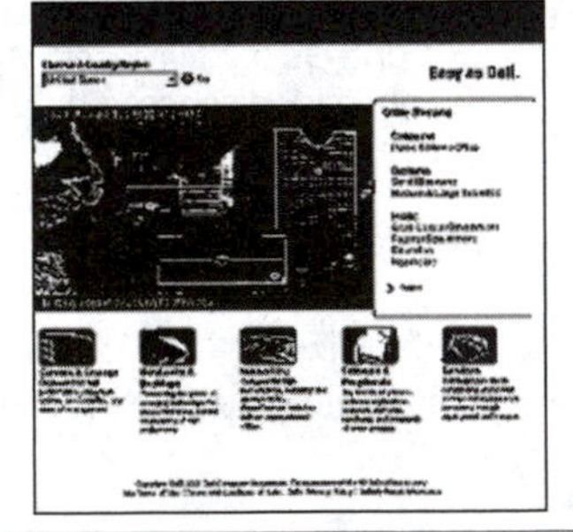
>

- Use breaks before and after lists and headings.
- To show paragraph spacing, use the following methods:
    - ☐ Indents.
    - ☐ Space between paragraphs. This technique makes paragraphs more skimmable, but makes pages longer.
- Keep line length short.
- Avoid the following:
    - ☐ Using so much white space that readers must scroll to find important information.
    - ☐ Using wasted space in the top third of the page–the most crucial area.

- ☐ Separating information so much that there is no obvious grouping/proximity of related items.
- ☐ Left-aligned text that contains wasted white space on the right.
- ☐ Centered graphics, which contain wasted space on either side.

▷ Wiki

A wiki (from the Hawaiian word "quick") is a site (or server software) used for collaborative authoring. Visitors can create content, add links, and edit content. Wikipedia is the most well-known example; wikis vary in purpose, audience, and scope. Writing should follow all the principles of good writing and editing in this handbook.

Tips

- ▸ Set up writing and file naming conventions, policies, and guidelines; if you are a contributor, follow those set up on the site you are contributing to.
- ▸ Determine whether it is best to create a new entry or add to another one.
- ▸ Use a neutral and objective point of view.
- ▸ Back up your information by using reliable sources.
- ▸ Use external links sparingly, but link to the most valuable updated resources.
- ▸ Be clear, concise, organized, and focused.
- ▸ Use specific information and examples, including visuals.

▷ Word Choice

Word choice is your selection of vocabulary. Your word choice affects clarity, tone, and how easily users can understand your Web pages.

Tips

Words To Use

- ▸ Use concrete, precise, and specific words.

> ☒ **Example**
>
> All or the servers and clients talk to each other through a bunch of straightforward (straightforward for computers, that is) "languages" called protocols. For example, that omnipresent little doohickey "http://" stands for "hypertext transfer protocol", which is how the World Wide Web works. There are other protocols for electronic mail, simple computer file transfers and many more.

- ▸ Use short, simple, one- and two-syllable words instead of long ones.

> ☒ **Examples**
>
> -Remember that, even though your visually-dense page may load quickly on your own machine, it is **a lumbering narcotized sloth** over the Net.
> -To make things really **obfuscated**, it should be noted that HTML is a *markup* language.
> **-Ubiquitous** <P> tag
> **-NEW!** ***Dr. Website Gets Avuncular***

-We can **utilize** words, pictures and sounds that you can provide to create a multi-dimensional color image of your product.

- Be aware that small words (such as *not, may, can, any, only*) can easily be missed online. If small words are important, emphasize them with bold or all caps.
- Use modern language.

☒ **Examples**

-The information herein is focused on courses that are taught primarily online.
-The average size of a web page is around 20k, and each picture therein is also around 20k.
-Hence what modem actually stands for (MODulation and DEModulation).

- Use words appropriate to the audience. Be especially careful in choosing words for an international audience.

☒ **Examples**

-Negotiate reciprocal links with other sites.
-We provide a number of site schemas to cater to our clients' individual needs.
-Web Design Extraordinaire!
-While our aim is to create a dynamic synergy between the client's vision and the implementation of their site, we will strive for an equilibrium between creative design, the client's current marketing philosophy and the realities of effective internet exposure.

☑ **Example**

Be a hip and happening teacher with Internet Explorer 4.0. Join our zippy, online tutorial and find out how IE4 makes learning fast, FUN and effective!

- Pick your words carefully. Be sure your words cannot be misinterpreted.
- When needed, use definitions and a glossary.
- Use words that will attract search engines. Choose the words in titles and headings carefully to describe both the type and use of the information that follows.
- Use consistent terminology.

☒ **Example**

Annotations begin with three different terms.
Accelerators and Instrumentation: This section includes . . .
Chemistry and Geology: In this category, you can find . . .
General and Power Engineering: This area encompasses . . .

Words To Avoid

- Avoid vague, abstract, fuzzy words, such as the following:
  - A lot (not *alot)*
  - Big
  - Bunch
  - Good
  - Nice
  - Pretty
  - Some
  - Stuff

- ☐ Thing

> ☒ **Examples**
>
> -Alot of people are learning how to program in Java and JavaScript.
> -Multimedia is a marketing term for stuff other than text.
> -The Internet is big. Really big.
> -There are a whole bunch of commands that start with & and end with ;
> -The first thing to understand about GIF and JPEG image formats is that they are both compression based formats.
> -To narrow your search, type some more words in the search box at the right.
> -This is a pretty neat utility.

- Avoid intensifying words (such as *really* and *very*).
- Avoid big words when a simpler one will do. *See the Reference section of this handbook for a list of words to avoid.*

> ☒ **Examples**
>
> In order to avoid the rocky shoals of copyright infringement, the veteran net surfer is well served by an understanding of the basic tenets of copyright law. This section covers the basics of the ubiquitous copyright notice, what copyright protects, and how long it lasts.
> • Linking Copyrightability of link-lists and the moral quagmire of deep linking.
> • Composite Pages Web aggregation pushes the envelope on linking.
>
> *Substitute a simpler word for the underlined words.*
> Fortunately, a number of tools are available to <u>facilitate</u> the task of locating and retrieving network resources, so that users anywhere can <u>utilize</u> texts, data, software and information for public access.

- Avoid buzzwords and acronyms.

> ☑ **Example**
>
> We specialize in Java Servlet Technology for dynamic database driven websites.

▷ Words: Errors to Watch for

Tips

- Check for omitted words and letters.

> ☒ **Examples**
>
> -The purpose of site is to provide information and news about the United States Navy to the Navy community.
> -You have to keep in mind that we will even re-design the whole site if are not satisfied with the work.
> -Suggestions for monitoring include combing the Web for references to your site, examining where your incoming traffic is linking in from, and asking other site to get you permission before linking to your site.

- Check for double words.

☒ **Examples**

-Technology Services recognizes that your your web site may have a very defined budget.
-The purpose of this site is is to present a small portion of the history of mathematics.

- Check for commonly misused words, such as *affect/effect.*

☒ **Example**

Keywords and a description can effect the number of hits you get...in a big way!

- Check for commonly misspelled words, such as *too/to* and *there/their/they're.*

☒ **Examples**

-List of search Engines we submit too
-Get there attention and then give them the links to get to the specific information they are looking for.

▷ "You" Orientation

"You" orientation is an emphasis on the reader rather than the writer. You orientation attracts readers to your pages and keeps them interested. It also gives your writing an informal, personal tone. "You" orientation is especially important in sales/persuasive writing.

Tips

- Analyze your audience to determine what will appeal to them.
- Focus on reader benefits, interests, and needs.

☑ **Examples**

-These pages are to provide YOU with more information, and to help you get more from your computer.
-We thank you for taking the time to stop by and sincerely wish for your visit here to be an informative one.
-Homestead: Your web site company.
Your complete web site building and hosting solution.

☒ **Examples**

-We specialize in personal attention to your Web Needs
-We have professional artists and page designers to create the best pages on the Web
-We can host your site on a high speed server for as little as $19.95 per month.

Approach Expertise Vision Results

- Emphasize what you can do for your readers, especially in the introduction of your Web page or site.

☒ **Examples**

-We bring the best to the web.
-We pride ourselves on being creative providers of online design for the digital community.
-These are what I've found to be the best resources available, but I am by no means omniscient. If you've found a site that you like, and I've overlooked it, please send me mail. **Please do not mail me for additional assistance with the WWW.** (Much as I might like to, I really don't have the time.) The links to the below resources are the only assistance I can provide you.
-I have designed this site to please myself, and if some other people like it then that's cool too!!

☑ **Example**

If you are developing or marketing a product, you should be aware of what usability testing can do for you.

- Use personal words and personal pronouns (you).

☒ **Examples**

-The interface is designed to promote advanced learning of Web page design. Only since the HTML 3.0 draft one is able to change the background by specifying an image.
-The author calls on an established body of theory related to the design of books and other print materials, along with the design of graphical user interfaces (GUIs).

Appendix

▷ Glossary

Accessibility: Characteristic of a Web site that allows anyone to obtain information on your site regardless of platform, browser, devices, and disabilities.

Advance organizer: A preview of what will be discussed. It can be in the form of text, list, links, or graphics.

Archives: Previous articles, postings, newsletters.

ASCII file: Plain text file in American Standard Code for Information Interchange format.

Attachment: An encoded file (binary or ASCII text) sent with an e-mail message rather than incorporated in the message itself.

Banner: A combination of text and graphics that appears at the top of many Web pages. It sometimes contains a logo, title, and navigational aids.

Blog: A "Web log;" an online journal that provides commentary on a particular subject.

Blurb: Summary or description of a page on your site. It is primarily used to attract readers to articles.

Branding: Creating both a visual and emotional image that readers associate with your site and thus your company or organization.

Breadcrumbs: Links (separated by >) that indicate the path a reader has taken in a site. This is a method of providing contextual clues and allowing quick navigation to topics.

Browse: Explore what is available on a Web page or site. It is different from searching.

Browser: Software program that displays a Web page by reading HTML codes and then formatting it.

Browser-safe palette: The 216 colors that look good on both PC and Macs and Netscape and Internet Explorer browsers.

Cascading Style Sheets (CSS): A W3C standard for defining Web page formatting.

Chat: Converse by computer in real-time with other participants.

Chunking: Breaking down information into separate topics, modules, or units.

Click through: Clicking a Web advertisement and going to the advertiser's Web site.

Community features: Elements of a Web site that help develop an online community: e-mail, forums, discussion groups, etc.

Contextual clues: Elements on a Web page that help readers determine their location in the site.

Deep linking: Linking to a page within an external site rather than the home page.

Disclaimer: Statement claiming you are not responsible for information on your site.

Discussion group: Online forum (e.g., newsgroup) in which people communicate about a common interest.

Document: Another name for a Web page. It is a single .htm/.html file.

Domain name: The unique name that identifies an Internet site.

Download: Transfer a file from a remote service or computer to a local computer.

Emoticons: Punctuation symbols used to show humor or emotions using ASCII characters.

External link: A connection between a location in one document and a document within the same or another Web site.

FAQ: Frequently Asked Questions. Collection of answers to the most commonly asked questions. FAQs are found most often in newsgroups or technical support.

Flaming: Insulting another person online.

Frame: A smaller window (or pane) within the browser window containing a separate HTML document. Frames can have separate backgrounds, sizes, or locations, and can scroll.

FTP: File Transfer Protocol. FTP software is used to transfer files to and from Web servers. For example, you use FTP software to upload your final Web pages to a Web server.

GIF: Graphics Interchange Format graphic; best for line drawings, screen captures, and logos.

Home page: The first page of a Web site; a one-page Web site.

HTML: Hypertext Markup Language. HTML files are indicated with an .htm or .html extension.

HTML Code: HTML tags that use brackets (see Tag).

Hyperlink: Connection between two locations. It may be text or a graphic. When you click a hyperlink, you can go to another location on the Web page, another location on the Web site, or an external Web page.

Hypertext: Text containing links to other pages, graphics, or multimedia files. These links allow you to read in a non-linear way.

Icon: A small graphical symbol that usually represents an object or abstract concept.

Image map: A clickable image; a graphic with hotspots that link to other pages or sites.

Information overload: Excess on a Web page: too much text, blinking, and animation; too many links and graphics; and overuse of devices of emphasis.

Internal link: A connection between two locations within the same document.

Internet: A global network of smaller networks and computers that communicate using a common language.

Inverted pyramid: A method of organization with the most important information (who, what, where, when, why, how) first and details and background last.

JPEG: Joint Photographic Experts Group graphic; best for photographs and continuous-tone images.

Layering: Technique of beginning with general information, then providing links to more detail and supplementary information.

Link: Connection between two nodes (locations in a document) that allows a user to jump from one location to another.

Listserv: Software that lets people subscribe/unsubscribe to an e-mail list.

Mailing list: An e-mail discussion on a specific topic directed to multiple recipients.

Mailto: A command that provides a link to your e-mail address from the Web page. The browser automatically opens an e-mail program.

Menu: A list of links to topics on a Web page or site. It is usually less detailed than a table of contents.

Navigation: The pattern through which users move around your site. Types of navigation are global (for the entire site), local (within subject or area), hierarchical, and sequential.

Navigation aids: Textual links or graphical buttons (or icons) used to move from the main contents or to jump between pages.

Netiquette: Net etiquette; good Internet manners.

PDF: Portable Document Format; the file format used for Adobe Acrobat documents.

Plug-in: A program that allows a Web browser to display inline multimedia.

Post (verb)/**Posting** (noun): Send a message to a newsgroup or other online forum.

Protocol: A set of rules and standards that let computers communicate with a minimum of errors.

RSS feed: (Really Simple Syndication) A method of delivering and publishing frequently updated syndicated digital content in XML format.

Screen: The amount of information displayed on the monitor without the need to scroll.

Scroll: Move a Web page displayed in the browser window vertically or horizontally.

Search engine: A program that lets readers search for keywords in a site or in database files or documents.

Signature: Information at the bottom of a Web page, including contact information. A signature is also a text file automatically attached to e-mail messages and newsgroup postings. It contains contact information and often a slogan or quote.

Site map: A representation of the Web site organization by using graphics or text.

Spamming: Inundating a Web page, e-mail, or online forum with unwanted information.

Splash page: A Web page that usually contains only a graphic, logo, quote, or brief text (e.g., Click Here to Enter) that links to the real home page.

Stickiness: Features of a site that make readers linger and return.

Tag: HTML markup that tells a browser how to display text. Tags are enclosed in brackets < > and usually are paired to indicate the beginning and end of the tag.

Template: A pre-designed document with placeholders for text and graphics. A Web template has HTML tags already inserted.

Thumbnail: A smaller version of a larger graphic usually used to link to the larger file.

Thread: Series of related postings to an online forum. A thread is usually a discussion or debate.

Trolling: Posting messages in a discussion group primarily to provoke a fight.

URL: Uniform Resource Locator, a unique address for every item on the Internet. The URL includes the protocol (transfer method), domain (server) name, directory path, filename, and type of site.

Usenet: Internet newsgroups that exchange messages, conversations, and debate.

W3C: World Wide Web Consortium; an international Web organization that develops open Web standards.

Web page: A single document or file in a Web site; ASCII text that includes HTML tags.

Web site: A collection of related Web pages. They are usually accessed through a home page and connected by links.

Web site: Group of related HTML documents and all related files.

Wiki: a site (or server software) that allows users to create and edit Web page content using any Web browser.

World Wide Web (WWW, W3): All the information and multimedia content available on the Internet. It is a subset of the Internet that uses a special set of protocols that allows multimedia information (such as text, graphics, audio, video, and animation) to be easily transmitted and viewed around the world.

WYSIWYG: What You See Is What You Get. What you see on the screen is exactly what will appear on a printed document. Web pages do not have this quality because readers have different monitors and browsers.

XML: Extensible Markup Language; a system for defining, validating, and sharing document formats that uses tags to distinguish document structures and attributes.

▷ Useful Lists

Words to Avoid

Substitute simpler words for the following, which are often difficult for average readers to understand.

abandon
abate
abbreviate
accelerate
accommodate
accompany
accomplish
accrue
accumulate
achieve
acknowledge
acquire
activate
actuate
adapt
additional
adequate
adhere
adjacent
adjust
administer
advance
advise
affix
alert
align
alleviate
allocate
allow
alter
amend
anticipate
apparent
appreciable
apprise
appropriate
ascertain
assist
attain
attempt
authorize
benefit
categorize
cease
cognizant
commence
components
comprehend
comprise
conclude
concur
conduct
consolidate
constitute
construct
consult
convey
correlate
corroborate
declare
decrease
deem
delegate
delete
delineate
demonstrate
depart
designate
desire
determine
difficult
diminish
disclose
disseminate
echelon
efficacy
eliminate
emphasize
encounter
endeavor
entire
enumerate
equivalent
establish
estimate
evaluate
evident
examine
exceed
excessive
execute
exemplify
exhibit
expedite
expend
expertise
express
exterminate
external
facilitate
familiarize
feasible
finalize
formalize
formulate
frequently
functional
fundamental
generate
hiatus
identical
illustrate
immediately
impacted
impair
implement
in lieu of
incapacitated
inception
increase
incumbent upon
indicate
initial
initiate
inordinate
input
inquire
insert
institute
interface
interoperability
issue
lethal
locate
magnitude
maintain
manifest
maximize
minimize
modify
monitor

necessitate
notify
numerous
objective
obligate
observe
obtain
obviate
occupy
occur
omit
operate
optimum
option
participate
perform
permit
pertain
possess
preclude
prioritize
probability
problematical
proceed
procure
productive
proficient
prohibit
promulgate
provide
proximate
purchase
purport
pursue
recapitulate
recipient
reduce
reflect
regarding
regimen
relocate
remain
remunerate
renovate
request
require
retain
select
separate
similar
solicit
submit
subsequent
substantial
terminate
transfer
transmit
transpire
ultimate
undertake
utilize
vacate
validate
verify
via
viable
visualize
warrant
wherewithal
witnessed

Redundant Phrases

____in color
____in number
____in shape
____in size

active consideration
actual fact
adequate enough
advance warning
arise up
at a cost of ____
at a distance of
at a time when
basic fundamentals
cease and desist
circle around
close proximity
completely correct
completely destroyed
completely unanimous
component parts
conclusive proof
connect up
consensus of opinion
cooperate together
couple together
definite decision
descend down
divide into two equal pieces
during the year
each and every
exactly identical/exact same
existing conditions
final conclusion
final outcome
finely divided
first and foremost
first began
group together
if and when
important essentials
in the city of/state of
in this day and age
intents and purposes
invisible to the eye
join together
mix together
modern, up to date
most unique
new innovation
one and only one
one and the same
one particular example

part and parcel
past experience
past history
plan in advance
rarely ever
repeat again
repeat back
revert back
revolutionary new
rough in texture
seesaw back and forth
sequential steps
serious danger
skeletal network
small in size
still remain
subject matter
sum total
tasteless to the tongue
totally complete
true fact
try and attempt
Xerox copy

Weak Verbs

arrive at a decision
ask the question
cause a change
cause an increase in
cause confusion
cause damage
come into contact with
come to a conclusion
come to a decision
come to an end
conduct a test
conduct an analysis
conduct an inspection
conduct an investigation
couple together
determination of
disclosure of
draw a conclusion
form a union between
get an understanding of
give aid to
give an explanation
give consideration to
give instructions to
give rise to
had an effect
has the capability of
has to
have a need to
have an interview with
have limitations on
have restrictions on
have the capability of
have the intention of
help in the removal of
hold a meeting
in a position to
in support of
in the process of
is cognizant of
is connected with
is engaged in
is in agreement with
is of the opinion
is required to
make a calculation
make a decision
make a determination
make a measurement
make a recommendation
make allowances
make an adjustment
make an announcement
make an approximation
make an evaluation
make an inquiry about
make an investigation
make application to
make changes to
make contact with
make every effort
make improvements on
make it a success
make provision for
make use of
occurrence of
offer a challenge to
perform an experiment
perform an investigation
provide an estimate of
provide assistance to
reach a conclusion
refusal of
reliance on
serve the function of
set limitations on
specification of
take a measurement
take action
take appropriate measures

take into account
take into consideration
take the initiative
undertake a study of
undertake the construction of
verification of
was a violation of

Wordy Phrases

a large number of
a majority/minimum of
a small number of
a variety of
absolutely essential
accompanied by
all of
as a consequence of
as a matter of fact
as of this date
as to whether
at a great/slow rate
at a later/future date
at a price of
at all times
at the present time
at this point in time
because of the fact that
by means of
by virtue of
by way of illustration
can be no doubt that
concerning the matter of
consider favorably
convert over to
despite the fact that
due to the fact that
during the time/course of
extent to which
few in number
for the most part
for the period of
for the purpose of
for the reason that
if and when
in a timely manner
in accordance with
in addition to
in an efficient manner
in association with
in back of
in case
in conformity with
in conjunction with
in connection with
in excess of
in favor of
in lieu of
in my opinion
in order to/that
in regard to
in short supply
in the absence of
in the amount of
in the city of
in the course of
in the event of/that
in the meantime
in the month of
in the near future
in the neighborhood of
in the process of
in the proximity of
in the same way as
in the vicinity of
in this location
in view of the fact that
inside of
is a case of
is apparent that
of an unusual nature
of the opinion
on a continual basis
on account of
on behalf of
on numerous occasions
on the basis of
on the grounds that
on the part of
one after the other
one by one
outside of
period of time
prior to
rarely ever
regardless of the fact that
subsequent to
sufficient enough
whether or not
with a view to
with the exception of

▷ Suggested Reading

Batschelet, Margaret W. *Web Writing/Web Designing.* Allyn & Bacon, 2000.

Clark, James L., and Lyn R. Clark. *Cyberstyle!: The Writer's Complete Desk Reference.* South-Western College Publishing, 2000.

Farkas, David K., and Jean B. Farkas. *Principles of Web Design* (Part of the Allyn & Bacon Series in Technical Communication), 2001.

Garrand, Timothy. *Writing for Multimedia and the Web: A Practical Guide to Content Development for Interactive Media*, 3rd. edition. Focal Press, 2006.

Hammerich, Irene, and Claire Harrison. *Developing Online Content: The Principles of Writing and Editing for the Web.* John Wiley & Sons, 2001.

Jeney, Cynthia L. *Writing for the Web: A Practical Guide.* Prentice Hall, 2006.

Kilian, Crawford. *Writing for the Web*, Self-Counsel Press, 2000.

Lynch, Patrick J., and Sarah Horton. *Web Style Guide: Basic Design Principles for Creating Web Sites*, 2nd. Ed. Yale University Press, 2002.

McAlpine, Rachel. *Web Word Wizardry: A Net-Savvy Writing Guide*, Ten Speed Press, 2001.

Maciuba-Koppel, Darlene. *Web Writer's Guide: Tips & Tools.* Focal Press, 2002.

McGovern, Gerry, Rob Norton, and Catherine O'Dowd. *The Web Content Style Guide*, Financial Times Prentice Hall, 2001.

Nielsen, Jakob. *Designing Web Usability: The Practice of Simplicity.* New Riders, 2000.

Price, Lisa, and Jonathan Price. *Hot Text: Web Writing that Works.* New Riders, 2002.

Scott, Karen. *The Internet Writer's Handbook 2001/2.* Allison & Busby, 2001.

▷ Review Checklist

General Characteristics

Standard Parts of a Home Page

- Header
  - ☐ Title
  - ☐ Logo or identity graphic
- Navigational aids
  - ☐ Site menu
  - ☐ Annotations
- Introduction
  - ☐ Purpose
  - ☐ Scope
- Footer
  - ☐ Author(s) and sponsor clearly identified
  - ☐ Contact Information
  - ☐ E-Mail Link
  - ☐ URL
  - ☐ Date Stamp
  - ☐ Copyright, disclaimer, privacy policy links

Accessibility

- ALT (alternative) text used for graphics, image maps, and multimedia
- Long descriptions provided to describe and summarize figures
- Plain text version of pages offered for pages that use special effects such as plug-ins

Audience Consideration

- Audience considered
- Definitions provided if appropriate
- International readers considered
- Pages platform-independent
- User options provided
  - ☐ Download menus
  - ☐ Other file formats
  - ☐ Printable version

Content

- Worthwhile content
- Geared to audience
- Scope clearly indicated
- Accurate

- Objective
- Current, updated, and maintained
- Links to external sites provided

Organization

- Important information on top 1/3 of page
- Inverted pyramid used
- Information chunked
- Information layered
- Information in logical order
- Structure made visible to reader

Design

General

- Readable
- Scannable
- Consistent
- Simple
- Uses appropriate theme/metaphor

Size

- Width and height of 800 by 600 pixels
- Page length appropriate
- Vertical scrolling minimal and horizontal scrolling avoided
- File size small; download time quick

Headings

- Headings used to aid skimming and show organization
- H1 to H6 tags used in sequence and only for headings
- Headings descriptive
- Headings parallel

Lists

- Lists used when appropriate
- Lists grouped into headings if long
- Lead-ins provided
- Numbered lists used when order is important
- Bulleted lists used when order is not important
- Graphical bullets small, consistent, fit design scheme
- Parallel construction used

Emphasis

- Devices of emphasis used sparingly
- Boxes used to highlight and separate information

- Bold used sparingly
- Italics used sparingly
- All caps used sparingly
- Blinking avoided
- Underlining avoided

TEXT

- Text left-justified
- Fonts large and legible
- Adequate white space used
- Line length short

RULES

- Horizontal rules used sparingly to separate information
- Graphic rules consistent, used sparingly, fit design scheme

COLOR

- Appropriate color scheme used
- Color used in moderation
- Colors used consistently
- Background color and pattern appropriate
- High contrast used between text and background
- Text color used sparingly
- Link colors standard
- Red/green combinations and blue text avoided
- Browser-safe color palette used

GRAPHICS

- Graphics used for a purpose
- Graphics appropriate
- Graphics file size small (<20 KB)
- Graphics aligned
- Limited number of graphics used per page
- Alternative text (<ALT> tag) used
- Long descriptions used
- Links to large graphics provided
- Graphics labeled and explained
- Drop caps and bitmap fonts used sparingly
- Icons legible and meaning obvious
- Related graphics and text close together

SPECIAL EFFECTS

- Multimedia and animation used for a purpose
- Alternative pages provided for readers who have slow connections or do not have required plug-ins

Positioning

- Grid used to show zones of information
- Most important information in focal point
- Pages look balanced

Navigation

Types

- Table of contents
- Menus
- Index
- Site map
- Internal links
- Search engine

Characteristics

- Information layered
- Navigational at top and bottom or side
- Navigation placed consistently
- Current location disabled
- If graphics used for navigation, text provided as well

Links

- Obvious where links lead
- Context provided for links
- Links annotated
- Links descriptive and informative
- Links used appropriately in both lists and as embedded text
- Links not overused
- Hotspots not too long, too short, or too close
- Internal vs. external links distinguished
- Internal links used for long pages
- Buttons distinguished from graphics
- Navigation to all pages provided
- *Click here* avoided
- Links checked; no dead links or under construction links

Writing

- Chunking used
- Layering used
- Advance organizers provided
- Contextual clues provided
- Paragraphs short
- Sentences short
- Text concise

- Active voice used
- Simple, precise words used
- No jargon and buzzwords used
- Writing “you” oriented
- Informal style used
- Humor avoided
- No spatial references used
- Transitions used carefully
- Correct spelling, grammar, punctuation used

▷ Index